370-4
35

THE COMPLETE PLYWOOD HANDBOOK

BY S. BLACKWELL DUNCAN

TAB **TAB BOOKS Inc.**
BLUE RIDGE SUMMIT, PA. 17214

DELTA COLLEGE
Learning Resources Center

DEC − 1981

FIRST EDITION

FIRST PRINTING

Copyright © 1981 by TAB BOOKS Inc.

Printed in the United States of America

Reproduction or publication of the content in any manner, without express permission of the publisher, is prohibited. No liability is assumed with respect to the use of the information herein.

Library of Congress Cataloging in Publication Data

Duncan, S Blackwell
 The complete plywood handbook.

 Includes index.
 1. Carpentry—Amateur's manuals. 2. Plywood—Amateurs' manuals. I. Title.
 TH5606.D78   674'.834   80-23567
 ISBN 0-8306-9671-7
 ISBN 0-8306-1304-8 (pbk.)

Contents

Introduction 6

1 What Is Plywood? 9
Origins of Plywood—The American Plywood Industry—How Plywood Is Made—The Advantages of Plywood—Plywood Benefits for the Do-It-Yourselfer—Disadvantages of Plywood

2 Plywood Properties 61
Understanding Plywood Characteristics—Softwoods and Hardwoods—Softwood Plywoods—Hardwood Plywoods—Miscellaneous Plywood Properties—Specialty Plywoods

3 Tools for Plywood Work 144
Selecting Tools—Tool Costs—Layout and Measuring Tools—Saws—Hammers—Screwdrivers—Chisels—Planes Bits and Cutters—Clamps—Forming and Finishing Tools—Router—Personal Protection Gear—Miscellaneous Tools—The Starting Toolbox—Keeping Tools Sharp

4 Basics of Working With Plywood 217
The Plywood Workshop—Layout and Planning—Handling and Storage of Plywood—Selecting and Buying Plywood—Workpiece Layout—Cutting Plywood With a Handsaw—Cutting Plywood With a Portable Circular Saw—Using a Saber Saw—Making Cuts With a Table Saw—Working With the Radial Arm Saw—Bandsaw Cuts—Squaring—Cutting Scrolls and Curves—Edge Finishing—Drilling—Mortising

5 Plywood Joining, Fastening and Finishing 291
Joints—Gluing Plywood—Fasteners—Builder's Hardware—Edging—Framing—Finishing—Surface Repairs—Sanding—Grain Filling—Application of Stain Finish—Color Toning—Paints—Varnish—Shellac—Oils—Exterior Finishes—Natural and Satin Finishes—Paint Finishes

6 Constructions and Projects 352
Conventional Plywood Subfloor—Plywood Underlayment—Combined Subfloor/Underlayment—Glued Plywood Floor—Cemented-Tile Subfloor—Plywood Corner Bracing—Conventional Plywood Wall Sheathing—Exterior Plywood Siding—Single-Layer Exterior Plywood Wall—Plywood Roof Decking—Plywood Soffits—Built-Up Thermal Roof—Fire-Rated Roof—Sound Control With Plywood—All Weather Wood Foundation—Interior Wall Paneling—Wall-Hung Utility Desk—Cube Storage—Fireside Bench—Doghouse—Planters

Index 401

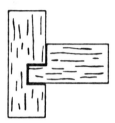

Introduction

Nowadays plywood is perhaps the most universal and utilitarian construction material available to us, not only for general building purposes but for literally thousands of different home shop and project applications as well. Oddly enough, to most do-it-yourselfers it seems also to be one of the least understood of the common building materials. The most prevalent misconception among many home mechanics with whom I've talked run along the lines of, "You see one sheet of plywood, you've seen 'em all." An oversimplified and understated comment, to be sure, and also unfortunate—nothing could be further from the truth.

Over the years I have been asked a good many questions about plywood, some general and some specific, often involving particular details or properties of the material. Most of them I was able to field readily enough, while a few required some research. But because of my own reasonable familiarity with plywood, I assumed that probably other do-it-yourselfers were also well acquainted with at least the principal facts concerning plywood. An over-the-coffee-cup discussion about plywood several months back with two men, one an "intermediate" do-it-yourselfer and the other a residential building contractor, disabused me of that notion. It developed that neither man, both of whom had years of experience in woodworking, construction, carpentry and associated areas, possessed more than a smattering, of solid knowledge about plywood. Some of that was misinformation. The culmination of the conversation appeared to revolve around the questions, "What's the big deal about plywood? What makes *it* so special?"

Those are good questions and ones that I, and evidently a lot of other folks, have never stopped to seriously consider. Some further investigation—very unscientific and done on an admittedly limited, localized basis—led to the strong hunch that probably a good many do-it-yourselfers are not realizing anywhere near the full potential and benefits of regularly utilizing plywood in their sundry projects, mostly because of a lack of concrete knowledge about plywood.

This book is an attempt to rectify that situation in some small degree by presenting some of the more important information about plywood, in layman's terms for the most part, and investigating how, under what conditions and for what purpose it can be used, as well as how to work with it. The text concentrates largely on gathering together and presenting the basic, essential information about plywood, much of which is presently scattered and difficult to find, except in bits and pieces, unless you happen to know just where to look and enjoy doing research.

You will learn of the origins of plywood, its history and how it is now manufactured. Plywood characteristics, properties and applications are covered in depth, along with some pertinent consumer information. The tools needed to work plywood are discussed, as are the basics of working and fabricating with plywood. This includes both home and shop projects, as well as numerous aspects of building construction. Techniques and materials employed in plywood joinery, fastening and finishing are included as well.

The last portion of the book contains a series of typical plywood constructions as used in house-building and similar endeavors, as well as a few representative home and workshop projects that use plywood as a basic material. They range from fairly simple to fairly complex, and at one stage or another incorporate much of the information contained in the preceeding text. These constructions and projects are intended primarily as idea-starters and to exemplify some of the practical results that can be gained from using plywood. For those of you who want to plan your own projects to exactly fit your own needs, this material will give you an idea of how it's done. For others who prefer preplanned projects and see nothing here that strikes your fancy, don't despair. There is a wide range of excellent project plans and instruction sets available in books like *Working With Plywood, including Indoor/Outdoor Projects* (TAB book No. 1144) and other sources.

In short, the book presents a rounded disclosure of plywood as a versatile, highly useful and extremely effective material for the do-it-yourselfer to work with in a myriad of applications. It will hopefully serve also as a springboard to many satisfying and worthwhile hours of constructive work around the home and in the shop, and lead to further investigation of the field and more advanced and sophisticated levels of skills, projects and constructions.

I must also add that the American Plywood Association (APA) and the Hardwood Plywood Manufacturers Association (HPMA), without whose help this book would have been very difficult if not impossible to compile, are fine sources of information, especially of the technical sort, about plywood and its uses. The APA can be contacted at 7011 So. 19th St., P.O. Box 11700, Tacoma, WA 98411. The HPMA address is 1824 Michael Faraday Drive, P.O. Box 2789, Reston, VA 22090.

S. Blackwell Duncan

Chapter 1
What is Plywood?

Plywood is one of those basic, primary products that we see every day in one form or another and to which we seldom pay any attention. This fundamental construction and fabrication material seems to (and perhaps does) have a million uses, but we take its presence entirely for granted. Plywood surrounds us, sometimes literally. There is plywood in our floors, in or on our walls and in our roofs. There is plywood in our furniture, our kitchen cabinets, our bookshelves and our stereo sets. Plywood is found in railroad boxcars, tractor-trailer rigs, campers and mobile homes. There is plywood in saloons and salons, boutiques and pawn shops, stockbrokers' offices and jails. Plywood is on the farms, in suburbia, and throughout the cities and industrial parks. It is everywhere. And yet, what do we know about plywood? We know very little, for the most part, unless we happen to be directly associated with some phase of its use, or a part of the plywood industry itself. What is this universal and ubiquitous material, anyway? Where did it come from, and why is it so useful that it impinges upon our lives daily?

ORIGINS OF PLYWOOD

Plywood is neither a "wonder material" developed of space age components and technology; nor is it (in basic form) by any stretch of the imagination a new material. As we shall see, modern technology has indeed made plywood the valuable asset that it is today. But the origin of plywood goes back a long, long time—3500 years. About the earliest evidence we have of the process of *veneering* dates back to approximately 1500 B.C., in the form of a mural that depicts workmen going through the actual steps of cutting thin strips of wood, mixing and applying glue, covering a base material with the veneer strips, and weighting them with stones while they cured.

Early Veneering Uses

Veneering was used in artistic and beautiful fashion on various early furnishings; among the first we know of is a cedar wood casket veneered with ebony and inlaid with ivory, found in the tomb of King Tut-Ankh-Amon. Veneering and what we now call plywood are known to have been used in Roman times, and to some extent during the Classical Greek period. The Greeks are generally credited with having devised the table, the couch, the day bed and the folding bed, all of which frequently were embellished with beautiful veneers. Pliny, a Roman historian, goes into considerable detail on veneering in Book XVI of his *Natural History*. At that time, veneering and the laying up of plywood (not known by that term back then) was not done for purposes of strength or economy. Veneering provided the best use of the most beautiful pieces of wood, and offered the greatest opportunity for artistic effects, including the cutting and inlaying of symbols, figures and such in the pictographic fashion common to those times. But Pliny does say, "In order to make a single tree sell many times over, laminae of veneer have been devised." So perhaps there was a conservation effort even in those days. The most likely reason, however, was to increase profits, just as popular a concern in those days as in these.

With the demise of classical civilization, veneering (along with most other aspects of early cultures) disappeared for a good many centuries. Not until the 17th and 18th centuries did veneering reappear, and once again it was used as an art form in the making of fine furniture. One of the best-known pieces from this period is an incredibly ornate and handsome desk, the prototype of today's familiar roll-top desk, called the *Bureau du Roi*. This piece is acknowledged generally as quite probably the plywood masterpiece of all time; it was made (by hand, of course) in France for Louis XV and was nine years in the making. Classic furniture made wide use of the veneering process, and of plywoods, during the period from approximately

1700 to 1820, particularly in England and Europe, and to some extent in the American Colonies as well. Walnut and mahogany veneers were the most popular choice, though other woods were sometimes used as well. Of course, many of these pieces are still extant today.

By the turn of the 19th century, fertile minds were beginning to explore possible uses of veneer other than in furniture making. One of the great drawbacks, even in those times when handwork and craftmanship were the order of the day, was in the cutting of veneers. This was a difficult, slow and arduous process when done by hand. The quantities of veneers that could be turned out in this fashion simply were not enough to allow any widespread usage. Though the first mechanically operated saw, a reciprocating blade type, is known to have been in use around 1650 and was undoubtedly used in cutting veneers, for some reason or other the process never came into widespread use. The circular saw was developed in the late 1700s but was of little practical import until the steam engine became better developed. It came into general use around 1840 and was of some help in veneer production, but it was also very wasteful. The saw kerf was a good deal wider than the veneers produced, so most of a veneer log was reduced to sawdust! A veneer shaver, a complicated and peculiar-looking device that was intended to shave slices of veneer from solid wood stock, appeared in the early 1800s, but apparently was unsuccessful and little-used.

Machinery

Though from a practical standpoint the machinery of those early days was a flop, the proper thoughts and intentions were indeed right on target. In 1797, an Englishman by the name of Sir Samuel Bentham set about developing a series of woodworking machines, and subsequently applied for patents. The patents are comprised of many sections, most of which are of little interest to the plywood historian. But two sections in particular looked directly ahead for two full centuries, squarely at a plywood industry that was then completely unknown. The first section concerns the forming of laminated wood from shavings (Bentham's term for veneers, which were to be cut on a machine that he invented for the purpose). Part of the patent claim describes in considerable detail not only plywood panels made in much the same fashion as they finally were when the plywood industry came into being, but also a method for winding veneers to make hollow plywood tubes. The second section had to do with the idea of giving curvature to a piece by bending it, which involved making a thick section from thin plies that could be easily bent and worked. This process, of course, is one that is widely used today in forming plywood shapes.

Bentham's was not the only mind at work in this area. A resident of Stockbridge, Massachusetts, one John Dresser, is credited with the first machine that was distinctly intended, and reasonably effective from an operational standpoint, for cutting veneer. The machine actually was a modified turning lathe fitted with a knife that would slice the veneer from a cylindrical block of wood revolving between the centers of the lathe. Over a number of decades, various pieces of machinery were developed and employed (with horrendous inefficiency by our present-day standards) in the manufacture of veneers, thus constituting a small but reasonable beginning for an unknown industry.

Mayo's Patents

While some were thinking about machinery, others were pondering various possibilities for practical use of veneers. John K. Mayo, a native of Portland, Maine, exercised his Yankee ingenuity to portend the future with incredible accuracy. He took out a series of patents over a short period of time that could well be construed as the birth, in tangible terms, of the plywood industry. Two paragraphs in particular are worth quoting from those papers, which are still in the Patent Office records, because of their farsightedness at that time and their timeliness now. Mayo used the term "scale" or "scale-boards" as his definition of a thin sheet of wood or veneer: "The invention consists of cementing or otherwise fastening together a number of these scales or sheets, with the grain of the successive pieces, or some of them, running crosswise or diversely from that of the others . . . The crossing or diversification of the direction of the grain is of great importance to impart strength and tenacity to the material, protect it against splitting, and at the same time preserve it from liability to expansion or contraction."

Some while later, another document was issued that said in part, "The invention consists in the formation of various structures used in civil engineering of a plurality of thin sheets or veneers of wood, cemented or otherwise firmly connected together, with the grain of the several scales or thicknesses crossed or diversified, so that they will afford to each other mutual strength, support, and protection against checking and splitting, shrinking or swelling, expanding or contracting." Mr. Mayo deserves a hearty round of applause from today's building contractors.

Commercial Uses

The fact is, though, that the piano industry (of all things) was one jump ahead of Mayo. The old adage that necessity is the mother of invention was at work. There was a problem with keeping stationary the pins that held and were used to tune the piano wires. If the pins were simply inserted in holes drilled in the pin plank, there was insufficient friction to keep the pins from continually turning slightly under the great pressure imparted by the tightened piano wires, and they quickly loosened. Someone, we know not who, devised the idea of laying thin strips of maple crosswise to one another and glued together. This allowed the pins to be gripped from all directions by the ends of the wood fibers, so they did not slip. This cross-laid maple construction was the first definite example of plywood as we know it today being used for a commercial purpose.

Some four decades later, the *Wheeler & Wilson Sewing Machine Company* started another operation called the *Sewing Machine Cabinet Company* of Indianapolis. The sole purpose of this company was to provide plywood parts for sewing machine cabinets. Thus began an era of commercial manufacture and end-use of plywood as a product, even though the two companies were dependent upon one another. Shortly thereafter, Gardner and Company of New York City began manufacturing plywood bench-type seats that were widely used in ferry and railroad stations, as well as assembly halls. This was the first known use of cross-ply curved plywood that was layered up in forms and cured to form solid-piece seats and backs. Most of these benches, which came in a variety of forms, were decorated with perforated designs and mottos ("friendship, love & truth"), and all of those holes were hand-drilled by child labor.

Not long afterward, plywood began to appear in cabinet ends, bellows, and various parts of organ cases, especially the fretwork. In 1883, the plywood desk top was developed and immediately hailed by the manufacturers of office furniture as the best thing to come along since the invention of the desk itself. Plywood tops soon appeared on virtually every desk manufactured. From this point, the jump from desk tops to furniture of all kinds was but a short and easy one. By the last decade of the 18th century, the tops, fronts and various parts of all kinds of furniture were being made of plywood. (Some parts, such as drawer bottoms, dust bottoms, backings and the like, had long been made from thick single thicknesses of veneer.) Even then, the fledgling industry was not without its problems, because the artisans and craftsmen of that time resisted this innovation with passionate fervor, just as they often do today. Calling the new material nothing more than "pasted wood," they refused to have anything to do with it—or at least they tried not to, but the bosses prevailed. By 1900 that particular flap died down, and the worthiness of plywood began to become generally accepted.

Doors and Barrel-Top Trunks

Around 1890 plywood doors began to appear. At first, the plywood was employed only as door panels, primarily because of its strength and stability, which was far greater than comparable solid wood panels. Shortly afterward the stiles and rails of panel doors also were fashioned of plywood, for the same reasons of strength and stability. At about the same time, what was popularly called the slab door and is now widely known as the flush door came into being; it was the use of plywood that made this door design a practical possibility. Sleigh and carriage makers adopted plywood rather quickly because the material could be bent with relative ease. This was particularly important to the sleighmakers, because they could easily bend the plywood to form the familiar S-shape of the curved dashboard, and also the gently curving back area. The curved shape of the plywood could easily be maintained, and the body of the sleigh itself strengthened, by installing iron straps bent to the proper contours and securely fastened to the body.

The same principles were employed by the makers of the popular barrel-top trunks; in this instance, the plywood not only could be easily curved to shape the trunk top, but the trunk had far greater strength and resistance to crushing than a similar one made of solid wood. In short, the use of plywood offered the consumer a far better product than he had previously been able to obtain, and the idea was a complete commercial success. As the 20th century dawned, plywood and its burgeoning use was essentially a *fait accompli*. But as yet there was no plywood industry, as such. Nor had the term "plywood" yet been coined; the process was still called *veneering* and the products were called *veneered*.

THE AMERICAN PLYWOOD INDUSTRY

The door started the whole business. That ordinary, commonplace, pedestrian bit of building material was the impetus, the initial charge, for the explosion of the United States plywood industry that we enjoy today. That's not so strange when you stop to think about it. Just before the turn of the century (and in fact, for a few years afterwards) there really was no such thing as mass production on any appreciable scale. And yet, particularly in the case of doors, the basic ingredient for mass production was there—demand. In those days, virtually all rooms in buildings were built with doorways. The open-design concept, especially in housing, was not unknown but certainly was not in vogue. If somebody built a doorway, he or she automatically installed a door to go with it. Practically every new room built in the land needed one door at least, and frequently several. Population at that time was beginning to increase more rapidly than it had in years past. The demand for dwelling units, offices, commercial space and other things was rising swiftly as the nation expanded. A lot of folks needed a lot of doors, and plywood was the best material for doors.

Those earliest doors were, practically speaking, handmade. There was some machinery in use, but it was not particularly efficient, effective or time-saving, and certainly not capable of true mass production in the usual sense of the term. As has been shown in countless occasions in the past, where a demand exists, but the means to meet the demands does not, the means will soon be found. In this case, the product and the demand for it existed. The necessary manpower, both skilled and unskilled, was readily available and the raw materials were at hand. All that was lacking was special machinery, techniques through which the machinery could be employed, and places to set up shop.

As we have seen, the rotary veneer slicer was around at that time—it had been for more than 30 years—and was in regular use. Through the years, various improvements had been made and techniques worked out to the point where the machine was a fairly useful one. But a principal problem had not yet been conquered, that of properly handling the cut, green veneer and drying it so it could be used for manufacturing. In 1902, the Coe Veneer Dryer appeared in answer to this problem. The veneer sheets coming from the rotary lathe could now be run between driven rollers that automatically propelled the veneer along, to be force-dried and fed out of the dryer onto an offbearing table. Air-drying in racks placed in yards or lofts, a slow and cumbersome proposition that required plenty of elapsed time and a lot of handling (and cost), was eliminated. At about the same time, various methods for applying glue to the veneers and then pressing the sheets together for curing were developed; they were rudimentary, of course, but they worked. The fledging plywood industry had stepped to the edge of the nest. Only one good shove was needed for a launch.

Paine Lumber Company

The shove actually came in the form of two separate, but healthy, nudges. The first was when the *Paine Lumber Company* of Oshkosh, Wisconsin, found that the demand for plywood doors had grown to such an extent that it could no longer be ignored. Nor were the old methods of building these doors sufficient to get the job done. In 1904 Paine constructed a separate plywood manufacturing facility to make the panels for the doors they manufactured, much in the way that the Sewing Machine Cabinet Company produced plywood parts for Wheeler & Wilson's sewing machine cabinets. So while the idea of a plywood manufacturing facility was not a new one, the scale and scope of the operation was, and the potential possibility of such an operation did not escape notice.

Portland Manufacturing Company

The second nudge was in the guise of the *Lewis and Clark Exposition*, held in Portland, Oregon, in 1905. At that time there was a concern in that city, the *Portland Manufacturing Company*, which produced drums, crates and baskets for various purposes. The company was asked by the directors of the Exposition to make something unusual and different to display at the Exposition that would catch peoples' fancy, preferably something that the collective public eye had never really perceived before. The plant manager and one of his best men, a skilled lathe operator, did just that—they decided to manufacture some plywood panels for exhibit.

The task would be a large one, for the only equipment available was a big veneer lathe, an ordinary kiln dryer, and a crew of eight men. The glue used was an animal glue, or hide glue, which had to be kept hot and stank abominably. The men frequently had to step outdoors for a while to revive themselves. The glue was applied with paintbrushes. An impromptu press was cobbled up from odds and ends, and pressure was applied to the press quite unscientifically and unevenly with the aid of a number of large house jacks. The crew could only make one set of panels in a full day's worktime, and the panels had to set in the press overnight to cure. The process was long and laborious. But the chore was not without its rewards, for the plywood panel exhibit, the first of its kind seen anywhere, caught not only the collective eye of the general public. The exhibit also attracted considerable attention from several door manufacturers. The interest was soon translated into firm orders, and the Portland Manufacturing Company found itself in the business of manufacturing plywood for doors.

Door, Furniture, Automotive and Aeronautical Uses

Some of this plywood was purchased by Wheeler-Osgood, a company that quickly came to a two-part conclusion. They saw that there was a rapidly increasing demand for plywood, and that their having only one source of supply (Portland) was ridiculous. By 1909, Wheeler-Osgood had its own mill in operation. The idea had spread to California where the Long Bell Lumber Company established another plywood production operation. The first Canadian mill went into production in New Westminister, British Columbia, in 1913. All of these plants, as well as those that were established for the next few years, were essentially captive operations, merely part of a door manufacturing operation. That situation changed in 1919.

The *Elliot Bay Mill Company* built a separate plant in Seattle for the express purpose of manufacturing plywood for whatever purpose and for whomever had a need. The mill was complete, equipped with the latest machinery, and produced quality plywood panels with the latest and best techniques. It included warehousing and shipping facilities, the latter being an innovation in the industry. Previously, production of plywood was scheduled only to keep up with and properly meet the demands for door production of the parent company, in a more or less continuous and hand-in-hand operation.

This is not to say that plywood was used only for doors during that time; there were other uses as well. All of the items mentioned earlier had increased their usage of plywood, except for the carriage and sleigh making trades, but the automobile and truck industry took up much of the slack in that respect. The furniture manufacturing concerns consumed quantities of plywood, and the material began to find its way into trolley cars, railroad cars and buses, not to mention phonograph and radio cabinets. During World War I, there was great interest in plywood as it could be employed in airplanes. A good deal of research went on in aeronautical plywood applications, and the material was in fact used extensively in many different models and in numerous ways.

Development of Better Glues

Plywood of those days had a particular problem in that it unfortunately tended to come apart. The glues then in use simply could not hold up under many adverse conditions—they did not provide a good enough bond. The glues were of the casein-lime type, and the one most commonly in use was a formula devised by the U.S. Forest Products Laboratory. Blood-albumin glues also appeared then, in conjunction with the first tentative tries of hot-pressing the veneer sheets and glue layers together in an attempt to form a more effective bond. The Wheeler-Osgood mill, mentioned earlier, was successful in developing a new type of glue made from peanut meal that was better than the casein glue and replaced it for a short while. Then soybean glue was developed by I.F. Laucks, Inc., of Seattle, Washington. This glue provided a bond superior to anything yet developed, and was also inexpensive to make.

A New Name

Veneering had been a highly respected art for decades. Prior to the advent of what we now call plywood, it was generally associated with fine furniture and decorative craftmanship. But somehow or other, as cross-laid veneers began to be used for various purposes, the tern "veneering" generated some unfavorable connotations. You will recall the sarcastic comment about "pasted woods" that was commonly used by many craftsmen of the day. For no good or apparent reason, cross-laid veneer construction began to earn an undeserved adverse reputation that was serious enough to be detrimental to the growth of the industry. A great potential for the application of plywoods in many fields was certainly there, and that potential was increasing yearly. But this unfounded reputation preceded the material. There were few standards of manufacture. Evaluation of the material and its advantages was difficult, and many potential users simply backed away without making a thorough investigation.

During the World War I years, when plywood suddenly became important from an industrial and war effort standpoint, industry leaders determined that something must be done to reverse the situation and convert the public conception of veneered wood from a negative to a positive one. One possibility for doing this was to change the name of the product and to proceed as if this were a brand-new material, in conjunction with heavily promoting its advantages (which were many). Disassociation, at least in the public view, from the term "veneering" and all its connotations was deemed advisable. Meeting after meeting took place over a period of many months, and finally the word "plywood" was settled upon. This term had been around for some while, used only in scientific and technical journals, but seemed to be a natural choice in the final analysis. It was a fortuitous choice, because the name stuck in the public mind (with a bit of help from the manufacturers) and is now a household word.

Post World War I Changes

After World War I, the interest in plywood for airplanes began to taper off as the aluminum industry virtually took command of the field and the plywood companies focused their attention on other matters. Foremost among those concerns was further and marked improvement of production methods and techniques, since demand was once again outrunning supply. The increase was gradual but insistent. Also, the separation of interests between the softwood and the hardwood plywood interests was now becoming apparent. In 1921 the *Plywood Manufacturers Association* was established in Chicago, and in 1922 the *National Veneer and Plywood Association* was formed; the industry was beginning to organize itself. The automobile industry, which was enjoying a phenomenal heyday during the Roaring '20s, consumed incredible quantities (for those times) of plywood for seat bottoms, seat backs, trunks, running boards, floorboards, dashboards and numerous other odds and ends.

Railroad passenger cars and boxcars needed plywood. The big boys of the radio world—RCA, Zenith, Atwater—Kent—used plywood by the carloads for radio cabinetry. An increase was needed in plywood production, along with greater efficiency, lower costs and better production schedules.

At the same time, another problem arose in that the supply of prime, old-growth "peelers"— the logs from which the veneer was cut—was rapidly dwindling. This meant more manufacturing problems to overcome and also that smaller, lower-quality logs with various defects and imperfections had to be used in production. The latter situation occasioned even more problems with machinery and techniques.

Pacific Coast Plywood Manufacturers, Inc.

However, the industry rose to the challenge with continued success. In 1928 a new product was brought forth by the *Washington Veneer Company*—the now familiar ¼-inch wall paneling in standard 4-foot by 8-foot size. This panel, not only for wall paneling usage but also for a tremendous variety of other purposes, quickly became an industry standard, as manufactured in numerous grades and types. In this same year, an organization called the Pacific Coast Plywood Manufacturers, Inc., was formed and headquartered in Seattle. Its purpose primarily involved softwood plywood sales and promotion activities in a cooperative effort, and essentially marked the beginning of a period of industry-wide growth. In 1938 the hardwood plywood interests formed another organization in Cleveland, Ohio, called the Plywood Manufacturers Institute. By this time, the plywood industry was a tremendous force in the country. It had organized itself, established certain standards and codes not only in manufacture, but also in sales, promotion, applications and such, and had diversified geographically. At the same time, the industry had without conscious effort gathered about itself a substantial core of forward-looking, energetic, capable leaders.

When the *Pacific Coast Plywood Manufacturers* was established in 1928, the members formulated what was called the Rule, a relatively simple set of standards by which their products would be manufactured. In 1933, the Douglas Fir Plywood Association was formed, and the original Rule was developed into the first published commercial plywood manufacturing standard in that same year. When the Great Depression struck during the 1930s, the plywood industry was seriously affected, along with most American industries. The volume of plywood being manufactured now exceeded the demand. Many complaints were being fielded by the plywood companies, because of the adhesive used to make the plywood panels (primarily soybean glue) was not good enough to hold up under many of the more strenuous applications. During this decade, improved adhesives were top priority in research and experimentation departments.

Advances in the 1930s

In the early 1930s a *phenol-formaldehyde* resin dry-film glue was developed in Germany and subsequently introduced in this country. This was a waterproof glue, something desperately needed by the American plywood industry, but had to be applied with care and cured at about 300°F in a special hot-press. From a practical standpoint, hot-presses were hardly even in use in this country. However, the industry went to work on the problem, and by late 1935 this glue was readily available and hot-presses were in operation, despite substantial costs and other problems. Because the cost of producing these panels was higher than those made with soybean glue, a need arose for an adhesive that would be satisfactory in applications where a waterproof glue was not really needed, but something better than the soybean glue was required. This led to the development of *urea-formaldehyde* resin glues which were effective but allowed a certain grade of panels to be manufactured for less cost and sold for a lower price. Soon afterward, *melamine-formaldehyde* resin glues came along, and about 1944 *resorcinol-formaldehyde* resin glues were introduced, with additional beneficial effects for the plywood industry.

In the late 1930s, the availability of quality timber had become an even more serious problem. The supply in Washington had dwindled to a low level, so far as conveniently available resources were concerned. There was a shift to Oregon, where excellent timber was readily available. As the years went by, the shift continued, though Washington continued to be a center of activity. Oregon, California and western Canada all became very important to the industry's overall growth and production capabilities.

The advances made during the 1930s turned out to be of great importance to the plywood industry when the United States entered World War II. Suddenly there was a tremendous demand for plywood to be used in all types of applications. Exterior coverings for both aircraft and boats required huge quantities of the material. The American troop-gliders (CG-4 and CG-4a) were made of plywood, as was the famous Mosquito used by Canada and England, and the American Air Force *Fairchild AT-21* all-wood training plane, made using the *Duramold* plywood molding process. Rescue boats were made with molded plywood hulls, as were small Army attack boats, while landing craft were made of flat plywood stock. Many of the latter, in the smaller sizes, were transported in kit form and assembled at the battle site while the action was going on.

Plywood was also employed in the famed PT boats, in aircraft carrier decking and a variety of other marine applications. Of course, there was a host of additional applications for plywood, such as in communications equipment (cabinets, antenna masts, radar domes, etc.), surgical equipment (splints, medicine chests, surgical supply cabinets, etc.), structures (from tiny communication shelters to huge aircraft hangers), and various plywood sandwich constructions that could be used to quickly slap together general purpose fabrications from doors and partitions to floors in aircraft.

Southern Plywood Manufacturers Association

During World War II production facilities were increased, research and development efforts redoubled, and plywood as a viable construction and fabrication material really came into its own. At the conclusion of the war, the plywood industry was able to convert rapidly from wartime to peacetime needs. In the decade that followed, great technical and technological advances were achieved and growth of the industry was very rapid. The *Southern Plywood Manufacturers Association* came into being in 1946. This association was hardwood plywood oriented and had a vested interest in the development of southern forests and plywood facilities. Methods were devised so that lower-grade and smaller logs could be effectively used. Demand for the product continued to increase, and for the first time true mass production was a reality.

1950s and 1960s Developments

In the 1950s and 1960s, tremendous strides were made in the way of equipment, machinery, manufacturing techniques and the like. The southern pine forests could now be harvested and manufactured into plywood, something that previously had been impossible because of the small size of the timber. The Southern Plywood Manufacturers Association, which had merged with the Hardwood Plywood Institute (formerly the old Plywood Manufacturers Association of Chicago) in 1953, became the *Hardwood Plywood Manufacturers Association (HPMA)* in 1964. This consolidation made the HPMA the voice of the hardwood plywood interests, which it remains today. The housing industry demanded great quantities of plywood. When that demand slackened in the late 1950s, the industry promoted new applications in institutional and industrial areas, agricultural buildings, concrete forms and so on. Speciality products, like exterior sidings, panels overlaid with other materials and underlayments were developed. Specialists in the manufacture of certain types of plywood began to appear. The old Douglas Fir Plywood Association became the American Plywood Association in 1964, and greatly increased its scope and efforts in the development and promotion of softwood plywood products over the ensuing years.

Today's Industry

Today, the plywood industry is a huge one in a constant state of flux and evolution. New products, techniques, machinery and applications are turned out so rapidly that keeping up is a virtual impossibility, even for those directly involved in industry affairs. The demand for plywood in a host of different forms is incredible and increasing. Today, the only constant factor in the plywood industry is change. There are many serious challenges to come: tight supply of raw materials, rising costs, increasing regulation, continued essential research and development leading to new end uses, improvements in manufacturing techniques and conservation of raw materials. The list is seemingly endless but doubtless the plywood industry will cope, just as it has in the past.

Production of plywood back around the turn of the century was a mere few thousand feet here and there. By 1925, for instance, the production of Douglas fir plywood alone had risen to some 153,000,000 square feet on a ⅜-inch-thick basis. In 1943 there was a temporary peak of 1,800,000,000 square feet. The overall United States consumption of softwood plywood for various end uses in 1963 was some 10 billion square feet. Over the span of a mere decade, that nearly doubled to over 18 billion square feet. In 1972 over 5 billion square feet of prefinished hardwood plywood alone were shipped in this country. Today's consumption continues to expand and shows little sign of diminishing in the near future, provided that the supply can keep up with the demand.

HOW PLYWOOD IS MADE

The manufacture of plywood begins in the forests. Back in the early days, those were "natural" forests—unmanaged and wild, with no hint of silviculture imposed upon them by the hand of man—comprised of virgin and some second-growth trees. Loggers established their camps, sharpened their saws and axes, and set about bringing out of the forest some logs of high quality. Trunks 3 feet thick were everywhere, 4-foot and 5-foot trunks were commonplace. Once the undercut "scarf" was cut from an average fair-sized Douglas fir, the cutter could comfortably lie down in the cut and take his ease in the shade created thereby. There were enough 8-footers and 10-footers (thick, that is) to keep crews busy for months on end. One fir in particular that was recorded on film measured 417 feet high, 25 feet in diameter at the butt, and 9 feet in diameter 207 feet above the ground; the bark was 16 inches thick at the butt end of the log. This tree was by no means of record size. Remember, all of these trees were felled and bucked (sawed into lengths) by sturdy individuals armed only with one-man and two-man saws 8 to 12 feet long, double-bitted axes, wedges and mauls, and the inevitable bottles of kerosene (for cleaning pitch from the sawblades).

By the time the plywood industry was beginning to show definite signs of life, not long after the turn of the century, much of the big timber had been cut, at least that which was easily accessible. Not many years went by before big, prime timber was in short supply. Smaller, less desirable logs were being substituted on a regular basis. That situation has compounded itself over the years. Of course, there is no such thing today as big, virgin timber, if one discounts certain federally protected stands like the redwoods in the Redwood National Park. Though some timber does occasionally come from lands that have not been previously cut over both for plywood purposes and for other applications as well, most of the timber taken for plywood production these days comes from *tree farms*.

A forest could be considered as a large expanse or stand of trees growing wild—a natural phenomenon. Once that forest is cut over or laid waste by fire, many years go by before regeneration is complete. The second forest growth is likely to be considerably different than the first. But a tree is just a plant, though a pretty big one, and the raising of trees can be fundamentally approached in the same manner as raising corn or raspberry canes. Under the proper circumstances, trees can become a crop, providing a certain yield on a constantly rotating yearly basis.

Furthermore, those trees can be planted, the stands maintained, the growth directed, and the harvesting done specifically with an eye toward commercial usage. Trees are a renewable resource. When properly farmed and utilized they can be raised specifically to fulfill our various needs.

Tree Farming

Modern tree farming incorporates all the latest scientific technology, advances and techniques. Soil is analyzed. Forest pathology is studied. Tree genetics and physiology play an important part. Soil analysis is essential. A tremendous list of important factors goes into the raising of trees for commercial purposes, just as it does for wheat, milo, sugar beets or roses. As little as possible is left to chance.

After mature trees are harvested, new ones are planted to take their place. These will likely be specialized breeds that have been painstakingly developed to exhibit certain characteristics and properties that will be advantageous in light of the particular anticipated end use of the tree as a commercial product, as well as rapid growth, straighter trunks, greater heights and similar traits. After the seedlings are planted, they are not just left to fend for themselves, but are carefully nurtured and brought along under constant surveillance. They are protected from wildlife and insects, forest fires and other natural hazards insofar as is possible. The new trees are thinned periodically, and weak or damaged ones are removed. This opens up additional space where the remaining trees can develop fully. The trees that are removed are immediately put to some use, even if only for chips needed for paper pulp, instead of being left to rot away as would happen in a natural forest. If necessary, the growing strands of young trees are fertilized and perhaps periodically sprayed with insecticides.

The species that are planted are suitable to the prevailing soil conditions, climatic conditions and other natural aspects of the area. When harvesting time comes, only those methods which are best suited to the topography and terrain, as well as to the species itself, are employed. This can range anywhere from clear-cutting blocks of timber with huge, specialized machines to hand-cutting particular individual trees and snaking the logs out of the stand with a twitch horse—the four-legged, hay-burning kind, not a mechanical monster with a cute nickname.

Demand for Unusual Grain Patterns and Figures

The harvesting process results in logs of various diameters ranging from as little as 5 inches or so to perhaps 2 feet or a bit more in diameter. Under modern forestry management, the remaining parts of the trees—the top and limbs that were formerly left as "slash"—are largely recovered and utilized for one purpose or another. The logs themselves are the useful part of the tree as far as plywood production is concerned (and lumber, too, of course). The straighter and cleaner-grained these logs are and the fewer defects they have, the better they are for most plywood manufacturing processes.

However, we must note here that such is not always the case. There is a great demand in the hardwood plywood industry for unusual *grain* patterns and *figures*, those that are often produced as random freaks of nature through uncontrolled growth, or those that may be found in parts of a tree that cannot be used in other forms of manufacture.

For instance, the large, misshapen bulges or knobs that sometimes protrude from a tree, called *burls*, cannot be used in lumber production. But they contain weird and wonderful grain patterns that when properly cut as veneers are incredibly handsome. They are invariably unique; no two are ever alike. It is this property that adds distinctive flair and beauty to hardwood panels. Similarly, the stumps of trees whose rather ordinary trunks have gone into some form of standard lumber or plywood manufacture are often uprooted and sliced for the same reason—unusual grain and figure patterns form where the bole of the tree expands and branches out to form the root system. Deformed tree trunks—trees that are twisted, gnarled or lightning-struck, or have suffered some sort of damage that has caused an abnormal growth pattern—are sometimes sought out to provide unusual veneer panels.

By comparison, of course, the quantity of wood that is used for these purposes is very small beside the incredible number of square feet of veneer that is produced for ordinary plywood panels. But this is a distinctive and important part of the industry, and the source of those eye-catching and unusual hardwood plywood panels that find their way into fine furniture and superb wall coverings. Unlike the logs that are used for most plywood production, which are handled in bulk by the thousands of tons and the millions of board feet, these special pieces are handled virtually on an individual basis. The time and effort involved, of course, is in part responsible for the high costs often associated with the end products made from them.

But it is not just the odd-shaped or strange-grained pieces that are in great demand. Many hardwood species that have relatively regular but handsome grain and figure patterns are in very short supply. Prime individual trees of those species, when they become available, command intense attention and spirited bidding from various commercial concerns. One interesting case involves a particular American black walnut tree.

The Black Walnut Tree

This tree took root naturally and began to grow just a few feet from the banks of the St. Joseph River in northwestern Ohio, a short distance from the new homestead of one James Hays, just after the turn of the 19th century. By 1900 or so, the tree was locally famous as a marvelous source for walnuts. It grew straight and strong and became acknowledged, finally, as one of the finest walnut trees standing anywhere. The Hays family continued to live on the family homestead, and in 1939 Lloyd Hays received his first offer for the tree—$300 for the lumber it would make. Many more offers came in subsequently over the years, and all were refused. Lloyd Hays had more regard for that majestic tree than he did for the money it represented. But finally he passed away, and the only feasible way for the heirs to retain possession of the family land and still settle the estate was to sell what had become known as "the Hays tree," along with several other merchantable trees. The timber was put up for bid, and of the $80,000.01 paid by *Atlantic Veneer Corporation* of Beaufort, North Carolina, $30,000 was attributable to the Hays tree alone.

The tree was taken down with careful, lengthy planning and extreme care, so as little damage as possible would result. The trunk measured 57 feet from the cut to the first limb and was over 38 inches in diameter at breast height—a perfect specimen.

After the tree reached the veneer plant, numerous knowledgeable persons examined, pondered, measured and calculated until they were satisfied as to just how the log could be made to yield up the greatest possible quantity of absolutely top-quality veneer. The log was cut into five pieces. The peeler blocks were processed into enough very thin and absolutely beautiful veneer to almost cover three full acres of ground.

The cost in dollars was enormous for one log, and the cost in time, effort and care was astonishing, especially by comparison to the normal procedures used in lopping down and slicing up ordinary softwood logs. But the venture was commercially successful because of the scarcity of such material. It was also successful in that the famous walnut tree, which had reached its peak and would shortly have gone into decline anyway, was put to use in the best possible fashion. Eventually the veneer sheets were shipped to Germany, where they were transformed into handsome furniture of lasting quality.

Debarking Logs

After harvesting, in due course the logs reach the mill where they are first separated by wood species (Fig. 1-1). At the same time, they may be sorted by length and perhaps also by rough grade quality. Then they are stacked to await the production process (Fig. 1-2). As logs are needed they are taken from the piles, debarked and cleaned down to the bare wood surface beneath. This job is done automatically, of course, by running the log against knurled wheels (Fig. 1-3), by forcing the bark off with a high-pressure water jet, or by peeling the bark off with knives. The bark, incidentally, is not wasted, but is packaged for use as *garden mulch* or may be burned as fuel to provide power for the plant itself. For instance, when the fuel crunch occurred in 1974, of the 192 plywood mills then opperating, 71 were self-sufficient in creating their own power and electricity by burning bark and other wood waste products.

The debarked logs, or *peelers* as they are called, are next transported into the mill (Fig. 1-4) and cut into lengths of between approximately 8 feet 4 inches to 8 feet 6 inches by passing them under a guillotine-type cutoff saw (Fig. 1-5). This particular length allows the resulting ribbon of veneer to be cut to standard 8-foot-wide sections with enough left over for edge trimming. These short logs are called *peeler blocks*, even though

Fig. 1-1. These redwood logs are being loaded for their journey to the mill (courtesy of the California Redwood Association).

they are still round and are by no means block-shaped. Then the blocks are *conditioned*.

Heating Peeler Blocks

Early on it was discovered that a block of hot wood cuts much better than a cold block. An analogy can be made to butter, which slices very easily when it is warm but is difficult to cut when very cold. The peeler blocks must be heated by one means

Fig. 1-2. At the mill, logs are separated and stacked according to species and sometimes length and quality as well (courtesy of the American Plywood Association).

Fig. 1-3. The first step in the manufacturing process is to remove the bark from the logs, here being done by knurled chipping wheels (courtesy of the American Plywood Association).

or another, clear through to the center and with relative uniformity, in order to facilitate the second phase of production. Since wood is a natural thermal insulator, dry-heating the peeler blocks is virtually an impossibility; they must be wet. A higher yield can be gained from a properly heated block because

Fig. 1-4. After debarking, the logs travel to the saws (courtesy of the California Redwood Association).

Fig. 1-5. Here a log is at the cutoff saw, being lopped into short lengths called peeler blocks (courtesy of the American Plywood Association).

veneer-weakening stresses are greatly lessened, compression tearing is minimized, and tension tears are reduced. There is also less breakage and splitting of hot-peeled veneer.

There are numerous methods and techniques used for heating peeler blocks. There is a whole range of possibilities and combinations running from simple immersion in hot water to application of high-temperature steam under pressure. Combination methods are most often used today, because live-steam heating is an uneven process. Hot-water immersion in tubs has safety and cost disadvantages, as well as little flexibility where several different kinds and sizes of peeler blocks are processed. In any event, the peeler blocks are preheated and then heated to whatever temperature is most advantageous for peeling that particular species. The conditioning times are variable, depending upon the method used, the species of wood and other factors. Conditioning may take place for only 12 hours or less, or for more than 40 hours. For example, a particular steam vat operation might involve: load 30,000 board feet of pine peeler blocks, 1.5 hours; heat blocks with steam, 6 hours; allow blocks to steep, 2.5 hours; unload blocks from vat, 2 hours. Thus, one production batch in this particular instance is cycled through the heat process every 12 hours.

Peeling Lathes

As soon as the peeler blocks emerge from the conditioning process, they are transported to the mill. This may be done by some sort of conveyor belt system, or the blocks may simply be picked up with the forks of a tractor-type loader and driven to the peeling lathes. Various methods and types of machinery are used here, but basically each peeler block is picked up by a charger, geometrically centered, and shifted into position to be chucked into the lathe (Fig. 1-6). There are many variations in lathes and the way they operate, too. Once the peeler block is properly secured and the cutting knife in the lathe comes into position, the manufacture of veneer begins. The lathe chucks spin the log (at speeds that are widely variable, depending upon numerous factors) against a full-length knife that peels the wood away from the block in a continuous strip (Fig. 1-7). This process is so rapid that in many instances as much as 600 linear feet or more of veneer may issue from the lathe *per minute*. For softwood veneers, the thickness of this ribbon of wood may range anywhere from 1/10-inch to as much as ¼-inch. Hardwood veneers are frequently much thinner, as little as 5/1000-inch (0.005). The peeler block is quickly reduced to what is, comparatively speaking, little more than a thin spindle of wood.

The diameter of this *core*, as it is called, varies—typically 3 or 4 inches—and no more veneer can be peeled from the log after a certain point because of the flexibility and springiness of the core. The knife blade simply can no longer make a proper cut. The core is automatically released and drops through the lathe bed onto a conveyor. A new peeler block is quickly swung

Fig. 1-6. A lathe spotter aligns the peeler block, and the charger moves the block into the lathe chucks for peeling (courtesy of the American Plywood Association).

Fig. 1-7. Peeler block revolves against the lathe knife, which peels off a continuous ribbon of veneer (courtesy of the American Plywood Association).

into place and the lathe is recharged, while the core goes on its way to a chipper or to be used for some other purpose.

Three Lathe-to-Clipper Process

Meantime, the stream of veneer is continuously conveyed away from the lathe and into any one of three different handling processes. The veneer may go directly into a machine called a *green clipper* (Fig. 1-8), which cuts the veneer ribbon into usable

Fig. 1-8. Veneer moves to the clipper, where it is cut into pieces of various sizes (courtesy of the Hardwood Plywood Manufacturers Association).

widths of as much as 54 inches. At the same time the green clipper slices out sections that have defects which cannot be repaired. This is called a *direct-coupled* or *close-coupled* method, and is often used with peeler blocks of small diameter. Another possibility is the *tray storage* system, where the veneer is sliced only in long lengths, typically 120 feet, and stored in trays until it can be conveyed and fed through the green clipper. This system may be used with large peeler blocks or with high-defect logs.

The continuous flow of veneer can be run onto reels, where the veneer ribbon is wound up into a huge roll, much like a roll of paper. The filled reels are then sequentially stored until they can be sent along to the green clipper. This system is often used with hardwoods, which are generally peeled thinner and at lower speeds than softwoods. The method offers the advantage to some production lines of allowing only one or two peeling shifts to produce a sufficient backlog of veneer to keep the green clippers operating around the clock. There are many variables and variations in these lathe-to-clipper processes, of course, and each individual plywood manufacturing plant operates under whatever system and with whatever techniques are best suited to its particular raw materials and end products.

Slicing Veneer

As was mentioned earlier, not all veneers are peeled on a rotary lathe; some are sliced instead. This process is used in particular for face veneers where the angles of the cuts to be made depend upon the grain and growth of the wood. The object is to cut the veneer in such a way that the most attractive grain and figure—the greatest or most emphatic beauty—of the wood will be revealed. Appearance is the key factor. The process is frequently used with walnut, oak, mahogany and particularly with a great many expensive and valuable imported woods like *rosewood, zebrawood, paldoa or wenge,* to name just a few. Veneer that has been sliced is, of course, cut flat so that it has the advantage of not having a tendency to curl as does rotary-cut veneer.

The preparation of the raw material for slicing is different than that used for peeling. Here, the log, or the burl, crotch or whatever, is first opened up so that decisions can be made on exactly which angle-cuts will produce the finest figure. The piece may be cut into segments or shapes and trimmed in various ways. The chunks that remain are called *flitches*. Once cut, the stack of veneer sheets that represents the original flitch is called a *flitch of veneer*. The flitches are then *tenderized* by cooking or steaming them, for the same reason that peeler blocks are. In this case the process is even more important, so that the veneer sheets can be cut cleanly from the flitch and the irregular grain will not lead to tearing or other damage. What little bark remains on the flitches after sawing from the original piece comes off during the cooking, or is removed manually afterwards.

The prepared flitch is then mounted on the bed of a *veneer slicing machine* (Fig. 1-9), which is adjusted to make the necessary thickness of cut and set up in a way to best handle the particular flitch. In operation, the bed moves back and forth or up and down, driving the flitch against the blade, which shears off a sheet of veneer. The veneer sheets are taken from the machine in sequence and carefully kept that way so that proper matching can be done when the time comes.

Sorting and Stacking Methods

Once the green veneer goes through the clipper, something must be done at once with the resulting sheets and strips. This is a tremendous job because the sorting and stacking must be done at a relentless pace. The job may be done manually, with various mechanical assists, or by automatic machinery, depending upon which process is best adapted to a particular manufacturer's production line and product. In a manual system, for

Fig. 1-9. Rather than being peeled on a rotary lathe, hardwoods are often sliced, as shown here (courtesy of the Hardwood Plywood Manufacturers Association).

instance, a man must watch the flow of veneer sheets coming from the clipper, sort them into various categories as they come along, and stack the sheets carefully on waiting carts. The loads must be well-stacked and squared on edges and sides for proper further handling and storage. Pulling the pieces from the moving conveyor is in itself a tricky task.

A run of veneer coming through the clipper might be handled by four men, for instance, each responsible for four individual categories of separation. Those categories might consist, for example, of heartwood random sheets, heartwood half-sheets, sapwood full sheets, heartwood full sheets and so on. Mechanical assists may aid in the back-breaking labor of slinging heavy wet sheets by substituting *pinch rolls* and short *feed lines*, *mechanical stackers* and the like. Automated stacking can be done by mechanical or vacuum processes. On the other hand, one man can handle the sheets being sliced from a flitch, especially a small one; he just drops the pieces sequentially on a cart. The systems used in any given production line depend primarily upon two factors: labor costs and potential veneer yield considerations.

The Drying Process

The next stop along the production line for both rotary-cut and sliced veneers is a critical one. The veneer sheets must be properly dried to reduce moisture content to the order of about 5% before they can be worked further. There are three general methods used, each comprised of a variety of techniques, handling methods and drying procedures. The simplest method, of course, is simply to stack the veneer outdoors in good weather so that air can circulate freely over all the surfaces. Obviously, this method is time-consuming, requires much handling, is inefficient and also chancy. The second method is *kiln* drying, where the veneer sheets are piled and then stacked in a conditioning room. Much better control of the drying process can be gained with this method. The third method makes use of *mechanical veneer dryers*, whereby the veneer sheets are conveyed continuously through a superheated drying chamber (Fig. 1-10).

There are many problems involved with the drying process, well outside the scope of this book. The moisture content of the veneer must be nicely calculated. Overdrying is a possibility. Underdrying with subsequent redrying is sometimes a problem.

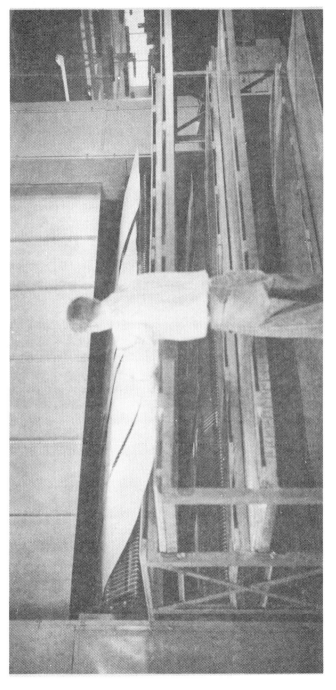

Fig. 1-10. After the veneer sheets are cut, they must be dried; here a mechanical dryer is used for the purpose (courtesy of the Hardwood Plywood Manufacturers Association).

Shrinkage must be controlled. Dryers tend to break down and sometimes catch on fire. Emissions from the veneer dryers must be controlled. The list goes on and on. But tricky or not, the job has to be done and is frequently the major bottleneck in a mill's production line.

At this point in the overall process, stacks of dry, untrimmed veneer eventually emerge from the dryers, representing only about half of the total raw materials that first appeared in the yard of the mill. The quantity varies, but only about 45 to 58% of the original green log is eventually realized at the dry, untrimmed veneer stacks. All the rest is lost, though a portion of the waste can be turned to other purposes. The core, of course, cannot be used, and a substantial 18 to 26% or so is lost at the clipper. There is some shrinkage, about 3%, and some veneer is below-grade and of no value. The total amount of loss, of course, depends upon the wood species as well as the specific manufacturing methods and their relative efficiency.

Layup Process

Once out of the dryer, the veneer sheets can be prepared for assembling into plywood panels. Very few of the sheets of veneer that reach the *layup* department nowadays are perfect. The quality of timber required for manufacturing quantities of virtually perfect veneer no longer exists to any great degree, at least from a mass-production standpoint. Of course, certain veneers are chosen and handled with great care, especially among the valuable hardwoods and fine-figure specimens, and well-nigh perfect sheets do appear. Obviously a great deal more attention is paid to the selection of virtually perfect sheets to be applied to the faces, and sometimes the backs as well, of plywood panels whose principal quality lies in appearance, than to veneers that will become interior plies or faces and backs for structural plywood. But the net result is that for one reason or another a large percentage of the veneer must be repaired in one way or another (Fig. 1-11). Some deflects like *wormholes, knots and pinholes* are removed by cutting out the bad piece and filling the holes with patches. Splits and gaps can also be repaired. Frequently a series of smaller sheets must be spliced together by taping them into one larger sheet (Fig. 1-12).

After repairs are made and full sheets are made up as necessary, the veneer sheets move on into the layup process. The veneers are separated according to the needs of the particu-

Fig. 1-11. Small defects in the veneer sheets can be repaired by diecutting to remove the defects and then patching (courtesy of the American Plywood Association).

lar kinds of plywood panels being assembled, in the way of core stock, cross-bands, face and back panels. Further sorting is done as well, depending upon the panel requirements. A *spreader crew*, surrounded by the necessary stacks of veneers and the various sorts of equipment and machinery necessary to lay up the panels, goes to work (Fig. 1-13). The *core feeder* moves a piece of stock to the *core layer*, and the *sheet turners* work in concert with the core layer to glue and lay up the plies. This job is not necessarily done entirely by hand, but may be mechanically assisted by devices such as *sheet carriers* and *whirlybirds*

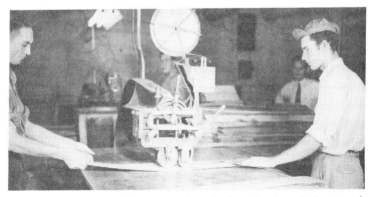

Fig. 1-12. Small pieces of veneer can be spliced together to make larger and more useful ones. Here the job is being done with a splicing machine (courtesy of the Hardwood Plywood Manufacturers Association).

Fig. 1-13. Sheets of veneer are being laid up to form softwood plywood panels (courtesy of the American Plywood Association).

that pick up and set the sheets. In some cases, the entire process may be automated to a relatively high degree—machinery sets a veneer back sheet, which moves under a glue applicator, another layer of veneer is dropped on, the two pass through another glue applicator and so on.

The Hot Press

From the layup process the assembled panels, which are merely slapped together with their edges more or less aligned but still in the rough, are moved to a hot press. Here the assemblies—they are not yet panels—are slipped into the slots in the press (Fig. 1-14). Again, there are many variables involved, but by and large the heat range of the press runs from about 230°F to 315°F. The pressure applied to the veneer assemblies runs from 175 to 200 pounds per square inch. Care must be used to cure the glue properly, to not compress the assemblies too much, and to maintain the thickness tolerances specified for that particular type of panel. Wood becomes more pliable as pressure and heat increase, so the amount of pressure, the temperature and the length of time the assemblies are left in the press all work together to determine the amount of panel compression. Of course, a certain length of time is required for the glue to cure.

Trimming and Sanding Panels

When the veneer assemblies are removed from the presses, they have become panels, but they are not yet ready to be put into service. These are the rough plywood panels, which now travel along the line to undergo further processes. The sheets must be trimmed to proper size, with all dimensions being held within certain tolerances. Certain types of sanding may also be required, depending upon the end product. This is taken care of as the panels are automatically fed into huge drum or widebelt sanders. For certain grades of plywood only touch-sanding is required, rather than sanding the entire face and/or back of the panel. Some plywoods intended for application as exterior siding must have grooves cut lengthwise of the panels—303 Texture 1-11, for example. These panels are automatically fed through special saws that cut the grooves. Other kinds of siding panels may have surface textures or patterns cut in them with roughening saws, brushes or other equipment. Prefinishing may be required, such as a clear finish on hardwood wall paneling, or a paint or stain on exterior siding material. All of these various chores are performed largely by special machinery devised for the purpose.

Inspecting and Grading Panels

The panels that have been trimmed to size must be inspected. Defects in the faces or backs of the panels can frequently be repaired so that a given panel may be upgraded. There are certain parameters that govern the type and extent of repair vis-a-vis the final grade. Some of the defects which appear at this stage are natural ones like pitch pockets, burls and splits.

Fig. 1-14. Glued-up veneer assemblies are placed in a hot press, where heat and pressure will bond the veneers into plywood (courtesy of the Hardwood Plywood Manufacturers Association).

Many, however, result from the manufacturing process itself—gaps at joints, indentations, scrapes or chipped-out places, broken corners, defective plugs and so on. Panels can be patched by puttying small defects and splits, and by sawing out larger defects and filling with wood shims. Large defects are routed out, and the holes are filled with wood patches. Synthetic patching with epoxy is frequently employed to fill knotholes and other defects, or a polyurethane substance may be foamed directly into a void and heated to cure it. All of this work is done manually. Amazingly enough, one experienced hand can make all necessary repairs to one 4-foot by 8-foot panel face in a minute or less—60 to 85 panels per hour.

The repaired panels, as well as those needing no repair, are then further inspected and graded. The grading system for softwood plywoods is entirely different than that used by the hardwood plywood industry, and we shall investigate both in considerable depth in the next chapter. Once the grade is established, the panels are stamped. Then they are stacked in standard bundles, strapped together, and crated or wrapped if and as necessary. They are warehoused temporarily on the mill premises (Fig. 1-15) and moved to waiting trucks and railroad cars as required, to be shipped off to building supply wholesalers and distributors, local warehouses and manufacturers who require various kinds of plywood for their own operations.

Stay Log Cutting

We must note here that the foregoing is only a rather general discussion of the way in which plywood is made. There are a great many variations in the manufacturing process, which depend upon many different factors. There are numerous specialty plywood products that require different techniques and machinery. Of course, there are hundreds of end products, all of which require their own special remanufacturing processes.

For instance, besides the two methods of veneer cutting that were previously discussed, rotary lathes and slicing, there is a third method that is sometimes used. This is called *stay log cutting*. In this process, a flitch of wood is secured to either a metal beam or a specially shaped, large chunk of wood called a *stay log*. The whole affair is mounted off-center in a lathe, and the rotary cutting produces a half-round veneer cut that is most often used as a decorative face on hardwood panels.

Fig. 1-15. After trimming, sanding, grading or whatever other operations may be necessary, the plywood panels are bundled and warehoused for distribution (courtesy of the American Plywood Association).

Cold Press Method

To take another example, not all plywood panels are made up in hot presses. Instead, some veneers are bonded together by the *cold press* method, which uses only pressure and no heat. This particular process is often used for making thick plywood or specialty panels where stability and strength of the material is paramount. Also, frequently when curved plywood panels are being manufactured, neither a cold nor a hot press is used. The job is done in a high-frequency press, where the glue is cured by means of high-frequency radiation.

Specialty Manufacturing

The specialty manufacturers, of which there are many, require some manufacturing processes that are yet again different. Various hardwood and softwood veneers may be produced in many stock sizes ranging from 1/100 inch to 3/16 inch thick. Plywood business cards may be made up of two pieces of sliced veneer that are each 0.005 inches thick, bonded together to make a card 10/1000 inch thick, complete with imprinting of the buyer's choice. Very thick stock, up to 5 inches or more, is produced for such uses as airplane propeller blocks, ski blanks, decks for pinball machines, stock for canoe paddles and a host of other items. For other purposes, such as *hardwood plywoc platens*, *mold stock*, *die boards*, *caul boards* and the like, there

Fig. 1-15. After trimming, sanding, grading or whatever other operations may be necessary, the plywood panels are bundled and warehoused for distribution (courtesy of the American Plywood Association).

Cold Press Method

To take another example, not all plywood panels are made up in hot presses. Instead, some veneers are bonded together by the *cold press* method, which uses only pressure and no heat. This particular process is often used for making thick plywood or specialty panels where stability and strength of the material is paramount. Also, frequently when curved plywood panels are being manufactured, neither a cold nor a hot press is used. The job is done in a high-frequency press, where the glue is cured by means of high-frequency radiation.

Specialty Manufacturing

The specialty manufacturers, of which there are many, require some manufacturing processes that are yet again different. Various hardwood and softwood veneers may be produced in many stock sizes ranging from 1/100 inch to 3/16 inch thick. Plywood business cards may be made up of two pieces of sliced veneer that are each 0.005 inches thick, bonded together to make a card 10/1000 inch thick, complete with imprinting of the buyer's choice. Very thick stock, up to 5 inches or more, is produced for such uses as airplane propeller blocks, ski blanks, decks for pinball machines, stock for canoe paddles and a host of other items. For other purposes, such as *hardwood plywood platens*, *mold stock*, *die boards*, *caul boards* and the like, there is a tremendous range of thicknesses and standard sheet sizes. Plywoods can be special-ordered to fulfill just about any end-product manufacturing demand. To meet those demands, frequently special manufacturing processes must be set up or occasionally devised from scratch to transform the raw material, logs, into the required plywood pieces or panels.

THE ADVANTAGES OF PLYWOOD

It is perfectly obvious, after considering all of the complexities involved in manufacturing plywood, that the whole affair is a rather expensive process. The price of the final product must be substantially higher than ordinary wood stock—boards, planks or whatever. While the retail price of nominal 1-inch boards, say, fir planed only on three sides, might run to 50 cents or 60 cents per board foot, a ¾-inch plywood panel might well cost $1 or more per square foot. In this instance, board feet and square feet are equivalent in terms of amount of material. Therefore, in order to justify the difference in cost, plywood must have something more going for it than ordinary, plain wood. And it does, in a great many different respects.

Distributive Strength

One of the most important advantages of plywood is its characteristic of *distributive strength*. Consider first an ordinary board, or a plank, with the grain running lengthwise. If the piece is reasonably straight-grained, splitting the piece lengthwise is not much of a chore. The split will follow right along the grain line. Some kinds of wood, like *cedar*, will split with remarkable ease, especially when dry. Other kinds, such as *elm* or *cottonwood*, are much more difficult to split, but nonetheless the job can be done. But if you try to split that board or plank crosswise, across the grain, you will be unsuccessful. The only recourse is to saw it. Likewise, if a good deal of pressure is placed upon a board along the line of the grain, it will readily break in half lengthwise, or a piece will shear off. If the same amount of pressure is applied across the grain of the board, only a relatively minor amount of deflection will take place. A great deal more pressure will be required to break the board in half, across the grain. Thus, while most woods are very strong across the grain, they are quite weak along the grain line.

If you take two pieces of board of the same size, lay one on top of the other with the grain lines running at right angles to one another, and fasten them together, you will be unable to split the piece in either direction. By crossing the grains, you have distributed the strength of the material so that it is roughly equal in all directions. This is the principle of distributive strength in plywood panels. By alternating the grain directions of the successive plies, the strength characteristics can be directionally balanced.

Furthermore, this balancing process can be carried out to whatever degree is necessary for the particular end use of the plywood. For instance, if one were to make up a plywood panel 4 feet square, it would be possible to make the panel with equal strength in all directions by using an even number of plies made up of veneers with identical characteristics. The panel could be made stronger in one direction than in the other by using a center ply at right angles to back and face plies of equal properties. The center ply might be made of a different kind of veneer, one with greater strength than either the back or face plies, or with greater thickness, to make a 3-ply panel of strength equal in all directions.

As you can see, by varying grain directions, veneer types and thicknesses and by changing grades and qualities, the strength characteristics can be balanced in whatever fashion is desired. The bottom line, however, is simply that pound for pound and square foot for square foot of coverage, plywood is far stronger than plain wood where the thicknesses are identical. In fact, plywood is likely to be considerably stronger than plain wood, even when thinner. That strength is more or less evenly distributed in all directions, depending upon the specific plywood panel in question.

Stiffness

Another important factor is *stiffness*. There is a wide variation in the stiffness of different species of woods when cut into boards or planks. In fact, some lumber can be stiffer than plywood. On the other hand, the reverse is also true. If stiffness is a requirement for a particular job application or a product end-use, plywood is the answer. Unlike boards, the stiffness in plywood can be equally distributed, and the degree of stiffness can be "built in" as required. Furthermore, that degree of stiffness will be very much uniform in a given production run of plywood, while similar uniformity is unlikely to occur in boards or lumber. Note, too, that often less thickness is required in plywood than in boards for an identical stiffness.

Added Strength

Because plywood is a manufactured product that is made up from several layers of raw material (the veneer), there is opportunity to add strength to the material that would otherwise be impossible from a practical standpoint. If you were to secure two pieces of ¼-inch wood stock together with nails or screws, with the grains running at right angles to one another, you would have a stronger piece of wood stock than a ½-inch piece of the same size and wood species. But the mechanical bond provided by the screws or nails is not an especially good one under many uses, and would not perform well over a long period of time. However, if you were to glue the two pieces together, with a complete bond that is stronger than the wood itself, the resulting piece would be much stronger than the one with screws or nails, and far stronger than the single piece of solid stock. The glue itself has added a substantial amount of strength.

This is exactly what happens in plywood, on a large scale. The thin layer of glue between each sheet of veneer adds a considerable amount of strength to the panel that could not be achieved in any other way. The only way that such a process can be practically carried out is in a large-scale manufacturing process. Thus, the addition of a small amount of glue adds a substantial amount of strength to the finished product, and the whole becomes greater than the sum of the parts.

Split Resistance

Earlier, in discussing the characteristic of distributive strength, we noted that while plain wood stock can be easily split, two pieces of wood stock placed at right angles to one another cannot be split, but must be sawn. This is another important characteristic, and a great advantage, of plywood. Splitting a plywood panel with an axe is a virtual impossibility; after repeated blows, the panel will finally give way and shatter, but it will not split. By the same token, plywood panels will not split while they are in service, from either natural or mechanical causes (unless extreme). In some kind of plywood, minor splits or checks may occur in face or back plies, but they are of little consequence and do not impair the overall integrity of the panel. Usually such occurrences have to do with weathering of unprotected plywood panels or perhaps improper application leading to unwarranted damage. Cracking, splitting and checking, while they may from time to time show up (if top-quality panels are used in the proper application, they seldom do), are greatly reduced and in some cases negated completely where plywood is used instead of plain wood.

But this is only part of the story. Because plywood is so resistant to splitting, it will accept fasteners with greater ease and has less chance of splitting. At the same time it provides more holding power than plain boards. For instance, if you drive a 10d common nail through a dry pine board about an inch from the end, the chances are reasonably good that the board will split. If it does not split immediately, it likely will at some later date. Of course, if the board happens to be a dry hardwood, the chances are excellent that the wood will split even if you drive the nail 2 inches back from the end, depending upon the species. Such is not the case with plywood. You could drive that 10d nail through the plywood as close as ½ inch from the edge at any point, and no split is likely to appear now or later. If you try to pry up a board that has been nailed down at the end, the chances are good that it will split as you are trying to lever it upwards. Plywood will not. A pinch bar will force both the plywood and the nail upward sufficiently in order to pull the nail, with little or no damage to the plywood itself.

Warp Resistance

As you know, raw wood has a definite propensity for warping. Boards can *crook* (the board looks like a rocker when stood on edge), and they frequently *bow*, or curl from end to end. Many boards, and even heavy beams, will *twist* about their lengthwise axis. There are few boards indeed, or planks either, that do not exhibit some degree of *cupping*, where the long edges tend to curl upward toward one another. Once this same wood is shaped into the form of plywood panels, warpage of whatever type is not longer of much concern. Because of the manner in which the veneer sheets are cut, the extent and the type of warping that takes place exhibits somewhat different characteristics than with ordinary boards or planks. There is indeed considerable warpage in the dried veneer sheets, but when they are glued and pressed the warpage disappears much as do the wrinkles from a shirt after ironing.

Furthermore, because the veneer sheets are solidly glued together with the grain lines in opposing directions, the possibilities for further warping are virtually eliminated. Warpage in a plywood panel is usually not a factor to be reckoned with. One frequently sees a plywood panel that is somewhat curled from end to end or from side to side, but it is a panel that is loose and not solidly attached to anything. The panel may have been left standing on edge in the weather and accumulated a bow, or it may have gotten soaked on one side and not on the other. But that same bowed piece of plywood can be nailed down tight to a floor frame or roof rafters, showing not a sign of warpage. Plywood panels that are flat when installed, are properly secured and are the correct panels for the installation at hand, will seldom, if ever, warp after being installed. The same certainly cannot be said for plain lumber.

Dimensional Stability

One tremendous advantage of plywood over raw wood, especially in such applications as furniture building, cabinetry or interior trimwork for houses, is its *dimensional stability*. As the plywood panels are manufactured, the veneers as well as the finished sheets are held to close tolerances with regard to moisture content. In addition, the way in which the panels are made insures that little addition or loss of moisture in the wood will occur. The result is that plywood panels will shrink or swell only to a minimal degree. The exact amount depends upon the specific kind of plywood panel, the degree of exposure to moisture and also the application in which it is used. The dimensions of the sheet remain remarkably constant. The sheets themselves are manufactured to close size tolerances to begin with, and there is very little variation, even in large sheets, over the life of the product. Naturally, the smaller the plywood piece, the smaller the overall size variation due to shrinking or swelling as moisture evaporates or is absorbed by the wood.

Weathering, sunlight or other similar conditions have little effect. Thus, unlike ordinary wood that can easily expand and is almost invariably afflicted with a certain amount of shrinkage, plywood panels can be cut to size, fitted in place and secured. There will be little, if any, problem later on with joints opening up, buckling seams or the like. This is true regardless of the conditions under which the plywood is used, provided that the proper type of plywood is used for the specific application to begin with, and the installation is correctly made.

Coverage

In today's lumber market, about the widest board one will find, except upon special order, is likely to be 12 inches nominal (about 11¼ inches actual). Those that are 10 inches or 8 inches wide, and smaller, are far more common and widely used. Sheathing a large roof, for example, with 8-inch boards is a long task. A good many boards have to be handled, cut and nailed. But plywood panels have the advantage of covering large areas with relative ease and speed. Roughly speaking, just as much nailing is required, but one standard 4-foot by 8-foot panel covers 32 square feet of area in one shot, as opposed to a bit more than 10 8-foot boards that are 8 inches wide (actually about 7½inches) for the same area. Thus, the job goes faster and easier. The finished roof decking is also both tighter and stronger. The same situation applies, of course, to plywood used in other ways as well.

There is a concomitant advantage, too. A 32-square foot area covered with 8-inch boards has seven joint lines between the boards, each 8 feet long. Even though the boards may be tightly edge-butted together at the outset, eventually they are bound to shrink and leave a series of fairly substantial cracks. Plywood panels, on the other hand, are seamless over the same expanse. So, by using plywood panels the number of joints is greatly reduced, making a stronger and more weatherproof surface. This in turn adds strength, since joints are a weak point in any wood structure or assembly (unless they happen to be securely glued). There are many occasions, such as in cabinetry fabrication, where the absence of any joint lines is an advantage from the standpoint of appearance.

Strength-To-Weight Ratio

Another of the important characteristics of plywood panels is its favorable *strength-to-weight ratio*. In other words, the material is quite strong for its weight. Though this factor is seldom actively considered by the do-it-yourselfer, nonetheless the advantage is there and is part of what makes plywood the tough and rugged material that it is. For example, a normal birch plywood might have a specific gravity of 0.67 and an ultimate tensile strength of 13,200. This results in a tensile strength/weight factor of 19,700. Now, standing alone, those figures do not mean much. But compare them with a particular heat-treated steel with a specific gravity of 7.75 and an ultimate tensile strength of 100,000. The tensile strength/weight factor is 12,900, some 7,000 less than the plywood! Consider an aluminum alloy with a specific gravity of 2.81 and an ultimate tensile strength of 40,000, for a tensile strength/weight factor of 14,200. Pound for pound, the plywood is even stronger than the aluminum. That is one of the reasons why plywood was widely used in the aircraft industry during the earlier days of aviation.

The same characteristics translate into excellent stiffness factors, or *EI factors*, where E is the *modulus of elasticity*. This is a complex subject, but to give a brief example, a particular spruce plywood might have an EI factor of 416. Birch has an EI factor of 178, while aluminum checks in with an EI factor of 22. The EI factor of steel is only 3.1. It is obvious that plywood is a very stiff material for its weight.

Increased Availability of Woods

For many years, standard industry practice has been to cut boards to a nominal thickness of 1 inch. In the early days, rough-sawn boards were actually 1 inch thick, or even a bit more. Today the thickness of a planed board may run from ⅝-inch to ¾-inch. But this can only be done with wood species that are in ample supply. If the same practice were followed with many of the rare hardwoods that are employed chiefly for their beauty of grain and figure, we would soon be left with no supplies upon which to draw. A particular butt, stump or crotch of wood with an exceptionally handsome grain or figure would hardly go far if cut into 1-inch thick slices.

By slicing this wood into very thin sheets of veneer, far more "mileage" can be gained from the same amount of wood. The face veneers are simply glued to cores of more readily available and less consequential wood, in the form of plywood. This makes possible, from both economical and practical standpoints, the manufacture of quantities of handsome book-matched, multiple book-matched, 4-piece butt-matched and other types of matched sheets of beautiful plywoods. Of course, this process also makes available substantial quantities of veneers that can be used in making up geometric shapes, marquetry work and inlaying.

By using plywood instead of solid wood, even substantial quantities of rare and unusual woods are generally available; otherwise, they would not be. A side benefit of this process is that because of the nature of plywood, face veneers which are very thin and fragile—indeed, some of the rare woods that are fragile even when cut relatively thick—can be reinforced and given considerable strength because of the plywood backing to which they are glued and by the glue itself.

Another advantage that accrues as a result of cutting veneer rather than cutting boards or planks lies in the fact that many species of wood, or certain sections of many species, cannot be successfully cut into boards. In the normal course of lumber production, for instance, stumps are of no value. Burls or crotches are merely troublesome chunks of wood to be discarded. Some species have grain patterns that are so twisted and gnarled that they cannot be properly cut and cured, or are virtually unusable when they are cut, because they twist or split or whatever. But all woods can be cut as face veneers for plywoods where appearance is the principal factor. For instance, *bastard cutting* can be readily accomplished. This involves a directional cut that is neither plain-cut nor quarter-cut, so that the face of the veneer shows the grain in its best or most attractive light. Burls, stumps and crotches are often cut; American black walnut is widely used for this purpose. Twisty grain, bird's-eye and other odd patterns are of prime value here, whereas in lumber they are merely defects. Sweet gum is often used, quarter-cut or sawn to show the heartwood figure. Thus, quantities of wood that might otherwise be commercially unimportant or unusable are of great value as veneer.

Bending and Molding

Another notable aspect of plywood is that it can be readily curved and molded. A number of special processes are used. From a commercial standpoint, special machinery is required. A good example of curved and/or molded plywood is in one-piece plywood furniture such as modernistic chairs and small tables. Though certain types of plain woods can also be bent or curved—the ash frames for snowshoes used to be made that way—by steaming and bending with jigs, plywood is the only wood product that really lends itself to this process. Practically any kind of curved and molded shape can be constructed with the proper equipment. Thin plywood panels lend themselves very nicely to fabrication of simple curves right in the home workshop without recourse to any heavy machinery or specialized techniques.

Conservation

Last, but hardly least, is the fact that plywood is highly conservative of timber. Timber is one of the few natural resources we have that is more or less holding its own as far as supply is concerned. It is a renewable resource. If wisely managed, the supply will likely keep pace, at least to a reasonable degree, with the demand for years to come. A much greater yield can be obtained from our timber resources by cutting veneers than by cutting lumber. We have already noted that in certain instances some species of wood, or parts of trees like stumps and burls, which otherwise would be either left alone or left to rot and would be of no commercial value, can be readily cut into veneers of one sort or another. Thus, better use of more wood can be made.

In addition, substantially greater yield can be obtained from a given log when it is rotary-cut into veneer than if it was sawn into lumber. Thus, fewer logs are used to achieve the same or greater purpose. In these days of dwindling resources and increasing demands upon them, we must make the best possible use of them in all possible respects. The manufacture and use of plywood instead of plain wood wherever possible helps greatly in fulfilling that purpose.

PLYWOOD BENEFITS FOR THE DO-IT-YOURSELFER

These, then, are some of the salient general advantages of plywood panels. There are others as well of a somewhat lesser nature, important primarily in specific applications. Some of these will be discussed in greater detail later on in the book, and some of the advantages just discussed will also be touched upon again, from a somewhat different angle. But there are also a few specific benefits in using plywood that are of interest to the home mechanic and do-it-yourselfer. There are a couple of drawbacks as well that deserve consideration.

Time

For most do-it-yourselfers, *time* is an important factor. One of the reasons for doing your own work is to avoid paying for someone else's labor, thus lessening the cost of the project at hand by a considerable amount. But just because that work time is not being paid for in cash doesn't mean that it is unimportant to the do-it-yourselfer. For most of us, there is little enough time to do everything that needs doing, and every minute is valuable. By using plywood panels, especially on large jobs such as decking a roof or sheathing exterior walls, plywood goes up much faster than board stock. With a helper or two, a worker can install an amazing number of plywood panels in a day's time. The same situation applies in projects of a different nature like furniture building or cabinetry. The need for time-consuming cutting, fitting and gluing up of smaller pieces is minimized and the major components for the project can be cut directly from plywood panels, in full size. For any kind of project where some sort of plywood can be employed for at least part of the fabrication, the job will go much faster.

Availability

Availability may also be an important consideration. There are few lumberyards, even in small towns, that do not stock a fair amount of different plywoods. Those most likely to be found are the more utilitarian grades commonly used in building construction, and these plywoods can be turned to many other purposes as well. Larger yards are likely to stock an even greater variety of plywood panels suitable for many different applications. In any case, it is always possible to order those types of plywood not available directly from local stock. Generally, this is not much of a problem. The material is usually available from the nearest warehousing center.

Larger lumberyards are also likely to stock a certain amount of hardwood plywood that can be used for a variety of decorative purposes. The plywoods most commonly stocked are generally wall panels, but these too can be turned to a variety of purposes, including cabinetry. On the other hand, the availability of many species of plain woods, with the exception of those few most commonly used in the building trades, is likely to be chancy. Special-ordering small quantities of plain woods that are not kept in stock, especially hardwoods, is a ticklish business at best. One might as well say that the available variety in plain woods is far less than that of plywoods.

Economics

For any do-it-yourselfer, the *economics* of a given project are most important. Often the object is to get the job done as inexpensively as possible. However, it is not a good idea to also get the job done as cheaply as possible. While super quality may not be important, reasonable quality always is. So is getting your money's worth. Do-it-yourselfers sometimes bypass plywood. They don't even stop to consider it, simply because plywood is a relatively expensive material on a square-foot basis. But this can be a matter of being penny-wise and pound-foolish. In fact, plywood is frequently less expensive on a per-project basis than a comparable amount of material needed in other forms.

Even though the plywood may be a bit more expensive than plain wood, for instance, the various advantages that plywood offers are frequently sufficient to offset that slight increase. By and large, using plywood instead of plain wood results in greater strength, greater durability, less waste, less time and so on. The net result can be a far better job for just a slight increase in cost, or perhaps even at less cost in the final analysis. When determining whether or not to use plywood on a project, be sure to compare it in all pertinent respects with the other materials that you might use—particle board, plain wood boards, hardboard or whatever.

Versatility

There is no question that wood is an extremely *versatile* material. But there is also no question that plywood is even more versatile. You can simply do more with plywood, and use it in more applications with greater ease and with greater expectation of longevity and durability, than you can with plain wood. This in itself is a good reason to opt for plywood rather than wood or some other building material, assuming that the application is a suitable one for plywood in the first place. Because plywood is available in so many different types and grades, sizes and thicknesses, there is a spot in virtually every project for some sort of plywood. There is some sort of plywood available that is particularly adaptable to practically all projects. In addition, the variety of different wood species as face plies is tremendous, compared with those readily available in plain wood. Decorative plywoods can be obtained to suit virtually any purpose and taste.

Durability

Strength, stiffness and durability are plywood characteristics that have already been discussed. These characteristics are of considerable importance to the do-it-yourselfer. Under many circumstances you can more easily achieve stronger and longer-lasting results by employing plywood than by using other materials. For instance, bookshelving can be made from particle board, which is nearly as easy to work with and quite a bit less expensive in identical sheet sizes. Unfortunately, particle board lacks both strength and stiffness, and shelves made from that material can easily break in half if not ruggedly supported. Even short shelf-lengths of particle board shelving will soon adopt a permanent sag if they are longer than about 20 inches (for ¾-inch thickness), even if fully supported by a rail across the rear shelf bottom.

Thus, this type of construction requires extra care and extra support, which in turn means a great deal of added work and more material. The same shelves built from ¾-inch plywood would be tough, sturdy and easily put together. The same situation is true of a great many other home building projects. Plywood is much stronger and sturdier when properly used in building a house, garage, addition to the house or the like, than would be the case with plain boards.

Workability

Because plywood is obtainable in standard sheets measuring 4 feet wide by 8 feet long (some types can be obtained in even larger sizes), it is easy for the do-it-yourselfer to work with. Large, monolithic pieces can be cut from a sheet, doing away with the chore of cutting out a series of small pieces from boards and then having to join them together. Plywood sheets will also cover large areas in short order, as when laying a subfloor or applying exterior siding material to a house. Though the thick sheets are relatively heavy, they are not particularly difficult to handle once you know how to go about it. With a helper or two, the job goes very quickly.

There are other factors that make plywood easy to work with, too. The fact that pilot holes need not be drilled for nails when nailing near the edge of the stock (since the plywood will not split) is a great help and saves a lot of time. Trimming pieces to size goes quickly with a circular saw because long, straight cuts can be made instead of a whole series of short cuts that require handling, measuring and marking a good many pieces. By the same token, a good deal of time is saved during the measuring and marking process in laying out workpieces, since again the job can be done on one large sheet rather than a series of small pieces.

The business of cutting, forming, shaping and sanding plywood is a simple matter once you learn how to accomplish those jobs properly, and the procedures and techniques are almost always invariable. One sheet of plywood can be worked in virtually identical fashion to any other sheet of plywood of the same general characteristics (there are some slightly different procedures for a few special types of plywood), and there are almost never any unexpected surprises. Raw woods, on the other hand, have different working and machining characteristics depending upon the species, and may even vary from workpiece to workpiece, requiring the use of somewhat different tools, techniques or procedures. While a good deal of time is required to become experienced in working many different kinds of woods under different conditions and circumstances, the basics of working with plywoods can be learned in a very short time and without much difficulty.

Stability

Perhaps one of the most beneficial aspects of plywood insofar as the do-it-yourselfer is concerned is its *stability*. There are few things more frustrating than making up a pretty set of cabinet doors from white pine only to have them warp and rack so that they do not close properly, and shrink so that they exhibit large gaps at the joints only two or three weeks after they have been installed. You might build a nice piece of furniture out of expensive solid birch only to discover a month or so later that the material really was not dry enough and all of the joints have pulled apart, leaving a wobbly and unsightly piece of work. There's all that time and effort, not to mention expense, gone down the drain. With plywood, this simply does not happen. Whether of full sheets or cut pieces, and irrespective of size and shape, the plywood will remain in its original form.

This does not mean that plywood does not absorb moisture from the air, or cannot dry out to a certain extent; it certainly can. But the extent of swelling or shrinking is so minor that it is virtually unnoticeable. If plywood joints are properly cut, glued and fastened and if the plywood is secured to a good framework (if and as necessary), the surfaces will remain plane and the joints will stay tight. If properly done, a plywood construction will remain stable and in exactly the same form as it was originally constructed, for an indefinite period of time. This assumes, of course, that the proper plywood for the job was chosen in the first place, that the construction was done with the proper procedures and techniques, and that there is no mechanical damage. Any time the job calls for fine joining, appearance is of consequence, and the component pieces of the project must remain stable, plywood is an excellent choice.

Other Characteristics

Depending upon the nature of the project at hand, there are various other properties and characteristics of plywood that may be of considerable value. Many of these properties will be discussed in greater detail in a later chapter. For example, plywood is a good insulator, with an approximate *R-value* of 0.93 in a ¾-inch thickness; in some applications this might be a consideration. Because plywood can be applied in large panels in building construction, there are fewer joints, and thus, less opportunity for air infiltration through the joints. Plywood can be used for its sound-deadening qualities in constructions where acoustics might be involved. Certain plywood constructions have a relatively low flame-spread rating. Other types are practically impervious to extreme moisture conditions and will even withstand agricultural acids found in ensilage, fertilizers and manure. A few types are made with tremendously hard surfaces that are impervious to practically anything, including rough treatment and mechanical damage. This means that when special applications crop up, there probably is a plywood suitable for the job. Frequently plywood is the *only* material worth considering.

DISADVANTAGES OF PLYWOOD

Lest you think that plywood is a miracle building material and a universal answer to all project problems, we must mention some of the drawbacks involved. Note that these are not blanket disadvantages that are part and parcel of the material itself, but rather disadvantages that arise due to particular conditions or circumstances.

Misapplication

For one thing, misapplication of plywood is a common happenstance. When plywood is misapplied—the wrong kind of plywood is used for a particular job—the results are fine when a higher quality level or grade level than is necessary is employed. In this case, about the only adverse factor lies in excessive cost. The job might have been done at less expense for materials had the quality or grade level, or type, of plywood been exactly commensurate with the job application. But if a lower grade or quality level of plywood was used than should have been, difficulties are bound to result. This is the situation that more often occurs, usually in an effort to cut corners, scrimp and save a few coins. The situation also arises from lack of knowledge and no realization that many kinds of plywood exist, all of which can be used for different purposes. Thus, the great variety of plywoods available can be a disadvantage to the unknowledgeable, but a tremendous advantage to those who fully understand these ramifications (which, incidentally, will be fully discussed in the next chapter).

Misapplication can occur in other ways, too. Although it is a fine material, plywood is not necessarily the best choice for every project. It is obvious, for instance, that one does not run out and buy expensive new plywood to take care of some temporary and low-key project around the grounds. If you need a temporary bridge over a trench, for instance, a couple of old planks will serve better than a fresh sheet of plywood. But there are other instances that are less obvious. The back panel of a bureau, for instance, could just as well be built from inexpensive ⅛-inch hardboard as ¼-inch plywood. The same purpose is served at less cost. There are occasions when particle board can serve equally well as plywood—as a backer for laminated plastic overlay in some applications, for instance. There are also cases where plywood simply does not serve as well as some other material entirely, perhaps not even a wood product. The point is that the best attributes of a particular material should be matched to the demands and conditions of the job or project, and the best material is not always plywood. The only way you can make a satisfactory judgment is to know and understand the materials under consideration so that you can make a fair comparison.

Handling Difficulty

Oddly enough, some of the very characteristics that are so advantageous with plywood can also be, at certain times, disadvantageous. For instance, the large size of plywood panels is generally a plus. But if you are working alone, that size can be a hindrance. A ¾-inch panel, for instance, is very heavy and awkward to handle. Though a single strong person can carry such a panel about, it is far more easily handled, and indeed generally must be, by two persons. Decking a roof, especially a fairly steep pitch in windy weather, can be an extremely difficult job for a single person, even when relatively lightweight ⅜-inch panels are being laid down. Making the first cuts in a full panel can also be difficult and awkward, especially in a small workshop. Sawhorses or other supports must be used. A good deal of fiddling around is generally necessary to get lined up for the first cut or two, particularly full-length cuts, and special supports must be used when working large sheets on a table saw or radial saw. Usually the extra effort is worthwhile, but the size problem must be recognized ahead of time and planned for.

As just mentioned, a 4-foot by 8-foot plywood panel is quite heavy. Even a ½-inch thick panel is plenty for one person to boost around. There are occasions when the do-it-yourselfer might opt to use plain wood boards instead of plywood panels, for reasons of weight and ease of handling alone. Hauling and positioning a whole stack of plywood panels is an arduous job, especially when done solo, but on the other hand one person can haul and position boards with little difficulty or effort.

Expense

Expense is an important factor for most do-it-yourselfers; that's one reason why they *are* do-it-yourselfers. And there is no question that plywood is an expensive material. There are times when the cost of plywood for a particular project is greater than the cost of other materials that will do the same job, although perhaps not quite as effectively. There are times, too, when the nature of the project is such that the additional cost for plywood may not be justifiable, at least in the eyes of the builder. This is especially true when other materials that might suffice are lying idle around the shop. It is true, also, when you must buy an entire sheet of plywood just to use only a few square feet of it. The question, then, is one of cost-effectiveness, rather than initial cost alone. This question is one that must be pondered by the do-it-yourselfer, and only he can make the decision as to whether the cost of plywood is justifiable or bearable for any given project.

Working Characteristics

We have mentioned earlier that plywood is a very easy material to work with. That statement is true. Many do-it-yourselfers, however, tend to avoid plywood because in their experience (usually limited) plywood is in fact difficult to work with. Conditionally, that statement is also true. The condition is that in many respects plywood is not worked in the same fashion that many of the common plain woods, such as pine, are handled. Some techniques are the same, to be sure, and most of the same tools can be employed as well. But there are numerous different techniques for using the tools, and different ways of working and handling plywood that make the job quite easy. However, just because you can readily work with pine does not necessarily mean that you can readily work with plywood. New techniques and procedures have to be learned first; this is especially true of joinery. Once these hurdles are overcome, and it is by no means a difficult chore, you will find that plywood on the whole can be more easily worked, and worked with, than practically any other material. Only if you don't know what you are about are the working characteristics of plywood a disadvantage.

Chapter 2
Plywood Properties

As anyone who has had even a little bit of experience with plywood realizes, a sheet of plywood is not just a sheet of plywood. There are a number of obvious differences, such as the stock length and width of a full panel, the thickness of the panel, the number of layers of wood that are joined to make a single panel and others. A sheet of plywood that you might use for subflooring, or roof decking, in addition to your house has an entirely different appearance than the plywood paneling you might choose to put up on your rec room walls. Both of those plywoods in turn look entirely different from the plywood exterior siding that you would use when building a garage. But there are other differences as well, and many of them are not at all obvious. They can only be ascertained by close inspection, translating the coded stamped markings on the plywood sheets in question, or by consulting the specification sheets that pertain to that particular type of plywood.

UNDERSTANDING PLYWOOD CHARACTERISTICS

The various plywood properties may seem at first to be rather complex and involved. You might think to yourself, "Well, baloney, I don't have to wade through all of this nonsense just so I can choose a hunk of plywood for my next project." True enough, you don't. Nor do you have to compare various specifications and performance data before you buy your next new washing machine or television set. You don't have to investigate and compare the various clauses and benefits when you take out your next insurance policy. But doing so certainly makes good sense, if you want the right product or service for your needs at an appropriate cost. The properties and characteristics of the many different kinds of plywoods are really not difficult to comprehend once you understand how the system works. In any event, there certainly is no reason for you to memorize all the pertinent information so that you can call it forth upon instantaneous demand. There are printed tables, charts and specification sheets, not to mention this book, that contain all of that information. All that is necessary is that you be able to refer to them and understand them.

The reasons for being able to do so are simple enough. First, there are many different kinds of plywood, and just looking at the physical appearance of a particular panel will not tell you if that plywood is going to be satisfactory for your purposes. It may not be suitable for the application you have in mind and consequently will not perform well, or even fail, sometime after installation or fabrication. On the other hand, that particular plywood panel might be of a level, from one standpoint or another, far in excess of what you need to adequately do the job at hand. Since there is a certain degree of overlap of properties and characteristics between numerous different kinds of plywood, you might be lucky and just happen to hit upon the right combination of properties and characteristics, and experience no difficulties.

The second reason is just as important, perhaps even more so. That is *cost*. There is a wide range of costs for different kinds of plywoods, on a per-square-foot basis, for panels of the same width, length and thickness. That cost is dependent to a large degree upon the properties and characteristics of the plywood, but does not necessarily relate to the relative quality of the product. Though quality is always a consideration, if you purchase plywood manufactured by reputable companies, you can be largely assured that quality will be perfectly satisfactory. The quality level will be comparable from manufacturer to manufacturer for plywood of the same size and properties.

The point is that only by choosing plywoods of the proper characteristics for the applications you have in mind will you be able to make cost-effective purchases. If you buy a kind of plywood that is very expensive, figuring that because it costs a lot it must be better, you may just be fooling yourself. If you build a child's sandbox from the plywood, you have spent far more than perhaps necessary, to no benefit. A less expensive grade or type might have worked just as well. On the other hand, if you try to save a few pennies by employing an inexpensive plywood, primarily priced that way because of its lower level of performance characteristics, you are again fooling yourself. You will end up spending extra money for repair or replacement because you should have used a more expensive plywood of a higher performance level. So, in order to get the most for your money, you have to use the right plywood for the job at hand. If you don't, the ultimate cost will inevitably be higher than need be, in one way or another.

Plywoods as a general group of wood products are broken down into two large subgroups: *softwood plywoods* and *hardwood plywoods*. Though there is some degree of overlap in properties, characteristics and especially in applications and uses, there are substantial differences in the makeup of the two subgroups. They are considered independently of one another.

SOFTWOODS AND HARDWOODS

This seems a good time to clear up any possible misconceptions, which seem to be relatively widespread, regarding the differences between softwoods and hardwoods. Those who are familiar with some of the more common species of American woods will immediately recognize that the oaks, for instance, are commonly called hardwoods. The pines are commonly called softwoods. But what, exactly, establishes that classification? Much confusion arises from the fact that some softwoods are actually hardwoods, and some softwoods are actually quite hard.

First, let's discuss the difference between hardwoods and softwoods. These are botanical terms and really have nothing to do with how hard or how soft a particular species of wood may be. Rather, the difference lies in whether or not the species of tree bears exposed seeds. The softwoods are *gymnosperms*, which *do* bear exposed seeds, usually in cones. These species of trees are commonly called *conifers*, or even more commonly, "pine trees" (though only those conifers of the genus *Pinus* really are). They are also often called *evergreens*, because they keep their needles year round (but a few conifer species, such as *larch*, do not). Hardwoods, on the other hand, are all *angiosperms*, and have true flowers, broad leaves and seeds *enclosed* in a fruit. All but a very few of these so-called broad-leafs are *deciduous*, losing their leaves every fall and growing them back in the spring.

Now let's turn to the matter of softwoods and hardwoods. The softness or hardness of various species of wood is relative and is dependent upon the density, or mass of substance per unit volume, of a particular species. The relative hardness of a piece of wood is a measure of its resistance to denting, scratching and similar mechanical damage.

In practice, density comparisons of different species of woods are expressed in terms of specific gravity of the wood. A unit volume of a particular species is compared with a unit volume of water and given as a ratio for comparative purposes. The results are usually expressed in terms of pounds per cubic foot or grams per cubic centimeter and are averages only. The figures are based on tests made with wood of oven-dry weight and volume with a 12% moisture content. Thus, *Eastern hemlock*, a softwood, has a specific gravity of 0.40, while *American basswood*, a hardwood, has a specific gravity of 0.37. The softwood is harder than the hardwood! Table 2-1 lists typical specific gravities for several of the more common species of woods, as a matter of interest.

Table 2-1. Specific Gravity of Several Common Wood Species.

Wood	Specific Gravity
Ash, white	.60
Aspen, quaking	.38
Birch, paper	.55
Cedar, eastern red	.47
Fir, Douglas	.48
Hemlock, eastern	.40
Locust, black	.69
Maple, red	.54
Maple, sugar	.63
Pine, eastern white	.35
Pine, Ponderosa	.40
Pine, red	.46
Redwood	.40
Sweetgun	.52
Walnut, black	.55

SOFTWOOD PLYWOODS

Plywood is doubtless one of the most versatile fabrication materials we have, and it surrounds us everywhere we go. Of the two subgroups, softwood and hardwood, softwood plywood is utilized in by far the largest quantity and in the greatest diversification of application. Softwood plywood is often termed the "construction and industrial plywood," and the name implies that it is made up entirely of softwoods. To further confuse the issue, that is not the case; softwood plywoods are in fact made up of softwoods, hardwoods, and both hard and soft woods. It is true, however, that most of the softwood plywoods that the do-it-yourselfer is likely to encounter, especially in building houses or additions and other projects, are at least faced and backed, and often cored as well, with a softwood, predominantly Douglas fir.

American Plywood Association Plywood

A tremendous quantity of softwood plywood is manufactured every year, and the principal watchdog for the greater proportion of it is the *American Plywood Association* (APA). Formed as the *Douglas Fir Plywood Association* back in 1933, the organization changed its name to the present one in 1964 and has become a respected name in the softwood plywood industry. The APA grade-trademark on a sheet of plywood assures the buyer of a quality-engineered product upon which he can depend. Since the APA has a large staff of professionals, including engineers, foresters and wood scientists, who conduct an ongoing and extensive research and development program in virtually every conveivable area having to do with softwood plywood, their program and their expertise has become widely known and respected throughout the country. In fact, they have done more to foster not only the use of softwood plywoods in innumerable applications, but also to establish standards of quality and manufacture, as well as broad-scale requirements, for the principal types and grades of softwood plywoods. Since this organization is an association of softwood plywood manufacturers and is supported by them, they have also striven to increase total sales and distribution of their products. That they have been successful in doing so in no way reflects adversely upon the worthiness of their standards, which are excellent.

As a part of their overall effort, the APA some years ago voluntarily worked with the *Product Standards Section* of the *National Bureau of Standards*, a part of the *U.S. Department of Commerce*, to develop what is known as *Product Standard PS1*. This standard not only details the specific requirements for manufacture of softwood plywood, such as techniques to be used, veneer and panel grades, workmanship and dimensional tolerances, but also establishes preparation for and methods of testing the completed product. Much of the softwood plywood that is manufactured in this country is made according to Product Standard PS1. As most of the major manufacturers of softwood plywood in the country are members of the American Plywood Association, their products conform to those standards as well as additional and often more stringent standards used only by the APA, over and above PS1.

As a consequence, approximately 75% of the softwood plywoods marketed in the United States carries the APA trademark, stamped in some fashion upon each individual plywood piece. This trademark, an assurance of quality, is contained within a larger grade mark, as a rule, and more about that later.

If the piece of plywood that you are about to buy does not carry the APA trademark, but instead carries some other trademark or perhaps none at all, will you be purchasing plywood of reasonable quality? The answer is probably yes, but not necessarily. Of the remaining 25% or so of the softwood plywood production in this country, the greater part will carry some other trademark or grade mark.

There are other independent agencies that can certify plywood as being made to certain quality levels and dimensional tolerances, including that manufactured under Product Standard PS1. Independent testing laboratories approve these plywood products as conforming to whatever standards are established as criteria, and the product is so stamped. These independent marks usually indicate quality comparable, at least in large measure, to plywood products bearing the APA trademark. It should be noted that all APA-stamped products are immediately acceptable for use under local building codes and by the authorities who enforce them (provided that the particular specifications of the plywood are appropriate for the application in question). Most independent marks are also usually accepted, though the burden of proof that the plywood is indeed of a sufficiently high quality level may rest with the user.

Mill Controlled Plywood

Some plywood is not trademarked at all and, in fact, may carry no markings whatsoever. These mill-controlled products are manufactured under whatever quality standards and dimensional tolerances the individual mill wishes to establish. There is considerable difference in these factors from mill to mill, and there may even be differences in the same product from the same mill over a period of time. The buyer has little if any idea of what he really is buying in the way of quality, stability, longevity and other properties. He likely will not be able to find out from the mill, even if he should be able to determine which mill in particular produced the material in the first place. It is strictly up to the buyer to make the decision as to whether or not to purchase. Building code authorities often will not accept unmarked plywood for construction purposes.

Two further facts deserve consideration. The first is that mill-controlled plywood is not necessarily of poorer quality than trademarked plywood. The second is that there are occasions when the relative quality of the plywood being used is of secondary (or even less) importance, as when building a simple form for a small backyard concrete retaining wall, or knocking together a crude crate.

In sum, where certain qualities or specifications are of importance and have some bearing upon the particular job at hand, use only plywood that is trademarked and whose specifications are clearly indicated and can be relied upon. In other instances, don't worry about it. Be sure you know ahead of time just what qualities and specifications will serve best for that job.

Sizes and Thicknesses

The most common size of softwood plywood, and certainly the most familiar, is the panel measuring 4 feet by 8 feet. Panel sizes of 4 feet by 9 feet and 4 feet by 10 feet are also in widespread use, though they frequently must be specially ordered by smaller lumberyards that do not carry a wide range of plywood. In addition, both 3 feet and 5 feet are standard plywood panel widths, and lengths are available to 12 feet, in 1-foot increments. Small sizes are now being stocked by some lumberyards and stores, specifically for use by do-it-yourselfers or anyone else who has need for only limited amounts. Sections measuring 2 feet by 4 feet and 4 feet by 4 feet, are easily obtained. Some lumberyards will also sell odd-shaped pieces on a per-square-foot basis, or by the piece. Often these pieces have been cut from partly damaged full sheets, or may be leftovers too good to discard.

There are a number of standard plywood thicknesses. The thinnest normally available is ¼-inch; others are 5/16-inch, ⅜-inch, ½-inch, ⅝-inch and ¾-inch. There is also one special type that measures 1⅛ inches thick, and additional thicknesses of ⅞-inch, 19/32-inch and 23/32-inch are available in certain types of plywood. Not all of the standard thicknesses are available in all kinds of softwood plywoods, even the most common ones.

Plywoods manufactured under U.S. Product Standard PS1 are required to maintain certain size and thickness tolerances. Each panel can be no greater in either width or length than its specified dimension, but the panel can measure as much as 1/16 inch less than either the specified width or length. However, one sometimes encounters panel dimensions that are off by a bit more than this, so it is wise to check carefully before making layouts and cuts on a plywood sheet.

The thickness of a plywood panel is often of less consequence than size, but is important when two or more panels must be matched to one another. Consistency of thickness from panel to panel can be very important when an article, such as a cabinet, is being fabricated from pieces that have been cut from two or more different panels. The allowable thickness tolerance under PS1 is plus or minus 1/64 inch for specified thicknesses of ¾-inch, for sanded panels. Unsanded, touch-sanded and overlay panels must have a tolerance of plus or minus 1/32 inch for specified thicknesses of 3/16-inch or less, and plus or minus 5 percent of the specified thickness for those panels that are thicker than 13/16 inch.

Plywood sheets manufactured under PS1 must also adhere to certain tolerances with respect to squareness and straightness. All panels that measure 4 feet or greater in length and width must be square within 1/64 inch per lineal foot. Thus, a 4-foot by 8-foot panel must be square within 1/16 inch (4/64) across the width and ⅛ inch (8/64) along the length. Panels that are less than 4 feet in length or width must be square within 1/16 inch as measured along the short dimension. As to straightness, all of the panels must be manufactured so that if a straight line is extended from one corner to the next corner— that is, across one side or across one end of a panel—the edge of the panel must lean no more than 1/16 inch from the line.

Panel Constructions

A panel of plywood is made up of both plies and layers. A single ply consists of a single sheet of veneer. A layer may consist of a single ply or veneer sheet, or may be made of two or more plies laminated with their grain directions running parallel with one another. Plywood panels are always made up of an odd number of layers, but may contain either odd or even numbers of plies. The grain direction of each layer, be that a single sheet of veneer or two or more plies laminated, must always run at right angles to the grain direction of the adjacent layer (Fig. 2-1). The face layers of rectangular panels have their grain direction running parallel with the long axis of the panel. Thus, with the bottom layer grain direction running with the long axis, the next layer will have its grain direction lying in the short axis, at right

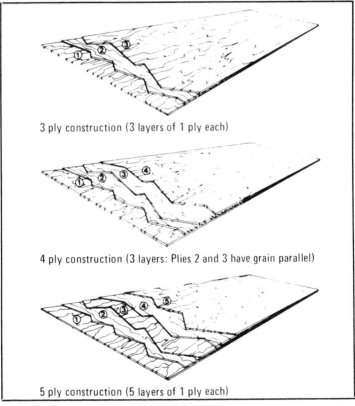

3 ply construction (3 layers of 1 ply each)

4 ply construction (3 layers: Plies 2 and 3 have grain parallel)

5 ply construction (5 layers of 1 ply each)

Fig. 2-1. Typical softwood plywood panel construction (courtesy of the American Plywood Association).

angles to the first. The third layer will have its grain direction running at right angles to the second, along the long axis again, and so on.

All of the layers are joined together with special glues that cover the entire area of each contacting surface, and the bonding is done mechanically under high pressure and (usually) high temperature. The result is a structural bond between the layers that is at least as strong as the wood itself, and frequently stronger. In the event that a layer consists of more than one ply, the plies are similarly bonded. Softwood plywood panels are constructed with a minimum number of three plies and a minimum of seven layers and/or seven plies, depending upon panel thickness, but more can be employed. Table 2-2 lists the panel constructions.

Exterior and Interior Types

Softwood plywood is manufactured in two general types: *exterior* and *interior*. The word "type" refers particularly to the overall durability of the glue used to laminate the layers in the panel, and to the quality and usefulness of the resulting bond that can be expected when the plywood is utilized in various applications. Veneer grade is also an important factor in type designation.

Interior plywood may be bonded with any of three different classes of glue, each with a differing durability level. Interior plywood bonded with *interior glue* is intended only for interior applications with normal atmospheric moisture conditions and no exposure to weather or excessive amounts of moisture. This kind of plywood is moisture resistant, however, so if it gets damp, or even wet, for a short period or two during the construction process, no harm is likely to be done (though objectionable staining or surface grain raising might result).

Interior plywood can also be bonded with *intermediate glue*. This kind is quite resistant to moisture, much more so than plywood bonded with interior glue, and also resists the growth of bacteria or mold. This kind is a good choice where the plywood might be exposed to weather for moderate periods of time during construction and cannot be adequately protected, or where the possibility of water leakage or abnormally high atmospheric moisture might be encountered.

Interior plywood bonded with *exterior glue* has the highest level of bond durability. The glue is completely waterproof, and

Table 2-2. PS1 Construction Specifications for Various Softwood Plywood Panels (courtesy of the American Plywood Association).

Panel Grades	Finished Panel Nominal Thickness Range (inch)	Minimum Number of Plys	Minimum Number of Layers
Exterior Marine Special Exterior (See para. 3.6.7) B-B concrete form High Density Overlay High Density concrete form overlay	Through 3/8 Over 3/8, through 3/4 Over 3/4	3 5 7	3 5 7
Interior N-N, N-A, N-B, N-D, A-A, A-B, A-D B-B, B-D Structural I (C-D, C-D Plugged and Underlayment) Structural II (C-D, C-D Plugged and Underlayment) Exterior A-A, A-B, A-C, B-B, B-C Structural I and Structural II (See para. 3.6.5) Medium Density and special overlays	Through 3/8 Over 3/8, through 1/2 Over 1/2, through 7/8 Over 7/8	3 4 5 6	3 3 5 5
Interior (including grades with exterior glue) Underlayment Exterior C-C Plugged	Through 1/2 Over 1/2, through 3/4 Over 3/4	3 4 5	3 3 5
Interior (including grades with exterior glue) C-D C-D Plugged Exterior C-C	Through 5/8 Over 5/8, through 3/4 Over 3/4	3 4 5	3 3 5

Note: The proportion of wood based on nominal finished panel thickness and dry veneer thickness before layup, as used, with grain running perpendicular to the panel face grain shall fall within the range of 33 percent to 70 percent. The combined thickness of all inner layers shall be not less than ½ of panel thickness based on nominal finished panel thickness and dry veneer thickness before layup, as used, for panels with 4 or more plys.

is the same as is used in the manufacture of exterior plywood. However, the inner plies and backs of the interior plywood are made with a lower grade of veneer than the exterior. Though this kind of plywood cannot be considered a substitute for exterior plywood, it is durable and can be used under virtually any weather and moisture conditions for substantial periods of time during the construction process. It must eventually be covered and protected, however, and should never be used in applications involving permanent outdoor exposure.

Most interior plywood is manufactured today with exterior glue and presents a waterproof glueline. The interior type made up with intermediate glue can be obtained, but it is manufactured and used to a much lesser degree. Interior plywood with interior glue is used in a variety of different applications and is available, but it may not be stocked by smaller lumberyards that maintain a supply of the more utilitarian interior type with exterior glue. As far as the do-it-yourselfer is concerned, the important factor is to know exactly which glueline you are purchasing, so that the plywood will stand up under the proposed job conditions.

The term "glueline," incidentally, refers to the glue joint that is formed between the plies and/or layers in a plywood panel. The gluelines in a factory-made panel are called primary gluelines. Those that are made during the process of fabricating something from plywood parts, or plywood and wood or lumber, are called secondary gluelines.

The second major type of plywood is the exterior type. To clear up a common misconception, the terms exterior plywood and *marine plywood* are not interchangeable. Marine plywood is a specific high-quality grade of exterior plywood, only one of several that make up the entire group. All exterior plywoods are made with very high quality veneers, much more so than the interior type. The gluelines and 100% waterproof. The adhesives used in assembling these panels, while not actually indestructible, are virtually so from a practical standpoint. Exterior plywood can be exposed indefinitely to weather and can be repeatedly wetted and dried with no ill effects. Of course, long-term exposure to weather will lead to some surface deterioration and discoloration, just as will occur with practically anything that is left outdoors. But the strength and integrity of the plywood will remain essentially unharmed. Exterior plywood is extremely durable, and age has no effect upon the glue bond.

Species Group Classification

Many do-it-yourselfers who are somewhat familiar with softwood plywoods and have used them in one project or another may be accustomed to hearing the material referred to as fir plywood, or Douglas fir plywood. Indeed, a large percentage of the softwood plywood panels produced in this country do consist in whole or at least in part of Douglas fir veneers. But in actuality there are more than 70 different species of woods employed in softwood plywood manufacture. As you might expect, many of these are conifers. In addition to Douglas fir, several other species of fir are used, along with numerous pines, such as *loblolly, red and western white*. Several species of spruce are used, and both cedar and redwood plywoods are widely employed as exterior siding material for residences, apartments and light-commercial buildings. But also included in these species are a surprising number of hardwoods or deciduous species. Some of them actually are quite hard, such as *yellow birch* and *sugar maple*. Others are less so, like *red alder*, and some are rather soft, such as *quaking aspen* or *basswood*. However, all of these various species and more, as listed in Table 2-3, can find their way into plywood panels.

The species are broken down into five groups. Those listed in Group 1 have the highest levels of strength and stiffness, while those in Group 5 are the weakest and limberest. The face ply and the back ply of a plywood panel can be made of any one of the species listed in the five groups. Whatever group number that species is listed under is also the classification of the panel. Thus, a panel faced and backed with western hemlock would be in the Species Group 2 classification. If for some reason either the face or the back ply must be made up of two separate pieces of veneer, they must be of the same species.

The inner plies or layers, however, of any plywoods in the first four groups can be made of any species listed in those groups, irrespective of the particular species of the face and back plies. Thus, a Group 1 panel with American beech face and back plies might contain inner plies or layers of jack pine (Group 3) or yellow poplar (Group 2). Just which materials are used for inner plies or layers depends upon numerous factors, such as price and availability.

Plywood panels that are classed in Group 5 can be made up with any of the entire group of species as inner plies or layers. In addition, the inner plies or layers of any plywood panels can also be made of any other (unlisted) hardwood or softwood species that has

Table 2-3. Classification of Species of Woods Used in the Manufacture of Softwood Plywood Panels (courtesy of the American Plywood Association).

Group 1	Group 2	Group 3	Group 4	Group 5
Apitong[a][b]	Cedar, Port Orford	Alder, Red	Aspen	Basswood
Beech, American	Cypress	Birch, Paper	Bigtooth	Fir, Balsam
Birch	Douglas Fir 2[c]	Cedar, Alaska	Quaking	Poplar, Balsam
Sweet	Fir	Fir, Subalpine	Cativo	
Yellow	California Red	Hemlock, Eastern	Cedar	
Douglas Fir 1[c]	Grand	Maple, Bigleaf	Incense	
Kapur[a]	Noble	Pine	Western Red	
Keruing[a][b]	Pacific Silver	Jack	Cottonwood	
Larch, Western	White	Lodgepole	Eastern	
Maple, Sugar	Hemlock, Western	Ponderosa	Black (Western Poplar)	
Pine	Lauan	Spruce	Pine	
Caribbean	Almon	Redwood	Eastern White	
Ocote	Bagtikan	Spruce	Sugar	
Pine, Southern	Mayapis	Black		
Loblolly	Red Lauan	Engelmann		
Longleaf	Tangile	White		
Shortleaf	White Lauan			
Slash	Maple, Black			
Tanoak	Mengkulang[a]			
	Meranti, Red[a][d]			
	Mersawa[a]			
	Pine			
	Pond			
	Red			
	Virginia			
	Western White			
	Spruce			
	Red			
	Sitka			
	Sweetgum			
	Tamarack			
	Yellow Poplar			

(a) Each of these names represents a trade group of woods consisting of a number of closely related species.

(b) Species from the genus Dipterocarpus are marketed collectively: Apitong if originating in the Philippines; Keruing if originating in Malaysia or Indonesia.

(c) Douglas fir from trees grown in the states of Washington, Oregon, California, Idaho, Montana, Wyoming, and the Canadian Provinces of Alberta and British Columbia shall be classed as Douglas fir No. 1. Douglas fir from trees grown in the states of Nevada, Utah, Colorado, Arizona and New Mexico shall be classed as Douglas fir No. 2.

(d) Red Meranti shall be limited to species having a specific gravity of 0.41 or more based on green volume and oven dry weight.

an average specific gravity of 0.41 or more, based upon oven-dry weight and green volume, as determined by the *Forest Products Laboratory* of the U.S. Department of Agriculture.

There is an exception, and this refers to sanded or decorative plywood panels that are ⅜ inch thick or thinner. In this case the panels are primarily used in applications where appearance is paramount and structural capabilities count for little or nothing. Here the group number refers to only the face ply, and the back ply can be any species. For instance, a ¼-inch panel faced with eastern white pine would not be used for structural purposes, but rather in cabinetry or as decorative paneling. Though the back ply might be made of Douglas fir (Group 2), for example, panel classification is made on the basis of the white pine face, Group 4.

In many cases, the particular species group classification holds little meaning for the do-it-yourselfer who is purchasing plywood for a particular project. This is especially true of plywood panels that have the APA grade marks affixed, because the strength and stiffness factors are automatically included in the overall specifications of the plywood panels and in an item known as the "identification index number," which we will investigate later. If you select plywood panels for a job on the basis of job conditions and panel specifications that fit job requirements, you will automatically get the best group classification as well. Of course, plywood is often selected not on the basis of performance characteristics but appearance, or for a face made of some particular species for decorative emphasis, such as redwood or mahogany, or for some particular grain or texture. In such instances, group classification has little import.

There is one exception where reference to a species group classification may be important, and this is in the selection of *plywood underlayment*. The underlayment produced by APA member manufacturers does not include an identification index number, so the only way to determine relative stiffness, an extremely important factor in flooring especially of the single-layer type, is to consult a species group classification chart.

Another exception is exterior siding panels, where both strength and appearance can be considerations. In constructions where the plywood siding panels are applied directly to studs, or where they must be depended upon for maximum structural strength, some compromise may be necessary between appearance and strength.

Veneer Grades

There are four areas of consideration with respect to plywood grades, all of which are important in the specification and selection of plywood for different purposes. These include *veneer grades, appearance grades, engineered grades* and the additional grading factors over and above appearance grading that are used with APA *303 siding*.

■ **Grade N.** Veneer grades are applicable to the sheets or pieces of veneer as they exist before being assembled into plywood panels. Grade N is of the highest quality, and is particularly intended for applications where a natural finish is required. The wood used is select. Each sheet much consist of all heartwood or all sapwood, never a mix. There can be no defects such as knots or knotholes, splits, stains, pinholes and such. Under Product Standard PS1, very small defects may be filled with synthetic fillers. Repairs can be made to the veneer sheets, too, but must be of wood, well-matched for grain and color and laid parallel to the grain of the sheet, so that the repairs are as unobtrusive as possible. No more than six repairs can be made per 4-foot by 8-foot panel. In short, Grade N veneer presents an excellent and attractive surface. Its use is limited to interior plywood. Although it may be applied to both face and back of a panel, it is generally used only as a face, with a lesser grade as a back. The inner plies are of either Grade C or D, and all panels made with N veneers are special order items.

■ **Grade A Veneer.** This is the next lowest category, and is widely used and readily obtainable in both exterior and interior types of the appearance grade plywoods. Grade A veneers are smooth and are suitable for painting, thought they are frequently used for a natural or a semitransparent finish. The veneer must be free of knots, pitch pockets and various kinds of open defects. When more than one piece of veneer is applied on a panel as a face or back, the pieces must be well-jointed. Synthetic fillers can be used to fill defects of larger size than is possible with Grade N, and as many as 18 well-made repairs of either wood or synthetic material, parallel to the grain, can appear on a 4-foot by 8-foot panel. The use of both sapwood and heartwood together is permissible, and slight discolorations or staining are also allowed. Plywood panels may be made with Grade A veneers on both faces and backs, but the backs are more often of a lesser grade. The inner ply grade is D for interior plywoods and C for exterior.

■ **Grade B Veneer.** This kind is solid and presents a reasonably good appearance, but it is generally used where the application is more utilitarian and the appearance does not matter as much. There may be a few minor sanding skips or defects, and the surface may be slightly rough. Minor splits are allowable, and so are tight knots up to 1 inch across the grain. Synthetic fillers are used to repair small splits or chipped areas or openings; their size limitations depend upon whether or not the panels are exterior or interior. Repairs can be made with either wood or synthetic material. Though they must be well-made, they may be quite obvious, as they need not be aligned with the the grain and may be made of circular plugs. Certain restrictions are placed upon the size of such repairs.

■ **Grade C-Plugged.** This is a utilitarian grade that is a slight improvement upon the next lower level in the grading system, Grade C. Splits may be as wide as ⅛-inch, and there may be open defects like knotholes or wood-borer holes as large as ¼-inch by ½-inch. There may be broken grain, and tight, sound knots up to 1½ inches wide across the grain. A variety of synthetic repairs in the way of plugs, patches and shims may be used. This grade may be identified as being fully sanded, in which case the sanding defects allowable are the same as for those of Grade B. Otherwise, in the so-called touch-sanded category there may be any degree of sander skips. Repairs are made with synthetic materials, not wood.

■ **Grade C.** This grade of somewhat lesser quality than Grade C-plugged. There may be substantial discoloration, and knotholes as much as 1½ inches across the grain may appear. Splits may run as wide as ⅜ inch to ½ inch, and voids that occur from missing veneer strips may be as wide as 1½ inches and as long as 6 inches. All manner of repairs can be made with either wood or synthetic patching material, though there are certain restrictions upon them as to size and placement. Grade C is used to some extent in exterior appearance grade plywood, and to a considerable extent in the engineered grades, both interior and exterior types. The inner ply grade may be either C or D.

■ **Grade D Veneer.** This one lies at the very bottom of the scale. As you might expect, all manner of defects are visible, but they cannot be so large or so numerous as to impair the serviceability or the strength of the completed plywood panels. Grade D is used as panel backs, and may have knots as large as 3 inches across the grain and knotholes equally as large under certain conditions. Splits are permissible, and so are voids. Grade D veneers are used only on interior plywood, in both appearance grade and engineered grade categories.

Grade Letters

An understanding of the veneer grades which are capsulized in Table 2-4 is important, because at least part of the selection process of plywood for various applications must be based upon them. An essential part of the specification terms that are used when discussing or obtaining plywood is these very same grade letters. One of the most commonly used plywoods in building construction and general purpose applications, for instance, is simply referred to as C-D. The term is so widely used that it has become generic and all-inclusive; mention the term "C-D"—you don't even have to say plywood—to anyone in the building trades and he will immediately take your meaning. But actually there are several types of C-D plywood. If you are referring to something other than C-D INT-APA or its equivalent, you had best be more specific.

Table 2-4. Softwood Plywood Veneers Are Categorized in These Grades by the APA (courtesy of the American Plywood Association).

N	Smooth surface "natural finish" veneer. Select, all heartwood or all sapwood. Free of open defects. Allows not more than 6 repairs, wood only, per 4x8 panel, made parallel to grain and well matched for grain and color.
A	Smooth, paintable. Not more than 18 neatly made repairs, boat, sled, or router type, and parallel to grain, permitted. May be used for natural finish in less demanding applications.
B	Solid surface. Shims, circular repair plugs and tight knots to 1 inch across grain permitted. Some minor splits permitted.
C Plugged	Improved C veneer with splits limited to 1/8 inch width and knotholes and borer holes limited to 1/4 x 1/2 inch. Admits some broken grain. Synthetic repairs permitted.
C	Tight knots to 1-1/2 inch. Knotholes to 1 inch across grain and some to 1-1/2 inch if total width of knots and knotholes is within specified limits. Synthetic or wood repairs. Discoloration and sanding defects that do not impair strength permitted. Limited splits allowed. Stitching permitted.
D	Knots and knotholes to 2-1/2 inch width across grain and 1/2 inch larger within specified limits. Limited splits allowed. Stitching permitted. Limited to Interior grades of plywood.

Note that a veneer of one of the lower grades as it comes from the cutting and drying machinery can frequently be raised to the next higher level, provided that suitable repairs can be made within the specifications laid down by Product Standard PS1. Such repairs are made immediately after the veneer appears from the drying process. Once repairs have been made and a grade established, the veneers are then processed into completed plywood panels.

Appearance Grades

Once the various veneer grades have been assembled into the several specified combinations for different kinds and classes of plywood, the resulting plywood panels are then further graded as unit pieces, divided into two major categories: *appearance* and *engineered*. These categories are based upon use of the plywood panels, and each category is further broken down into specific kinds of plywood.

As the name implies, plywood panels whose appearance is a primary consideration—even though other factors may be of considerable or even great importance—make up this category. Though the panels are selected, and a manufacturing process is geared toward a presentable and good looking panel face (and sometimes back as well), this does not mean that the plywood panels are lacking in the typical characteristics of strength, stiffness, durability and so on. In fact, appearance is an added factor. The panels are made in several different forms, but include those with smooth, sanded surfaces that are ready for finishing, textured panels like those used for exterior siding on houses, and various special overlaid surfaces. Veneer Grades N, A and B are used as faces, and the backs range from Grades N to C. Both interior and exterior types of plywood panels are available in the appearance grades, as well as a wide range of thicknesses. Table 2-5 lists those appearance grade plywood panels that are commonly available; use of this listing makes selection of plywood panels for various applications a simple matter. Note that the Grade N veneer is not available in many parts of the country, since this fine veneer is in very short supply. Before establishing final specifications for a project, check local availability first. You may be forced to substitute an A grade for an N grade.

Plywood panels used for exterior sidings are available in many different surface patterns, and they have their own grading system in addition to the grading already discussed. Overlaid plywoods

Table 2-5. The APA Guide to Appearance Grades of Softwood Plywood Panels (courtesy of the American Plywood Association).

		Typical Grade-trademarks	A	C	C			
Exterior Type	A-C EXT-APA	Use where the appearance of only one side is important. Soffits, fences, structural uses, boxcar and truck lining, farm buildings. Tanks, trays, commercial refrigerators. (4)	A-C GROUP EXTERIOR PS 1-74 000	B	C	1/4	3/8	1/2	5/8	3/4
	B-B EXT-APA	Utility panel with solid faces. (4)		B	C	1/4	3/8	1/2	5/8	3/4
	B-C EXT-APA	Utility panel for farm service and work buildings, boxcar and truck lining, containers, tanks, agricultural equipment. Also as base for exterior coatings for walls, roofs. (4)	B-C GROUP EXTERIOR PS 1-74 000	B	C	1/4	3/8	1/2	5/8	3/4
	HDO EXT-APA	High Density Overlay plywood. Has a hard, semi-opaque resin-fiber overlay both faces. Abrasion resistant. For concrete forms, cabinets, counter tops, signs, tanks. (4)	HDO A·A G1 EXT-APA PS 1-74	A or B	C or C plgd	A or B		3/8	1/2	5/8	3/4
	MDO EXT-APA	Medium Density Overlay with smooth, opaque, resin-fiber overlay one or both panel faces. Highly recommended for siding and other outdoor applications, built-ins, signs, displays. Ideal base for paint. (4)(6)	MDO B·B G1 EXT-APA PS 1-74 000	B	C	B or C		3/8	1/2	5/8	3/4
	303 SIDING EXT-APA	Proprietary plywood products for exterior siding, fencing, etc. Special surface treatment such as V-groove, channel groove, striated, brushed, rough-sawn and texture-embossed MDO. Stud spacing (Span Index) and face grade classification indicated on grade stamp.	303 SIDING 16 GROUP 2 EXTERIOR PS 1-74 000	(5)	C			3/8	1/2	5/8	
	T 1-11 EXT-APA	Special 303 panel having grooves 1/4" deep, 3/8" wide, spaced 4" or 8" o.c. Other spacing optional. Edges shiplapped. Available unsanded, textured and MDO.	303 SIDING 16 oc GROUP 2 EXTERIOR PS 1-74 000	C or btr.	C				19/32	5/8	
	PLYRON EXT-APA	Hardboard faces both sides, tempered, smooth or screened.	PLYRON EXT-APA 000		C				1/2	5/8	3/4
	MARINE EXT-APA	Ideal for boat hulls. Made only with Douglas fir or western larch. Special solid jointed core construction. Subject to special limitations on core gaps and number of face repairs. Also available with HDO or MDO faces.	MARINE A·A EXT-APA PS 1-74 000	A or B	B	A or B	1/4	3/8	1/2	5/8	3/4

Grade Designation (2)	Description and Most Common Uses	Typical (3) Grade-trademarks	Veneer Grade Face	Veneer Grade Inner plies	Veneer Grade Back	Most Common Thicknesses (inch)				
Interior Type										
N-N, N-A N-B INT-APA	Cabinet quality. For natural finish furniture, cabinet doors, built-ins, etc. Special order items.		N	C	N,A or B					3/4
N-D-INT-APA	For natural finish paneling. Special order item.		N	D	D	1/4				
A-A INT-APA	For applications with both sides on view, built-ins, cabinets, furniture, partitions. Smooth face, suitable for painting.		A	D	A	1/4	3/8	1/2	5/8	3/4
A-B INT-APA	Use where appearance of one side is less important but where two solid surfaces are necessary.		A	D	B	1/4	3/8	1/2	5/8	3/4
A-D INT-APA	Use where appearance of only one side is important. Paneling, built-ins, shelving, partitions, flow racks.	A-D GROUP 2 APA INTERIOR P.S. 1-74 000	A	D	D	1/4	3/8	1/2	5/8	3/4
B-B INT-APA	Utility panel with two solid sides. Permits circular plugs.		B	D	B	1/4	3/8	1/2	5/8	3/4
B-D INT-APA	Utility panel with one solid side. Good for backing, sides of built-ins, industry shelving, slip sheets, separator boards, bins.	B-D GROUP 2 APA INTERIOR P.S. 1-74 000	B	D	D	1/4	3/8	1/2	5/8	3/4
DECORATIVE PANELS—APA	Rough-sawn, brushed, grooved, or striated faces. For paneling, interior accent walls, built-ins, counter facing, displays, exhibits.	DECORATIVE B-D G1 INT-APA PS 1-74	C or btr.	D	D		5/16	3/8	1/2	5/8
PLYRON INT-APA	Hardboard face on both sides. For counter tops, shelving, cabinet doors, flooring. Faces tempered, untempered, smooth, or screened.	PLYRON INT-APA 000		C & D				1/2	5/8	3/4
A-A EXT-APA	Use where appearance of both sides is important. Fences, built-ins, signs, boats, cabinets, commercial refrigerators, shipping containers, tote boxes, tanks, ducts. (4)		A	C	A	1/4	3/8	1/2	5/8	3/4
A-B EXT-APA	Use where the appearance of one side is less important. (4)		A	C	B	1/4	3/8	1/2	5/8	3/4

(1) Sanded both sides except where decorative or other surfaces specified.
(2) Can be manufactured in Group 1, 2, 3, 4 or 5.
(3) The species groups, Identification Indexes and Span Indexes shown in the typical grade-trademarks are examples only. See "Group," "Identification Index" and "Span Index" for explanations and availability.
(4) Can also be manufactured in Structural I (all plies limited to Group 1 species) and Structural II (all plies limited to Group 1, 2, or 3 species).
(5) C or better for 5 plies, C Plugged or better for 3 plies.
(6) Also available as a 303 siding.

that have special surfaces also are considered somewhat differently. Both of these will be discussed a bit later on.

Choosing the right plywood for a particular job is simple once you become familiar with the listing in Table 2-5. All you need to do is first determine whether or not you need an exterior or interior type of plywood. If the plywood will be exposed to the weather, whether painted or not, or if severe moisture conditions will be encountered, you will need an exterior type. If the plywood will be permanently protected and under normal use subject to no more moisture than is found in average relative humidity conditions, an interior type of plywood will be satisfactory. Of course, there are a great many instances that are not particularly clear-cut.

For instance, a piece of lawn or patio furniture made from plywood might be normally kept indoors, or on a covered porch, but could occasionally be left outdoors and drenched in a sudden rainstorm. In this case, an exterior type of plywood would certainly do the job with no problems, but an interior type bonded with exterior glue would doubtless also serve, particularly if the piece of furniture were well-finished and the wetting or weathering were only occasional and short-term. So, some degree of judgment might be exercised in situations of this sort.

Once that decision is made, check the "common uses" column in the guide and find a use that is akin (or identical) to the project you have in mind. For instance, if you want to build a child's sandbox, obviously you should use an exterior plywood since the sandbox will doubtless be out in the weather more or less permanently, and will be filled with damp or wet sand most of the time. Looking under the use column for Exterior plywood, we find that there are several possibilities—outdoor built-ins, tanks, outdoor shelving or service buildings or bins, for instance. Again, some judgement must be exercised. Looking in the next column, which gives the grade of the face and back veneers, it is apparent that several possibilities exist. The A-A plywood could be used, but is probably a bit much (and a little expensive) for an ordinary sandbox. The A-B grade is next, and that would work well with the face on the outside and the B back on the inside. The A-C could also be used, but the C back is likely to be a bit rough and would perhaps be a little unsuitable, though it certainly would stand up all right. The best all-around combination might well be the B-B plywood, since it is a utility grade and will afford good service while at the same time presenting a decent appearance and being capable of taking a good paint job.

Engineered Grades

The engineered grades of plywood are those that are primarily used for construction purposes. Though most panels of this type are used by the building trades for such purposes as wall sheathing, subflooring, roof decking and things of this nature, the engineered grades are also widely used in making stressed-skin panels, gusset plates, bins, tanks. Other uses are for siding on farm buildings, as underlayment, and as backing for walls or shower tiles, in building concrete forms, and for a long list of miscellaneous purposes. Appearance is of little or no consequence in these plywoods; strength, serviceability and durability are. In short, these are the workhorses of the plywood field and can be depended upon to do just about anything (except present a nice appearance).

In this category of plywood panels, there are two important subcategories: *Structural I* and *Structural II*. In the guide to engineered grades of plywood shown in Table 2-6, only *C-D INT* and *C-C EXT* are shown. However, there are other Structural grades as well, such as *C-D Plugged Interior* and *Interior Underlayment*, both of which are available in both I and II grades. All of the Structural grades are made only with exterior glue, whether they are listed as interior types or not. The Structural I grades are made with Group 1 woods only, while the Structural II grades may include Group 2 or Group 3 woods. Only small quantities of Structural II grades of plywood panels are produced, and they are not readily available. Check local supplies first before specifying these panels.

Of all of the various plywoods listed in the guide, the C-D INT is doubtless the most widely used and can be employed for virtually any kind of job, from structural covering of roofs and floors to the making of crates and pallets. This plywood is manufactured with interior, intermediate and exterior glues, though the intermediate type is of limited production. Those with the exterior glue have come to be generally known as "CDX" plywood by dealers and building construction workmen alike. It should not, however, be confused with a fully exterior panel; the proper term for that type is C-C EXT, or one of the Structural grades.

The last category listed in the guide is *B-B PLYFORM*, which is an APA trade name for a special type of panel that has been developed especially for use in making concrete forms. It is a very sturdy and stiff kind of plywood, engineered to withstand the high pressures and hard knocks that are inevitable in concrete work. Properly cared for, the panels can be used again and again. Com-

Table 2-6. The APA Guide to Engineered Grades of Softwood Plywood Panels (courtesy of the American Plywood Association).

	Grade Designation	Description and Most Common Uses	Typical[1] Grade-trademarks	Veneer Grade Face	Veneer Grade Inner Plies	Veneer Grade Back	Most Common Thicknesses (inch)				
Interior Type	C-D INT-APA	For wall and roof sheathing, subflooring, industrial uses such as pallets. Most commonly available with exterior glue (CDX). Specify exterior glue where construction delays are anticipated and for treated-wood foundations. (7)	C-D 32/16 (APA) INTERIOR PS 1-000 / C-D 24/0 (APA) INTERIOR PS 1-000 EXTERIOR GLUE	C	D	D	5/16	3/8	1/2	5/8	3/4
	STRUCTURAL I C-D INT-APA and STRUCTURAL II C-D INT-APA	Unsanded structural grades where plywood strength properties are of maximum importance: structural diaphragms, box beams, gusset plates, stressed-skin panels, containers, pallet bins. Made only with exterior glue. See (6) for species group requirements. Structural I more commonly available. (7)	STRUCTURAL I C-D 24/0 (APA) INTERIOR PS 1-000 EXTERIOR GLUE	C[3]	D[3]	D[3]	5/16	3/8	1/2	5/8	3/4
	STURD-I-FLOOR INT-APA	For combination subfloor-underlayment. Provides smooth surface for application of resilient floor covering. Possesses high concentrated- and impact-load resistance during construction and occupancy. Manufactured with exterior glue only. Touch-sanded. Available square edge or tongue-and-groove. (7)	STURD-I-FLOOR 24oc T&G 23/32 INCH (APA) INTERIOR PS 1-000 EXTERIOR GLUE NRB-108	C Plugged	(4)	D				19/32 5/8	23/32 3/4
	STURD-I-FLOOR 48 O.C. (2-4-1) INT-APA	For combination subfloor-underlayment on 32- and 48-inch spans. Provides smooth surface for application of resilient floor coverings. Possesses high concentrated- and impact-load resistance during construction and occupancy. Manufactured with exterior glue only. Unsanded or touch-sanded. Available square edge or tongue-and-groove. (7)	STURD-I-FLOOR 48oc T&G 1-1/8 INCH (APA) INTERIOR PS 1-000 EXTERIOR GLUE NRB-108	C Plugged	C[5] & D	D			1-1/8		
	UNDERLAYMENT INT-APA	For application over structural subfloor. Provides smooth surface for application of resilient floor coverings. Touch-sanded. Also available with exterior glue.	UNDERLAYMENT GROUP 1 (APA) INTERIOR PS 1-000	C Plugged	C[5] & D	D		3/8	1/2	19/32 5/8	23/32 3/4
	C-D PLUGGED INT-APA	For built-ins, wall and ceiling tile backing, cable reels, walkways, separator boards. Not a substitute for Underlayment or Sturd-I-Floor as it lacks their indentation resistance. Touch-sanded. Also made with exterior glue. (2) (6)	C-D PLUGGED GROUP 2 (APA) INTERIOR PS 1-000	C Plugged	D	D		3/8	1/2	19/32 5/8	23/32 3/4

Exterior Type

					5/16	3/8	1/2	5/8	3/4
C-C EXT-APA	Unsanded grade with waterproof bond for subflooring and roof decking, siding on service and farm buildings, crating, pallets, pallet bins, cable reels, treated wood foundations. (7)	C-C 42/20 EXTERIOR APA 000	C	C	5/16	3/8	1/2	5/8	3/4
STRUCTURAL I C-C EXT-APA and STRUCTURAL II C-C EXT-APA	For engineered applications in construction and industry where full Exterior type panels are required. Unsanded. See (6) for species group requirements. (7)	STRUCTURAL I C-C 32/16 EXTERIOR APA 000	C	C	5/16	3/8	1/2	5/8	3/4
STURD-I-FLOOR EXT-APA	For combination subfloor-underlayment under resilient floor coverings where severe moisture conditions may be present, as in balcony decks. Possesses high concentrated and impact-load resistance during construction and occupancy. Touch-sanded. Available square edge or tongue-and-groove. (7)	STURD-I-FLOOR 20oc 5/8 INCH EXTERIOR APA 000	C Plugged	C(5)				19/32 5/8	23/32 3/4
UNDERLAYMENT C-C PLUGGED EXT-APA	For application over structural subfloor. Provides smooth surface for application of resilient floor coverings where severe moisture conditions may be present. Touch-sanded. (2)(6)	UNDERLAYMENT GROUP 2 EXTERIOR APA 000	C Plugged	C(5)		3/8	1/2	19/32 5/8	23/32 3/4
C-C PLUGGED EXT-APA	For use as tile backing where severe moisture conditions exist. For refrigerated or controlled atmosphere rooms, pallet fruit bins, tanks, box car and truck floors and linings, open soffits. Touch-sanded. (2)(6)	C-C PLUGGED GROUP 2 EXTERIOR APA 000	C Plugged	C		3/8	1/2	19/32 5/8	23/32 3/4
B-B PLYFORM CLASS I & CLASS II EXT-APA	Concrete form grades with high reuse factor. Sanded both sides. Mill-oiled unless otherwise specified. Special restrictions on species. Available in HDO and Structural I. Class I most commonly available. (8)	B-B PLYFORM CLASS I EXTERIOR APA 000	B	C				5/8	3/4

(1) The species groups, Identification Indexes and Span Indexes shown in the typical grade-trademarks are examples only. See "Group," "Identification Index" and "Span Index" for explanations and availability.
(2) Can be manufactured in Group 1, 2, 3, 4, or 5.
(3) Special improved grade for structural panels.
(4) Special veneer construction to resist indentation from concentrated loads, or other solid wood-base materials.
(5) Special construction to resist indentation from concentrated loads.
(6) Can also be manufactured in Structural I (all plies limited to Group 1 species) and Structural II (all plies limited to Group 1, 2, or 3 species).
(7) Specify by Identification Index for sheathing and Span Index for Sturd-I-Floor panels.
(8) Made only from certain wood species to conform to APA specifications.

parable types of plywood for concrete forming are manufactured by other concerns, without the APA trademark. Unless you plan to use and re-use this kind of plywood for its intended purpose, however, other and less expensive kinds of plywood can also be used for concrete forming. In fact, a fairly common practice is to use CDX for forming the foundation walls of a house on a one-time basis and then recycle the plywood panels into floor, wall or roof sheathing.

Siding Face Grades

A great many different kinds of plywood exterior sidings are made for application on houses and other buildings. Of the sidings, a great proportion is made by APA members under the trademark of *303 Series Specialty Siding*. This group of plywoods, which is listed in the guide in Table 2-5 simply as 303 Siding, actually has a fairly involved grading system of its own. The group is broken down into four basic classes: Special Series 303, 303-6, 303-18, and 303-30. Each of the four classes is further broken down into grades, four for the Special Series 303, and there each for the remaining classes. These sidings are manufactured under PS1, as are other plywoods, and also meet various proprietary specifications of the APA. The grading has to do with the appearance characteristics and the number and kind of repairs that can be made to the face veneers. The Special Series is of the highest quality, while the 303-30 panels are of the lowest quality of the group. Price, of course, is commensurate. Table 2-7 lists the various grades.

These plywoods are available in a great many different patterns and finishes, which include overlaying, striating, embossing, scratch-sanding, brushing, rough-texturing, grooving and the like. Among the most popular in this group of plywoods is the *Texture 1-11* (T 1-11), a designation given to a specific grooving detail used only on 5-ply panels with a nominal thickness of at least 19/32 inch. The spacing of the grooves is standard at either 4 or 8 inches center-to-center, but other spacings can be obtained.

Redwood Plywood Grades

One of the more popular kinds of finish exterior siding is *redwood plywood*. Much of this material, as well as redwood lumber, carries the trademark of the *California Redwood Association* (CRA). Member companies of the CRA manufacture redwood plywood siding under U.S. Product Standard PS1, and the material meets the APA 303 Siding Specification when APA grade marked.

Table 2-7. The APA 303 Plywood Siding's Special Grading System (courtesy of the American Plywood Association).

Class	Grade[1]	Patches	
		Wood	Synthetic
Special Series 303	303-OC[2][3]	Not permitted	Not permitted
	303-OL[4]	Not applicable for overlays	
	303-NR[5]	Not permitted	Not permitted
	303-SR[6]	Not permitted	Permitted as natural-defect shape only
303-6	303-6-W	Limit 6	Not permitted
	303-6-S	Not permitted	Limit 6
	303-6-S/W	Limit 6 — any combination	
303-18	303-18-W	Limit 18	Not permitted
	303-18-S	Not permitted	Limit 18
	303-18-S/W	Limit 18 — any combination	
303-30	303-30-W	Limit 30	Not permitted
	303-30-S	Not permitted	Limit 30
	303-30-S/W	Limit 30 — any combination	

(1) Limitations on grade characteristics are based on 4 ft. x 8 ft. panel size. Limits on other sizes vary in proportion. All panels except 303-NR allow restricted minor repairs such as shims. These and such other face appearance characteristics as knots, knotholes, splits, etc., are limited by both size and number in accordance with panel grades, 303-OC being most restrictive and 303-30 being least. Multiple repairs are permitted only on 303-18 and 303-30 panels. Patch size is restricted on all panel grades. For additional information, including finishing recommendations, see APA's *303 Plywood Siding—Grades and Finishing*.
(2) Check local availability.
(3) "Clear"
(4) "Overlaid" (e.g., Medium Density Overlay siding)
(5) "Natural Rustic"
(6) "Synthetic Rustic"

This plywood is manufactured with a saw-textured surface, and the face veneers are available in two grades. The top grade is "Clear All Heart," and the panels are identified by proprietary tradenames such as "Premium Rough-Sawn," and "Clear Heart Saw-Textured." The lower grade is made up of sapwood-containing veneers, which are identified by such proprietary names as "Custom Rough-Sawn," "Select Saw-Textured," and "Select Ruff-Sawn"

In the top grade, the veneers must be relatively closely matched. Pin knots up to ⅜-inch diameter are allowable, along with certain patches reasonably matched in. Synthetic repairs may be used. The economy grades do not allow knots, knotholes or open slits, but wood or synthetic shims, pin knots up to ⅜-inch diameter and various patches are allowable.

Redwood siding panels may contain inner plies of redwood, but other species may be used except in certain instances where the center ply of 5-ply types must be of redwood. The panels are made up with fully waterproof exterior glue, and are factory-coated with water repellant. Available thicknesses are ⅜-inch and ⅝-inch, 3-ply and 5-ply respectively, and the panels are made in standard lenghs of 8, 9 or 10 feet. Those panels that include a shiplap-edge for tightly joining abutting panels measure 48⅜ inches wide instead of the standard 4 feet. When they are lapped, the net coverage is 48 inches. As with other types of exterior plywood siding, the T 1-11 pattern is one of the most popular. Standard redwood profiles are shown in Table 2-8. Note that here the standard groove spacing for T 1-11 is 4 inches, 8 inches, or 4 inches and 12 inches.

Plywood Identification Index

As was explained earlier, most plywood panels are marked in one fashion or another with a trademark, or a combination grade-trademark. A small amount of plywood production is unmarked. Unless the prospective buyer has access to the specifications under which that plywood was manufactured, he has no real knowledge of exactly what he is about to buy. Likewise, purchasers of those plywood panels that carry only a company trademark are not much more aware of the product quality, except insofar as the company name itself may stand for a quality product. But there are two sets of markings in particular, one or the other or both of which are found on most of the softwood plywood panels manufactured in this country, that are of considerable importance to the prosepctive buyer. Knowing how to read them and what the numbers and symbols stand for is a great aid in specifying plywoods on building plans, selecting plywood panels at the lumberyard for various purposes, or for ascertaining that you have indeed been delivered the particular kind of plywood that you requested. The first of these is the *identification index*, and the second is the *grade-trademark*.

The identification index is found upon all plywood panels in the basic unsanded grades produced by APA members: C-C and C-D Interior, Exterior, and Structural I and II. There are two sets of numbers, separated by a slash, found in the grade-trademark or on the edge mark. For example, you might find the numbers 32/16, or 16/0. The number on the left has to do only with roofing, and is of consequence only when the plywood is intended for a roof sheathing. The number on the right is used only when the plywood is to be employed as subflooring.

Table 2-8. Identification Index Guide for Softwood Plywood Panels Used in Construction and Building (courtesy of the American Plywood Association).

Species of face and back	Grade		
Group 1	C-C, Str. I C-C, C-D, Str. II C-C, C-D [c], C-D — (b)		
Group 2 (d)	C-C, Str. II C-C, C-D, C-D	C-C, Str. II C-C, C-D, C-D — (b)	
Group 3		C-C, Str. II C-C, C-D, C-D	
Group 4		C-C, C-D — (d)	C-C, C-D — (b)
Nominal Thickness			
5/16	20/0	16/0	12/0
3/8	24/0	20/0	16/0
1/2	32/16	24/0	24/0
5/8	42/20	32/16	30/12
3/4	48/24	42/20	36/16
7/8		48/24	42/20
(e)			

(a) Identification Index refers to the numbers in the lower portion of the table which are used in the marking of sheathing grades of plywood. The numbers are related to the species of panel face and back veneers and panel thickness in a manner to describe the bending properties of a panel. They are particularly applicable where panels are used for subflooring and roof sheathing to describe recommended maximum spans in inches under normal use conditions and to correspond with commonly accepted criteria. The left hand number refers to spacing of roof framing with the right hand number relating to spacing of floor framing. Actual maximum spans are established by local building codes. See reference source given in section 2 for complete description and product use information.

(b) Panels of standard nominal thickness and construction.

(c) Panels manufactured with Group 1 faces but classified as Structural II by reason of Group 2 or Group 3 inner plys.

(d) Panels conforming to the special thickness and panel construction provisions of 3.8.6.

(e) Panels thicker than 7/8 inch shall be identified by group number.

Here's how the numbers work. Suppose the identification index is 32/16. This means that the maximum recommended spacing of the roof rafters is 32 inches for this particular kind of plywood. The rafters may be on less than 32-inch centers, in which case the roof will be just that much ruggeder, but the spacing can be no more than 32 inches. If the same plywood were to be used for subflooring, the maximum spacing on centers for the floor joists would be 16 inches. Less than that distance would be fine and would result in a stouter floor, but this particular plywood should not be placed upon wider-spaced joists.

In the case of the number, 16/0, this simply means that the plywood can be used as roof decking where the rafters are spaced at a maximum of 16 inches on centers. The 0 means that the plywood should not be used under any circumstances as subflooring. Note that plywood designated as underlayment does not carry an identification index number, even though it is used on floors. Here the important factor is the group number, because the stiffness of the wood makes the difference in the thickness of plywood required.

One important point should be made with respect to the use of identification indexes. The framing member spans, or center-to-center spacings, that are listed in Table 2-8 are based upon the fact that the face grain of the plywood panel will always be placed at right angles to the supporting framing members (that is, the principal members such as joists and rafters). In every case the panel will be continuous across two or more spans. A short length of plywood placed across a single span exhibits quite different deflection and other characteristics than does a full, or even a half, panel.

Span Index

The *span index* is a number that appears on some types of APA plywoods, in particular, *Sturd-I-Floor* and 303 Siding. The latter are special panels that are designed to be laid in one operation, as a single-layer floor combining subflooring and underlayment, with only a finish flooring (carpeting, sheet vinyl flooring) left to be applied. The span indexes may be 16, 20, 24 or 48. These numbers simply stand for the distance in inches between centers of the floor framing members over which the panels can be installed. Several thicknesses of panels may be manufactured which are acceptable over a single span width, or a single thickness of panel may be marked as suitable for two different spans. In any case, the panels should never be used over spans greater than the largest number indicated in the span index. The 303 Sidings are manufactured with span indexes of 16 and 24 inches. They may be secured over nonstructural sheathing, over lumber or plywood sheathing, or in many cases directly to the wall studs. The panel should be applied over stud spacings no greater than the span index indicated on the panel, if the panels are positioned vertically. However, if the panels are positioned horizontally, they can be applied directly to studs either 16 or 24 inches on center.

Note that both the identification index and the span index numbers are accepted by the major building codes. However, local interpretation or regulations can vary from these standards and should always be checked before proceeding on the basis of the imprinted indexes.

Grade-Trademarks

A grade-trademark, sometimes abbreviated GTM, is a marking generally stamped on the back of a plywood panel. It may also be embossed into the edge of a panel, in which case it is called an edge mark. The components of the marking give the basic information about the plywood panel.

With respect to the grade-trademark, at the top left or across the top of the mark appears either a pair of large capital letters, or a term. The letters such as A-C or C-D indicate the veneer grade used on the face (left-hand letter) and on the back (right-hand letter) of the panel. The nomenclature indicates the general type of panel, such as M.D. Overlay, or Structural I. In some instances, the nomenclature will be at the top of the grade-trademark, with a pair of veneer grade letters located below and to the left.

In the center of the grade-trademark, on the left side, the Species Group Number appears. Directly below this, the type of plywood, interior or exterior, is noted. Directly below in the lefthand corner, the Product Standard that governs the manufacture of the panel is stamped; PS1-74 is the standard currently in effect. Centered in the bottom boundary line is another number, and this designates the particular mill that produced the plywood. In some cases, the words "Exterior Glue" will also appear below the grade-trademark.

Edge marks are read in much the same way, except the information goes from left to right. The first set of letters indicates the veneer grades of the panel face and back, and the second set indicates the species group number. Then comes the type designation, followed by the product standard and the mill number. In the grade-trademarks, the APA logo appears at the right, while in the edge mark it appears directly after the type designation. Thus,

Sanded Grades

- Species Group number
- Designates the type of plywood
- Product Standard governing manufacture
- Grade of veneer on panel face
- Grade of veneer on panel back
- Mill number

(Also available in Groups 1, 3, and 4)

Unsanded Grades

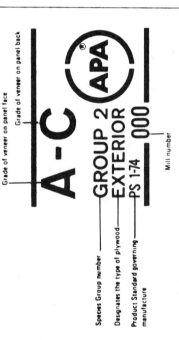

- Grade of veneer on panel face
- Grade of veneer on panel back
- Identification Index
- Designates the type of plywood
- Product Standard governing manufacture
- Type of glue used
- Mill number

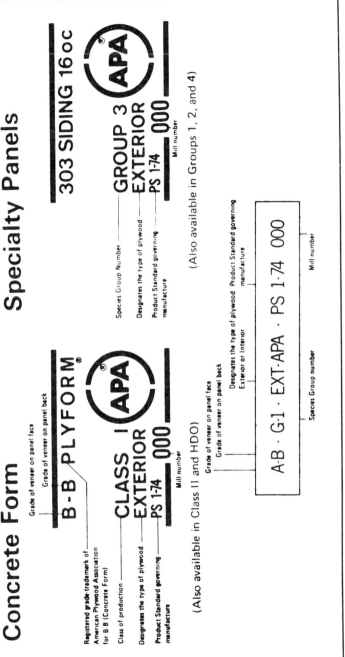

Fig. 2-2. Explanations of typical grade-trademarks (courtesy of the American Plywood Association).

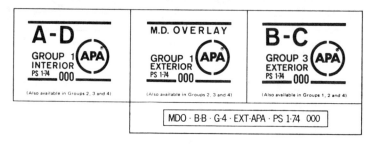

Fig. 2-3. These typical grade-trademarks are found on appearance grades of softwood plywood panels manufactured under PS1 (courtesy of the American Plywood Association).

while in many cases one panel of plywood looks much like another, by inspecting the grade-trademark or edge mark you can quickly ascertain exactly what the panel is. Examples of typical markings are shown in Fig. 2-2. Other typical grade-trademarks and edge marks are shown in Figs. 2-3 and 2-4.

Finishes

Most of the softwood plywood panels manufactured today are unfinished. There are a few exceptions. The various exterior siding plywoods mentioned earlier have textured or otherwise treated surfaces, and the special MDO/HDO panels are also treated with surfacing materials. But many of the plain panels are surfaced by sanding, and important factor in many do-it-yourself projects. There are three possibilities: *unsanded, touch-sanded*, and *fully sanded*. In fully sanded panels the surfaces are well-smoothed in a uniform fashion. Touch-sanding, however, consists of merely hitting the roughest spots with a small portable sander. Surface treatments for the various interior type grades are shown in Table 2-9, and those for exterior type grades are in Table 2-10.

Manufacturing Standards

We have previously mentioned that much of the softwood plywood manufactured in this country is made according to the basic standards set forth in U.S. Product Standard PS1. The one in current effect is PS1-74, which supercedes PS1-66; the standard is revised periodically as necessary to keep up with new developments and findings in the plywood field. Going into the specifics of this standard is not necessary. Some of the details have been and will be noted as the occasion warrants. Much of the information

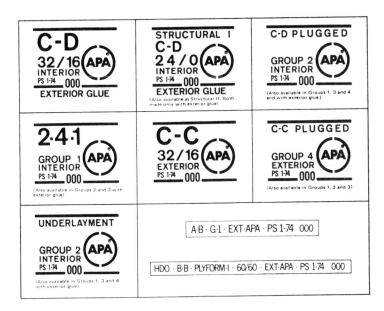

Fig. 2-4. These are some of the typical grade-trademarks that appear on engineered grades of softwood plywood panels (courtesy of the American Plywood Association).

Table 2-9. Panel Grades, Veneer Combinations and Surfacing Required Under PS1 for Interior Grades of Softwood Plywood Panels (courtesy of the American Plywood Association).

Panel Grade Designations	Minimum Veneer Quality Face	Back	Inner Plys	Surface
N-N	N	N	C	Sanded 2 sides
N-A	N	A	C	Sanded 2 sides
N-B	N	B	C	Sanded 2 sides
N-D	N	D	D	Sanded 2 sides
A-A	A	A	D	Sanded 2 sides
A-B	A	B	D	Sanded 2 sides
A-D	A	D	D	Sanded 2 sides
B-B	B	B	D	Sanded 2 sides
B-D	B	D	D	Sanded 2 sides
Underlayment[a]	C Plugged	D	C & D	Touch-sanded
C-D Plugged	C Plugged	D	D	Touch-sanded
Structural I C-D		See 3.6.5		Unsanded[b]
Structural I C-D Plugged, Underlayment		See 3.6.5		Touch-sanded
Structural II C-D		See 3.6.5		Unsanded[b]
Structural II C-D Plugged, Underlayment		See 3.6.5		Touch-sanded
C-D	C	D	D	Unsanded[b]
C-D with exterior glue (See para. 3.6.6)	C	D	D	Unsanded[b]

(a) See 3.6.3 and Table 5 for special limitations.

(b) Except for decorative grades, panels shall not be sanded, touch-sanded, surface textured, or thickness sized by any mechanical means.

Table 2-10. Panel Grades, Veneer Combinations and Surfacing Required Under PS1 for Exterior Grades of Softwood Plywood Panels (courtesy of the American Plywood Association).

Panel Grade Designations	Minimum Veneer Quality			Surface
	Face	Back	Inner Plys	
Marine, A-A, A-B, B-B, HDO, MDO	See 3.6.1			See regular grades
Special Exterior, A-A, A-B, B-B, HDO, MDO	See 3.6.7			See regular grades
A-A	A	A	C	Sanded 2 sides
A-B	A	B	C	Sanded 2 sides
A-C	A	C	C	Sanded 2 sides
B-B (concrete form)	See 3.6.4			
B-B	B	B	C	Sanded 2 sides
B-C	B	C	C	Sanded 2 sides
C-C Plugged[b]	C Plugged	C	C	Touch-sanded
C-C	C	C	C	Unsanded[c]
A-A High Density Overlay	A	A	C Plugged	—
B-B High Density Overlay	B	B	C Plugged[d]	—
B-B High Density Concrete Form Overlay (See para. 3.6.4)	B	B	C Plugged	—
B-B Medium Density Overlay	B	B	C	—
Special Overlays	C	C	C	—

(a) Available also in Structural I and Structural II classifications as provided in paragraph 3.6.5.
(b) See 3.6.3 and Table 5 for special limitations.
(c) Except for decorative grades, panels shall not be sanded, touch-sanded, surface textured, or thickness sized by any mechanical means.
(d) C centers may be used in panels of five or more plys.

contained therein is of little interest to the do-it-yourselfer or general consumer anyway. Suffice to say that this basic standard is quite all-inclusive. It covers not only the specifications to which the various kinds of plywood panels must be manufactured, but also establishes specific, definitive procedures and parameters for specimen preparation and testing. It insures that plywood panels coming from the various APA member mills, as well as others who manufacture their products to this standard, are indeed manufactured correctly. Various kinds of inspections and reinspections are made from time to time to insure conformance. This is turn means that the consumer is assured of getting a quality product upon which he can depend.

Standard PS1 has been promulgated by the softwood plywood manufacturing industry through the U.S. Department of Commerce, but that agency actually has no regulatory authority or enforcement power to stand behind the regulations set forth. In other words, from a governmental standpoint, Standard PS1 actually has no teeth in it. On the other hand, the regulations represent an industry consensus and the provisions in it have been established by trade custom. The effectiveness of Standard PS1 lies in this fact, plus the fact that it is used as a reference in federal specifications, building codes, sales and building contracts, specifications and advertising. Because of the various pressures exerted from several different directions, the standard is in effect a solid one.

It should be noted, too, that PS1 is actually a set of minimum standards. The fact is that the American Plywood Association has developed a number of standards that exceed those set forth in PS1, which the member companies of APA are obliged to follow. Likewise, those manufacturing concerns that are not members of APA may establish standard specifications that exceed those in PS1. In sum, the softwood plywood that you buy in today's marketplace is, for the major part, the best and most reliable plywood product on a widespread scale that has ever been made available in this country.

HARDWOOD PLYWOODS

Hardwood plywoods are used for all manner of purposes, and their manufacture constitutes a large and important part of the plywood industry in this country. Consumption hardwood plywoods has risen steadily over the years. It peaked in 1972 and has since dropped somewhat, but still represents a good many billions of square feet (in surface measure) of plywood (Fig. 2-5). Wall paneling constitutes a major end-use, particularly for mobile

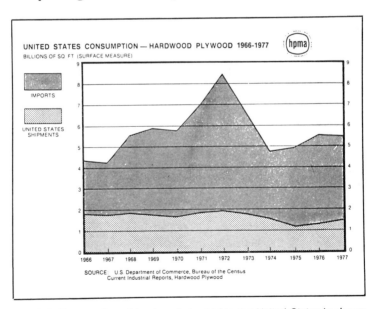

Fig. 2-5. Consumption of hardwood plywood in the United States is shown here, broken down into domestic and foreign supply categories (courtesy of the Hardwood Plywood Manufacturers Association).

homes, but there are many other uses as well. With respect to furniture, probably 90% of the visible portions of today's furnishings are of hardwood plywood construction. And there are many other industrial and miscellaneous end-uses as well, such as skis, wardrobe doors, hockey sticks, elevator cabs, boat/railroad/aircraft parts, saw and tool handles, toilet seats, shelving, etc. But the classification of hardwood plywood panels is considerably different than for softwood panels. Because of the tremendous variety that is available, classification can be somewhat more complex. Here, however, we will investigate only the basics of hardwood plywood properties.

Varieties

A good part of the hardwood plywood industry is devoted to manufacturing products that, while they have many of the usual characteristics and properties of all plywoods, are particularly made with appearance as a primary factor. For this reason, a great variety of different woods is used, especially as face veneers. This in turn leads to a tremendous number of possibilities for different appearing panels. At this time there are more than 150 different species of wood that are commercially available and in more or less common use. Each of these different wood species can be provided as veneer cut in any of three different ways: *rotary-cut, quartered* or *plain-sliced*. This leads to about 450 or so different choices for face veneers. Any of these plywoods can be obtained in four different core constructions; now we have some 1800 possibilities. Since there are 10 or 11 different possibilities for finishing the panels, that brings us to some 18,000 combinations.

To these must be added a host of special plywoods, rare and unusual woods, custom methods of cutting or slicing the veneers, special ways of manufacturing the panels, practically limitless possibilities in the way of curved or shaped panels other than the standard sizes and others. The overall picture is not only confusing, it can be mind-boggling. Then there are the different ways of positioning the veneers on the panel surface, various ways of arranging consecutive panels, and the increasing use of the so-called character grades of veneers where the possibilities are infinite. However, when viewed on the basis of what is available from stock at any given time or on what might be specified on an architectural basis for custom jobs, the entire process can be reduced to simple terms without much difficulty.

Manufacturing Standards

As with the softwood plywoods, the bulk of the hardwood plywoods produced in this country are made under a standard called the *NBS Voluntary Product Standard PS 51*. The standard currently in use is PS 51-71. This standard governs the manufacture of hardwood and decorative plywood. It was established by the Office of Engineering Standards Services of the National Bureau of Standards, a part of the Department of Commerce, in conjunction with scientific and trade associations and organizations, business firms, testing laboratories and other groups. This standard is followed by the members of the Hardwood Plywood Manufacturers' Association (HPMA) and is backed by this group.

In similar fashion to the PS1 Standard employed by the softwood plywood industry, this standard spells out the requirements for hardwood plywood wood species, veneer grading, panel core types, glue bonds, panel constructions, dimensions, moisture content, sanding and finishing, as well as sampling and testing provisions. As PS 51 itself states, "The standard is intended to provide producers, distributors, architects, contractors, builders and users with a basis for common understanding of the characteristics of this product." As far as the do-it-yourselfer is concerned, he can be assured of getting a quality product whose specifications are indeed as expressed, if they are made to this standard. Labeling with the HPMA trademark provides the same assurance since they were primarily responsible for the product standard in the first place. As with softwood plywoods, additional requirements, specifications and grade levels can be initiated and put in force by individual manufacturers at any time since these are minimum standards.

Note, however, that this is a *voluntary* standard, and manufacturers of hardwood plywood products do not have to conform to it. Members of the HPMA do and so do other companies, but there are always those who do not and who may consequently (but not necessarily) produce an inferior product. It is up to the purchaser of the product to determine whether or not it was made under PS 51, and if not, whether the product is an acceptable one for his purposes.

Wood Species

As mentioned earlier, as many as 150 different wood species are commonly used in hardwood plywoods, and others are often obtainable on special order or by architectural specifications. In

Table 2-11. The Categorization of Wood Species Most Commonly Used in the Manufacture of Hardwood Plywood Panels (courtesy of the Hardwood Plywood Manufacturers Association).

Category A species (0.56 or more specific gravity)	Category B species (0.43 through 0.55 specific gravity)	Category C species (0.42 or less specific gravity
Ash, Commercial White Breech, American Birch, Yellow, Sweet Bubinga Elm, Rock Madrone, Pacific Maple, Black (hard) Maple, Sugar (hard) Oak, Commercial Red Oak, Commercial White Oak, Oregon Palado Pecan, Commercial Rosewood Sapele Teak	Ash, Black Avodire Bay Cedar, Eastern Red[b] Cherry, black Chestnut, American Cypress[b] Elm, American (white, red, or gray) Fir, Douglas[b] Gun, Black Gun, Sweet Hackberry Lauan, (Phillippine Mahogany) Limba Magnolia Mahogany, African Mahogany, Honduras Maple, Red (soft) Maple, Silver (soft) Prima Vera Sycamore Tupelo, Water Walnut, American	Alder, Red Aspen Basswood, American Box Elder Cativo Cedar, Western Red[b] Ceiba Cottonwood, Black Cottonwood, Eastern Pine, White and Ponderosa[b] Poplar, Yellow Redwood[b] Willow, Black

[a]Based on ovendry weight and volume at 12 percent moisture content.
[b]Softwood.

fact, any hardwood species can be used as face veneers. If the panel is for decorative purposes only (or primarily), any softwood species can be used as well. Whatever wood species is used for the face veneer represents its identifying name; a hardwood panel faced with knotty white pine would be called knotty pine plywood (even though it is a softwood), while a panel faced with *hackberry* would simply be called hackberry panel. The back ply can be of any species, whether or softwood or hardwood, and so can the inner plies.

The species of wood that are most commonly used for hardwood plywood veneers are listed in Table 2-11, comprising some 52 species. They are categorized by specific gravity, which in turn determines the maximum thickness of the veneer. Any species that is unlisted can be simply categorized by determining its specific gravity and making a proper assignment. Where the specific gravity is unknown or questionable, the Forest Products Laboratory is the final arbiter of the data.

Veneer Grades

The veneers used in the manufacture of hardwood plywood panels are broken down into seven categories, six official and one unofficial. *Premium Grade (A)* veneers are the finest available. The sheets must be full-length, tight-cut and smooth. When used in two or more pieces to form a panel face, they must be carefully edge-matched by either book-matching or slip-matching (more about this later), with tight and practically invisible joints. Several species of woods have severe limitations placed upon them as to the defects that may be permitted in the veneer, as well as allowable characteristics and permissible types of matching. These specifications are found in Table 2-12. Those hardwood veneers not listed in this table have other and more general specifications to govern their selection for grade. For instance, they may contain occasional pin knots, usual characteristics that are found in the species, small burls, a few color streaks or spots and the like. However, wormholes, splits, decay, knots and such are not permitted.

The next grade in terms of quality is *Good Grade (1)*. This veneer is of only somewhat less quality than Premium. When used as a face in more than one piece, it need not be matched but must have tight edge joints. There cannot, however, be sharp contrast between adjacent pieces of veneer. The veneer characteristics and defects for these veneers can be found in Tables 2-12 and 2-13.

In terms of quality, the median level is *Sound Grade (2)*. These veneers must have a smooth surface, and must be free from any open defects. When two or more pieces of veneer are used to make up a panel face, neither color or grain matching is required. Practically any combination can be used. The various veneer characteristics and allowable defects for this grade of veneers are shown in Table 2-14.

Utility Grade (3) is a general purpose veneer in which open defects are allowable such as wormholes, splits that do not exceed 3/16-inch wide and extend more than half the length of the veneer panel, knotholes up to an inch in diameter and so on. Further characteristics and defects can be found in Table 2-14.

The lowest grade of hardwood plywood veneer in terms of quality is the *Backing Grade (4)*. In this veneer, which is similar to the Utility Grade, even larger defects are allowable, such as knotholes up to 3 inches in diameter and splits up to 1 inch wide. As you might imagine, this grade of veneer is used wherever appearance and relatively substantial defects will be of no consequence. Again, the permissible characteristics and nonpermissible defects for this grade of veneer are listed in Table 2-14.

Table 2-12. Summary of Veneer Characteristics and Defects of Premium Grade and Good Grade Hardwood Species Is a Part of PS 51 (courtesy of the Hardwood Plywood Manufacturers Association).

Characteristics	Rotary-Half round-Plain sliced birch						Rotary cativo		Plain sliced cherry	
	Natural[1]		Select white	Select red	Uniform light	Uniform dark			6	7
	Premium	Good	Premium	Premium	Premium	Premium	Premium	Good	Premium	Good
Sapwood	Yes	Yes	Yes	No	Yes	No	Yes	Yes	10%	20%
Heartwood	Yes	Yes	No	Yes	No	Yes	No	No	Yes	Yes
Color streaks or spots	Yes	Yes	Slight	Slight	Slight	Yes	Slight	Slight	Yes	Yes
Color variation	Yes	Yes	Slight	Slight	Slight	Yes	Slight	Slight	Yes	Yes
Mineral streaks	Slight	Slight	Slight	Slight	Slight	Slight	Slight	Slight	Slight	Slight
Small burls	Occ[2]	Yes	Occ	Occ	Occ	Occ	Yes	Yes	Yes	Yes
Pin knots	Occ	Yes	Occ	Occ	Occ	Occ	Occ	Yes	Occ	Yes
Knots (other than pin knots)	No	No	No	No	No	No	No	No	No	No
Worm holes	No	No	No	No	No	No	No	No	No	No
Open splits of joints	No	No	No	No	No	No	No	No	No	No
Shake or doze	No	No	No	No	No	No	No	No	No	No
Rough cut	No	No	No	No	No	No	No	No	No	No
Cross bars	No	No	No	No	No	No	No	No	No	No
Inconspicuous patches	Small	Yes	Small	Small	Small	Small	Small	Yes	Small	Yes
Type of matching	3	4	3	3	5	5	5	4	3	4

| Characteristics | Rotary gum-tupelo-magnolia-bay poplar ||||||| Quarter sliced gum || Rotary lauan ||
| | Natural || Select for white || Select for red || | | | |
	Premium	Good	Premium	Good	Premium	Good	Premium	Good	Premium	Good
Sapwood	Yes	Yes	Yes	Yes	No	20%	Yes	Yes	No	No
Heartwood	Yes	Yes	No	No	Yes	Yes	Yes	Yes	Yes	Yes
Color streaks or spots	Yes	Yes	Slight	Slight	Yes	Yes	Yes	Yes	Slight	Slight
Color variation	Yes	Yes	Slight	Slight	Slight	Yes	Yes	Yes	Slight	Slight
Mineral streaks	Yes	Yes	Slight	Yes	Slight	Yes	Slight	Slight	Slight	Slight
Small burls	Yes	Yes	Yes	Yes	No	Yes	Occ	Yes	Yes	Yes
Pin knots	Occ	Yes	Occ	Yes	Occ	Yes	Occ	Yes	Occ	Yes
Knots (other than pin knots)	No	Sound	No	No	No	No	No	No	No	No
Worm holes	No	No	No	No	No	No	No	No	No	No
Open splits of joints	No	No	No	No	No	No	No	No	No	No
Shake or doze	No	No	No	No	No	No	No	No	No	No
Rough cut	No	Small	No	Small	No	Small	No	No	No	No
Cross bars	No	No	No	No	No	No	Occ	Occ	No	No
Inconspicuous patches	Small	Yes	Small	Yes	Small	Yes	Small	Yes	Small	Yes
Type of matching	5	None	5	4	5	4	3 & 8	4	5	4

Table 2-13. More Characteristics of Premium Grade and Good Grade Hardwood Species (courtesy of the Hardwood Plywood Manufacturers Association).

Characteristics	Quarter sliced lauan		Quarter sliced limba	African and Honduras mahogany			Quarter sliced		Rotary-Half round-Plain sliced maple				
				Rotary, plain sliced, flat cut					Natural			Select white	Uniform light
	Premium	Good	Premium	Premium	Good	Premium	Good	Premium	Good	Premium	Premium	Premium
Sapwood	No	No	No	No	No	No	No	Yes	Yes	Yes	Yes
Heartwood	Yes	Yes	Yes	Yes	Yes	Yes	Yes	Yes	Yes	No	No
Color streaks or spots	Slight	Slight	Slight	Slight	Slight	Slight	Slight	Yes	Yes	Slight	Slight	Slight
Color variation	Slight	Slight	Slight	Slight	Slight	Slight	Slight	Yes	Yes	Slight	Slight	Slight
Mineral streaks	Slight	Slight	Slight	Slight	Slight	Slight	Slight	Small	Small	Slight	Slight	Slight
Small burls	Occ	Yes	Occ	Occ	Yes	Occ	Yes	Occ	Yes	Occ	Occ	Occ
Pin knots	Occ	Yes	Occ	Occ	Yes	Occ	Yes	Occ	Yes	Occ	Occ	Occ
Knots (other than pin knots)	No	No	No	No	No	No	No	No	No	No	No	No
Worm holes	No	No	No	No	No	No	No	No	No	No	No	No
Open splits or joints	No	No	No	No	No	No	No	No	No	No	No	No
Shake or doze	No	No	No	No	No	No	No	No	No	No	No	No
Rough cut	No	Yes	No	No	No	No	No	No	No	No	No	No
Cross bars	Occ	Yes	Occ	No	No.	Occ	Yes	Occ	No	No	No	No
Inconspicuous patches	Small	Yes	Small	Small	Yes	Small	Yes	Small	Yes	Small	Small	Small
Type of matching	3	4	3	3	None	3	None	3	4	3 & 8	5

1—Rotary ash, rotary basswood, rotary elm, rotary sycamore as birch grades (natural).
2—Occasional.
3—Book matched. Matched for color and grain at the joints (this can be furnished slip matched if customer so specifies).
4—Sharp contrast will not be permitted.
5—Matched for uniform color.
6—Occasional gum spot allowed.
7—Gum spots allowed.
8—Slip matched. Must be matched in sequence with tight side out.

Characteristics	Red and white oak						Pecan hickory		Walnut			
	Rotary		half round Plain sliced		Rift-comb grain				Half round Plain sliced	Rotary Half round Plain sliced	Quarter Sliced	Quarter Sliced
	Premium	Good	Premium	Good	Premium	Good	Premium	Good	Premium	Good	Premium	Good
Sapwood	Yes	Yes	No	Yes	No	Yes	Yes	Yes	10%	20%	10%	20%
Heartwood	Yes	Yes	Yes	Yes	Yes	Yes	Yes	Yes	Yes	Yes	Yes	Yes
Color streaks or spots	Slight	Yes	Slight	Yes	Slight	Yes	Yes	Yes	Slight	Slight	Slight	Slight
Color variation	Yes	Yes	Slight	Yes	Slight	Yes	Slight	Yes	Yes	Yes	Yes	Yes
Mineral streaks	Slight	Yes	Slight	Slight	Slight	Slight	Slight	Yes	Slight	Yes	Slight	Yes
Small burls	Occ	Yes	Occ	Yes	Occ	Yes	Occ	Yes	Yes	Yes	Yes	Yes
Pin knots	Occ	Yes	Occ	Yes	Occ	Yes	Yes	Yes	Occ	Yes	Occ	Yes
Knots (other than pin knots)	No	No	No	No	No	No	No	Sound	No	Sound	No	Sound
Worm holes	No	No	No	No	No	No	No	No	No	No	No	No
Open splits or joints	No	No	No	No	No	No	No	No	No	No	No	No
Shake or doze	No	No	No	No	No	No	No	No	No	No	No	No
Rough cut	No	Small	No	Small	No	No	No	No	No	No	No	No
Cross bars	No	No	No	No	No	No	No	No	No	No	Occ	Occ
Inconspicuous patches	Small	Yes	Small	Yes	Small	Yes	Small	Yes	Small	Yes	Small	4
Type of matching	3 & 5	4	3	4	3 & 8	4	3	3	3	4	3 & 8	Yes

Table 2-14. Summary of Veneer Characteristics and Allowable Defects of Sound, Utility and Backing Grades Used in the Manufacture of Hardwood Plywood Panels (courtesy of the Hardwood Plywood Manufacturers Association).

Defects	Sound Grade (2)[a]	Utility Grade (3)[a]	Backing Grade (4)[a]
Sapwood	Yes	Yes	Yes
Discoloration & Stain	Yes	Yes	Yes
Mineral Streaks	Yes	Yes	Yes
Sound Tight Burls	Max. diam. 1"	Yes	Yes
Sound Tight Knots	Max. diam. ¾"	Yes	Yes
Knotholes	No	Max. diam. 1"	Max. diam. 8"
Worm holes	Filled or Patched[b]	Yes	Yes
Open Splits or Joints	No	Yes; 3/16" for one-half length of panel	1" for one-fourth length of panel; ½" for one-half length of panel; ¼" for full length of panel
Doze & Decay	Firm areas of doze	Firm areas of doze in face. Areas of doze and decay in inner piles and backs provided serviceability of panel is not impaired.	Areas of doze and decay provided serviceability of panel is not impaired.
Rough Cut	Small area	Small area	Yes
Patches	Yes	Yes	Yes
Crossbreaks and Shake	No	Max. 1" in length	Yes
Bark Pockets	No	Yes	Yes
Brashness	No	No	Yes
Gum Spots	Yes	Yes	Yes
Laps	No	Yes	Yes

The sixth grade is known as *Speciality Grade (SP)*. This is a catch-all grade that covers those plywood panels or veneers that do not fit into the categories listed above. In general, the grade refers primarily to the veneer face, and the particularities are usually decide upon between the buyer and the seller of the plywood. Wall panel face veneer grades generally are in this category, and so are many of the specialty veneers that have unusual decorative features such as handsome coloration, unusual grain, distinctive figure and others. This includes such woods as English brown oak, bird's-eye maple, wormy chestnut and a host of others.

This brings us to the unofficial grade that was mentioned a bit earlier, which in fact is merely an extension of the Speciality Grades. There has been a trend toward the informal use of the term "character grade" over the past few years, and with good reason. This is simply because such veneers are now beginning to attract a considerable amount of attention by the general buying public, and perforce by the manufacturers of hardwood plywood products. These plywood panels have a distinctive appearance, and each veneer has individual character since it includes natural wood stains, knots, small splits, grain irregularities, pinholes and such. They are neither highly matched nor uniform, but rather are all quite different.

Table 2-15. Summary of Veneer Characteristics and Allowable Defects for Premium Grade and Good Grade Decorative Softwood Species Used in Making Hardwood Plywood Panels (courtesy of the Hardwood Plywood Manufacturers Association).

	Rotary—Sliced—Knotty Veneer				Sliced—Vertical grain		
	Western red cedar		White pine			Douglas fir	Redwood
Characteristics	Premium (A)	Good (1)	Premium (A)	Good (1)	Characteristics	Premium (A)	Premium (A)
Discoloration	No	Slight	No	Slight	Sapwood	Limited[d]	No
Burls	Yes	Yes	Yes	Yes	Heartwood	Yes[e]	Yes[e]
Knots	—	—	—	—	Color streaks	No	No
Pin knots	Yes	Yes	Yes	Yes	Color variation	Slight	Slight
Sound knots	Yes	Yes	Yes	Yes[b]	Mineral streaks	No	slight
Spike knots	Slight	Yes[b]	Slight	Yes[b]	Small burls	Yes	Yes
Filled knot holes[d]	¾ in.	1½ in	¾ in.	1½ in.	Pin knots	No	Yes
Worm holes	No	No	No	No	Knots	No	No
Open splits or joints	No	No	No	No	Worm holes	No	No
Doze	No	Inconspicuous	No	Inconspicuous	Open splits or joints	No	No
Rough cut	Small	Yes	Small	Yes	Shake or doze	No	No
Inconspicuous patches					Rough cut	No	No
					Crossbars	Yes	Yes
Pitch streaks	Small	Small	Small	Small	Inconspicuous patches	Small	No
Pitch pockets	No	Small	No	Small	Pitch streaks	Small	No
Crow's foot	Slight	Occasional	No	No	Pitch pockets	No	No
Matching	a or c	a or c	a or c	a or c	Matching	c or f	c or f

[a] Randomly spaced for pleasing appearance.
[b] Maximum 2 inches.
[c] Slip matched. Must be matched in sequence with tight side out.
[d] Bright sapwood not permitted.
[e] Six or more annual rings per inch.
[f] Book matched. Matched for color and grain at the joints.

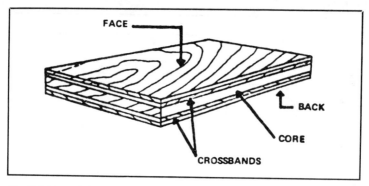

Fig. 2-6. Typical 5-ply veneer-core hardwood plywood construction (courtesy of the Hardwood Plywood Manufacturers Association).

In addition to the traditional hardwood species, many new woods that heretofore had no importance whatsoever are now being commercially used in this "grade," including domestic alder and aspen and a great many imported species that only a few years ago were never seen in this country. The idea of the character grade occurred first with wall panel products, but now has been adopted to some degree by manufacturers of cabinetry, both kitchen and otherwise, and is spreading into other areas of the hardwood plywood field, especially in furniture making. As time goes on, the use of character grades will be widespread. A series of custom character grades will likely be developed, thus affording ample opportunity for products of distinctive and individualistic appearance, as well as for the use of previously disregarded species of woods—an important factor when one considers the relative scarcity of top-grade traditional kinds.

As mentioned earlier, softwoods can also be used for the faces of hardwood panels, when appearance is the primary requisite. In this case, the face grade requirements for the most popular and commonly used species are listed in Table 2-15. Those softwoods not listed that are used as inner plies, faces or backs must meet the same grading requirements as those for hardwoods.

Panel Constructions

Panel constructions are basically the same for hardwood panels as they are for softwood, but there are several variations. The panels must be made up of an odd number of plies, all of which must be calculated to give a balanced panel. Except for the center core, the inner plies must occur in pairs of the same thickness and

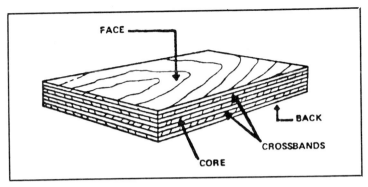

Fig. 2-7. Typical multi-ply veneer-core hardwood plywood panel construction (courtesy of the Hardwood Plywood Manufacturers Association).

placed in the same direction, one on each side of the core. The grain directions must be at right angles to one another, and to either the ends or edges of the panels.

There are four commonly used methods of coring hardwood panels. One possibility is to use a *veneer core*, the same as is done with softwood plywood. In a 3-ply or 5-ply panel, the core veneer is likely to be thicker than the face, back or crossband veneers. In a *multiply veneer core* construction, however, there may be seven sheets of veneer, all at right angles to one another and of the same thickness. A typical 5-ply veneer core construction is shown in Fig. 2-6 and a multi-ply construction in Fig. 2-7. The minimum thickness of veneers is not regulated, and can be whatever figure is

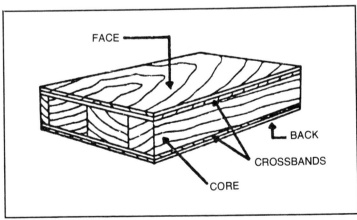

Fig. 2-8. Typical 5-ply lumber-core hardwood plywood panel construction (courtesy of the Hardwood Plywood Manufacturers Association).

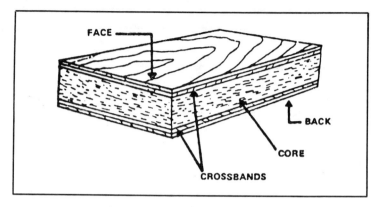

Fig. 2-9. Typical 5-ply particle board-core hardwood plywood construction (courtesy of the Hardwood Plywood Manufacturers Association).

agreed upon between the buyer and the seller of the plywood panels.

Another coring method commonly used with hardwood plywoods is the *lumber core* (Fig. 2-8). In this case, the core is relatively thick (or may be quite thick) and is made up of short, narrow strips of lumber of identical thickness that have been edge-glued together. Any species of wood can be used in making lumber cores, but each individual core for a panel must be made up of pieces of the same species; they cannot be interchanged. If the core wood is in Category A (Table 2-11), the maximum permissible width of the pieces is 2½ inches. The permissible width for Category B is 3 inches; for Category C, it is 4 inches.

There are four grades of lumber that are used in lumber cores. *Clear Grade* strips must be either full length of the panel or finger-jointed to make up the full-length strips, and must be free of knots and defects of any serious nature. Discolorations of the wood are permitted, as is the use of wood filler for patching, but wood patches or plugs cannot be used. *Sound Grade* lumber is the next lowest grade, wherein no defects of serious nature are allowable, but discoloration, small open defects if well patched or plugged with wood or wood filler, and sound knots are permitted. These strips, too, must be either full-length or finger-jointed. Regular grade lumber, however, need not be finger-jointed, but instead can be tightly end-butted to make the joints. Otherwise, the grade is much the same as Sound Grade. The Clear Edge Grade is a bit different, in that the wood strips must be of regular grade except the edge strips, along the edges of the panel, must have a 1½-inch or wider Clear Grade area. This is so the panel edge can be shaped or molded without running into any defects or patches.

Panel cores can also be made from particle board or hardboard (Fig. 2-9). Different kinds of each are used, and in several thicknesses, but in any case the core material must be manufactured in accordance with Commercial Standard CS 236, *mat-formed wood particle board*, or Commercial Standard CS 251, *hardboard*. This assures a certain amount of uniformity and a reasonable quality level in all hardwood plywood panels so cored.

Hardwood panels may also be made with a banded core, wherein one or more edges or ends are fitted to the panel between the face and back plies, or between the outer crossbands that lie just beneath the face and back plies (Fig. 2-10). The various designations of banding for hardwood plywood panels are as follows: banded one end (B1E); two ends (B2E); one side (B1S); two sides (B2S); two ends and one side (B2E 1S); two sides and one end (B2S 1E); and two sides and two ends (B4).

In addition, there are a good many special forms of hardwood plywood panels or constructions that can be made up for special applications. Different core materials may be used under certain circumstances, curved panels may be made and so forth. Despite these variations, the final product must come up to the general specifications laid out in PS 51.

Hardwood Plywood Types

As with the softwood plywoods, hardwood plywood panels are classified into types primarily depending upon the glue bond that is formed between the veneer sheets; other limiting criteria are

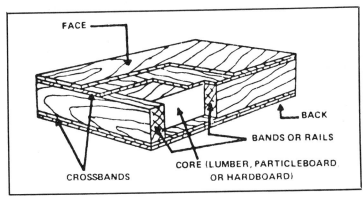

Fig. 2-10. Typical 5-ply banded-core hardwood plywood construction (courtesy of the Hardwood Plywood Manufacturers Association).

Table 2-16. Limiting Criteria for Hardwood Plywood, Including Veneer Thicknesses Relating to Table 2-11 Categorizations (courtesy of the Hardwood Plywood Manufacturers Association).

Limiting factors	Technical (Exterior)	Type I[a] (Exterior)	Type II (Interior)	Type III (Interior)
Glue bond Adhesive performance (3.9)	Fully waterproof Dry and cyclic boil shear	Fully waterproof Dry and cyclic boil chear	Water resistant 3-cycle soak	Moisture resistant 2-cycle soak
Species or specific gravity category of veneer (3.2)	Specify	Specify	Specify	Specify
Veneer edge joints (3.3)	No tape	No tape	Tape	Tape
Grade of faces or face and back (3.3)	Specify	Specify	Specify	Specify
Grade of hardwood inner plies adjacent to faces (3.3)[b]	2 under A or 1 3 under 2	2 or 3	2 or 3[c]	2 or 3[c]
Grade of softwood inner plies adjacent to faces (3.37)[b]	2	2 or 3	2 or 3[c] (4 under 3 or 4)	2 or 3[c]
Grade of other inner plies (3.3.1 through 3.3.7)	3 or 4	3 or 4	3 or 4	3 or 4
Grade of lumber core (3.5)	Not suitable	Specify	Specify	Specify
Particleboard or hardboard core (3.6)	Not suitable	Specify	Specify	Specify
Maximum veneer thickness in inches by specific gravity category (3.2.1):				
Category A	1/12	1/8	3/16	3/16
Category B	1/10	3/16	1/4	1/4
Category C	1/8	1/4	1/4	1/4
Percentage of wood in face direction	40 to 60	No limitation	No limitation	No limitation
Sanding (3.11)	Specify	Specify	Specify	Specify

[a] Not recommended for continuous exposure to moisture.
[b] Specify Grade 2 for solid inner piles.
[c] Where 1/16 inch or thicker faces are used, grade 4 or better inner plies are permitted.

Table 2-17. Wood Failure Requirements for Technical and Type I Hardwood Plywood Glue Bonds (courtesy of the Hardwood Plywood Manufacturers Association).

Average failing load	Minimum wood failure	
	Indiv. specimen	Test piece average
lb/sq. in. Under 250 250 to 350 Above 350	Percent[a] 25 10 10	Percent[a] 50 30 15

[a]These values are the percentage of wood area remaining adhered to the fractured surface in the test area.

shown in Table 2-16. The first of these is *Technical type plywood*, wherein the glue bond must be strong enough to meet certain wood failure requirements when the panels are tested. Those failure requirements are shown in Table 2-17. The tests are rather complicated and involve the following: shearing under pressure of a dry panel; a shear test after the specimen plywood has been boiled in water for 4 hours, dried for 20 hours and boiled again for 4 hours and cooled in water; and a 3-cycle soak test to see whether or not the panels will delaminate. All in all, the testing is rigorous. In addition, if this type of plywood is constructed with softwood inner plies and hardwood face veneers, it must also fulfill the requirements for the exterior type of bond that is specified in Product Standard PS 1, the same standard that is used in the manufacture of softwood plywoods.

Type I plywood must meet the same requirements as Technical type. *Type II* plywood need not meet standards that are quite so rigorous. The glue bond must be able to withstand only the 3-cycle soak test, which consists of 2-inch by 5-inch specimens taken from the test panels being submerged, in 75°F water for 4 hours and then dried at a temperature of 120 to 125°F for 19 hours, reducing the moisture content to a maximum of 8%. If the samples pass this test through three cycles without delamination beyond a certain set of parameters, and subject to certain other details as well, the plywood is suitable for Type II designation. Type III plywood must have a glue bond of sufficient quality to pass a similar 2-cycle soak test. Of the four types of plywood, Technical and Type I are categorized as exterior, while Types II and III are interior.

The Technical type plywood is a completely waterproof variety that can be used in continuous exposure to water, with no ill effects. Type I, while it is bonded with waterproof adhesives and can be used outdoors or in exposure to substantial moisture conditions, cannot withstand constant exposure to water. It will eventually fail. Type II hardwood panels are almost entirely bonded with urea-resin adhesives. The glueline is water resistant and will perform satisfactorily in conditions of high humidity or even very occasional wetting, but is unsuitable for exterior use. Type III plywood has less resistance to moisture and is a type not commonly made. It is generally utilized in certain commercial and industrial products such as containers.

Dimensions

Hardwood plywood panels are made in a wide variety of sizes and thicknesses as stock items, and a practically limitless number in the area of specialized and custom-ordered pieces. Most of the stock panels that are generally available around the country in lumberyards and home centers are wall panels (which can be used for many other purposes as well) and measure 4 feet wide by 8 feet long. This is considered a standard panel. However, hardwood panels can be made 5, 6 or 7 feet wide by as much as 12 feet long. In fact, individual panels can be manufactured as long as 40 feet or even more, by using special equipment and techniques. The principal problems with panels of such large sizes, however, are the limitations in handling and transportation.

There is likewise a wide range of panel thicknesses. Some aircraft plywoods measure only 3/64 inch thick, and mention was made earlier of the plywood business cards that are only 1/100 of an inch thick. On the other hand, standard flush doors may be as thick as 2¼ inches. Most of the hardwood plywood panels of interest to the do-it-yourselfer, however, will be the standard ¼-inch thickness, and in some cases ¾-inch. Others, of course, are available, though not commonly stocked by most lumberyards or the smaller plywood dealers.

Since there are so many specialty items in the hardwood plywood field, often nominal dimensions of the panels are agreed upon between the buyer and the seller. In this case, the tolerance in either length or width of the panel is plus or minus 1/32-inch. The thickness of an unsanded panel carries the same tolerance, but sanded panels must be made to plus 0, minus 1/32-inch dimensions unless the panel has a nominal thickness of ¼-inch or more, in which case the tolerances are plus 0, minus 3/64-inch.

If the panels are 4 feet by 4 feet or larger in size, they must be square within 3/32 inch. If the panels are smaller than that, they must be square within 1/16 inch. The standard method for determining the squareness of a panel is to measure and compare the two diagonals. The panels must also be straight enough along the edges that a straight line from one corner to the next falls within 1/16 inch of the panel edge, if the panel is less than 8 feet long. This dimension increases to 3/32 inch for panels 8 feet long and more.

Finishing

The first step in the finishing of hardwood plywood panels is sanding. There are four possibilities. The first is no sanding at all, and in this case tapes used to join veneer pieces need not be removed either. *Rough* sanding is merely a hit-or-miss method that somewhat smooths off the roughest portions of the panels; tape removal is not necessary in this case. *Regular* sanding requires that the entire surface of the panel face be sanded, cleaned and relatively smooth. All tapes must be removed. There may, however, be sander streaks, so this surface is not a suitable one on which to apply a fine finish, at least without further work. *Polish* sanding results in a perfectly smooth and clean panel face.

Many hardwood panels are prefinished at the factory. This is true of most of the plywood wall paneling, for instance. Several different kinds of finish are used, ranging in sheen from quite low to glossy. In addition, there are any number of special finishes that can be factory-applied to hardwood plywood panels. These decisions and specifications are generally agreed upon by the buyer and

Fig. 2-11. Book match is one of the most popular arrangements for decorative hardwood plywood face veneers (courtesy of the Hardwood Plywood Manufacturers Association).

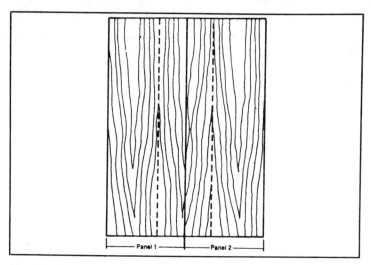

Fig. 2-12. Typical running match arrangement, where the veneer sheets are carried along from panel to panel (courtesy of the Hardwood Plywood Manufacturers Association).

the seller. Likewise, there are a great many different finishes that can be applied in the home workshop or at the job site. These procedures and techniques will be discussed in a later chapter.

Matching

Matching is a process whereby two or more pieces of veneer are arranged in a particular fashion to form the face of a plywood panel. The process can be used for either softwood or hardwood plywood, but it is common to hardwood plywood panels. Certain kinds of matching are either required or may be specified by architects where appearance is a keynote. If appearance is not a special requisite, or if there is no necessity for matching the grain, color, figure or other characteristics of the face veneer pieces, the usual practice is simply to apply the required veneer pieces in random fashion to cover the face of the plywood panel. In the matching process, however, considerable attention is paid to exactly how those pieces of veneer are set. There are two possibilities: matching of veneer pieces on each individual panel and matching the veneers from panel to panel to cover an entire project, which can be a large wall expanse like a comparatively small table top.

One of the most popular types of matching is called *book match*, an arrangement used on almost all architectural plywood.

The sheets of veneer are kept sequentially as they are cut. Then alternating sheets are turned over so that they face each other edge to edge, like the pages of a book. They are then arranged on the plywood panel so that the most emphatic grain lines and patterns match up as closely as possible (Fig. 2-11).

Book matching, or book-leaf matching as it is sometimes called, can be done in other ways as well. One possibility is the *running match*. In this case, the face of each panel is made up from as many veneer sheets as is necessary for the particular width of the panel; they are lined up side by side. The portion of a veneer sheet that is left over, and extends past the edge of the panel, is trimmed off flush with the panel edge and is then used as the starting piece for the next panel. This system is also known as *lotted in sequence* (Fig. 2-12).

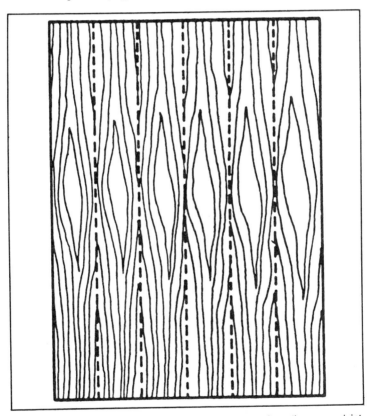

Fig. 2-13. Center match is a more costly arrangement where the veneer joint occurs in the center of the panel (courtesy of the Hardwood Plywood Manufacturers Association).

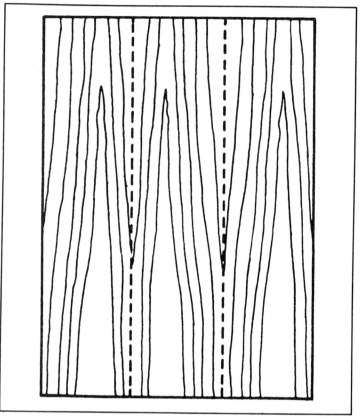

Fig. 2-14. The balance match is made up of veneer sheets positioned to balance the grain and figure across each panel (courtesy of the Hardwood Plywood Manufacturers Association).

A second possibility, and a very handsome one, is called *center match*. An even number of veneer sheets is used to make the panel face, and the full length sheets start with a pair joined edge to edge along the vertical center line of the panel. Then pairs of veneer sheets of the same width are placed on each side of the starting panels, until the width of the panel is filled out. This makes a nicely balanced and symmetrical panel (Fig. 2-13).

A system called *balance match* can be employed. Any number of veneer sheets, odd or even, but full length and clipped to approximately equal width, are arranged on the panel face to form a balanced arrangement (Fig. 2-14).

There is another general type of matching that can be used, too, and this is called *slip matching*. Here the grain, pattern or figure "reads" across the veneer sheets, from sheet to sheet in more or less of a continuum (Fig. 2-15).

There are other methods of matching as well, which are not normally employed in stock production panels but can be specified as desired. For instance, a *random* or "no-match" arrangement of veneer sheets can be used, wherein there is no particular pattern or effect. *Two-piece, four-piece* and *eight-piece* arrangements are commonly made up. Various designs can be made up in geometric shapes, such as *herringbones, checkerboard* or *diamonds* (Figs. 2-16A through 2-16D). In fact, veneer matching can be done in various grades of complexity up to the point where the final result is virtually a form of marquetry or inlay work, though this is seldom done because of the high cost involved. Obviously, extremely complex matching procedures are not suited to production runs of plywood panels, but are rather almost a handcrafting process.

Matching can also be done from panel to panel. Once the particular arrangement of veneer sheets on the panel faces have been determined, then a method of panel matching can be selected. Often, of course, there is no actual matching process. The panels are simply employed on a random basis and edge-butted to produce an interesting, or at least unobtrusive, pattern from whatever

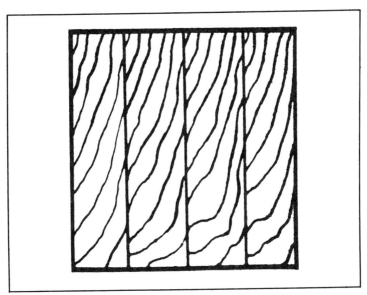

Fig. 2-15. In the slip match, the grain pattern follows along from sheet to sheet of veneer (courtesy of the Hardwood Plywood Manufacturers Association).

Fig. 2-16. A variety of different and unusual matches can be made with hardwood veneers (A) Diamond match. (B). Checkerboard match. (C). Four-piece match. (D) Reverse diamond match.

panels are at hand. But where specific matching is desired, there are two possibilities.

One is called *blueprint matching*. In this process, every space where a panel is to be employed is blueprinted in elevation view. For wall paneling, for instance, each wall would be drawn in elevation, showing all of the doors and windows, trimwork or any other parts of the wall. Then each plywood panel is individually fabricated to exact size and in a particular sequence to fit each individual panel location exactly. The panels are numbered to match the blueprints, and each is installed in its assigned location. This is a complex and expensive proposition, but results in a appearance that is superior to any other kind of installation. The grain, figure or pattern is closely matched from panel to panel and extends unbroken around doors and windows. It may even include the door face paneling itself.

The second method that is often used is called *sequence match*. In this case, the required number of panels, whose within-panel veneer arrangement has already been determined, is manufactured and the panels are matched as closely as possible. All of the panels are numbered in sequence from left to right. They are installed in that sequence, being cut and fitted on the site as job conditions dictate.

MISCELLANEOUS PLYWOOD PROPERTIES

In addition to the foregoing, there are various miscellaneous properties of plywood panels that can be of considerable importance to a given project. While they are not always of consequence by any means, the consumer and the do-it-yourselfer should have some knowledge of them.

Acoustical Properties

In general, plywood can be considered as a reflector of sound. Sound transmitted through the air is simply energy in motion, and thus can be either absorbed or reflected by a surface. Sound absorption is the usual measure by which various materials are compared for their acoustical properties, by means of a figure called the *sound absorption coefficient*. A material that absorbs all sound energy directed at it is rated as 1.0. The basis of comparison often used is an open window, which passes all sound waves. If the material is a perfect reflector of sound, its coefficient of sound absorption would be 0.0. A look at Table 2-18 will give you an idea of how plywood compares to other materials. Note particularly the substantial difference between plywood and the various special acoustic materials.

Even though plywood is a reflector of sound, rather than an absorber, it can be effectively used in sound control in residential applications (and others as well). There are various methods of making sound control walls and partitions which will be investigated later in this book.

Table 2-18. Sound Absorption Coefficients of Various Building Materials (courtesy of the American Plywood Association).

Material	Coefficient
Open Window	1.0
Special acoustic materials	0.2 - 0.8
Brick	0.03
Window glass	0.03
Plywood	0.04
Varnished wood	0.03

Table 2-19. Average Thermal Conductivity, K, for Softwood Plywood Species Groups at 12% Moisture Content (courtesy of the American Plywood Association).

Species Group	k (Btu/hour/sq ft/degree Fahrenheit/inch thickness)
1	1.02
2	0.89
3	0.86
4	0.76

Thermal Properties

Temperature has a number of important effects upon plywood. In the normal temperature ranges, plywood expands slightly as it becomes warmer, just as any wood and practically all other solids do. However, thermal expansion in plywood is very small—so small, in fact, that it is difficult to measure even under laboratory conditions. For all practical purposes, thermal expansion is negligible and can be discounted insofar as ordinary uses of the kinds that are likely to be found around the home are concerned.

Like all other materials, plywood has the ability to conduct heat. This capability is called *thermal conductivity* and is represented by the symbol k. The higher the k of any given material, the greater its ability to conduct heat. Table 2-19 shows the average thermal conductivity, rated in Btus of heat that will pass through 1 square foot of material 1 inch thick in 1 hour, per degree Farenheit difference in temperature between the cool side and the warm side. The figures given are average for the various wood species of each species group and are not totally accurate. Complete accuracy could only be achieved where specific parameters are designated, but these figures give a good idea of how plywoods compare to other materials. The k for glass wool thermal insulating material, for instance, is 0.27. Copper, on the other hand, which is an excellent heat conductor, has a k of 2,700.

Of more interest in most circumstances is the resistance that plywood exhibits against the transfer of heat. Plywood does have some thermal resistance, sufficient to warrant its inclusion when calculating the overall thermal resistance of a material to the passage of heat is called R and is the reciprocal of its ability to conduct heat, or k. The usual k used in determining the resistance of plywood to heat transfer is 0.80, a single value that works well enough and simplifies calculations (other k values based upon specific wood species/panel constructions can be used, however). The R value of various plywood panel thicknesses is shown in Table 2-20.

When plywood is exposed to extreme heat, degradation in various degrees takes place. When the temperature to which the plywood is subjected goes above approximately 212°F, exothermic decomposition takes place over a period of time. How fast this happens depends upon temperature, time length of exposure, repetitiveness of exposure and air circulation. When plywood is exposed to temperatures between 230 and 302°F, the wood will char very slowly over a long period of time and charcoal will form on the surface. If the heat is dissipated from the wood, this is all that will happen; otherwise, spontaneous combustion may take place. One test showed a 60% weight loss and appearance of charcoal after 320 days at 284°F. If the exposure is to heat in the range of 302 to 392°F, charring will take place more rapidly. From 392 to 536°F, charcoal formation takes place rapidly and spontaneous combustion is likely. At 532°F or more, spontaneous combustion will inevitably occur after a very short period of exposure.

The specific ignition temperature of plywood, or wood, is very difficult to determine and is dependent upon a number of contributing factors, most of which are variable. This ignition point

Table 2-20. Thermal Resistance, R, of Various Panel Thickness of Softwood Plywood Panels, Based Upon an Average K for Softwood of 0.80. as Listed by the American Society of Heating, Refrigerating and Air-Conditioning Engineers (ASHRAE) (courtesy of the American Plywood Association).

Panel Thickness	Thermal Resistance R
1/4"	0.31
5/16"	0.39
3/8"	0.47
7/16"	0.55
1/2"	0.62
5/8"	0.78
3/4"	0.94
1"	1.25
1-1/8"	1.41

can range anywhere from 510 to 932°F, but no particular point is an absolute for all plywoods or woods. From a practical standpoint, unprotected plywood should not normally be applied where ambient temperatures are greater than 230°F. For instance, if used on a wall in relatively close proximity to a wood stove, especially one that is used on a regular basis for comfort-heating purposes, it is obvious that the wood will be exposed to temperatures in excess (often greatly in excess) of 230°F. The result in due course will be decomposition of the wood. It is entirely possible that at some point, albeit only after a fairly long period of time, spontaneous combustion of the panel surface could result. If the paneling happens to be a relatively thin commercial wall paneling with a factory-applied finish, many of which are quite combustible, degradation and possible combustion is likely to take place much more rapidly.

While high temperatures degrade plywood, interestingly enough low temperatures have either a null or a positive effect. Tests have been made upon wood in temperatures as low as −300°F that show an increase in strength of as much as three times over that normally found at room temperature. Variables include the initial strength property of the wood and also its moisture content. Freeze-thaw cycles seem to have little or no effect upon wood, but these cycles can reduce the strength and holding capability of many fasteners by 10% or more. Thus, in practical application any increase in strength is likely to be offset by losses in other areas. Plywood, however, can be successfully used under the proper circumstances in temperatures down to −300°F, with no ill effects upon the plywood. Some studies have been performed that show gluelines made with urea, phenolic and casein glues are not affected by temperatures of −68°F. From a practical standpoint, experience has shown that plywood panels used in forming insulation jackets for the hulls of ships that transport liquid natural gas and serve at temperatures of about −150°F perform very well.

Equilibrium Moisture Content

The presence of moisture has certain definite effects upon plywood panels but, as we have mentioned earlier, plywood is a particularly stable material. Assuming that the right plywood has been specified and properly installed commensurate with prevailing moisture conditions, be that water vapor in the air or free water, moisture really has little practical effect upon plywood.

Wood is *hygroscopic*, like a blotter, and thus can absorb moisture as either vapor or water. No matter how dry a sheet of plywood is, it almost invariably has some moisture content. If plywood is exposed to an essentially constant relative humidity, such as might be found inside a house during the winter heating season, in a short period of time it will reach a state of balance with the surrounding atmosphere. This balance point is called the *equilibrium moisture content* of the plywood. While the moisture content dependent upon relative humidity, it is quite independent of ambient temperatures ranging between 32°F and 100°F.

This moisture content (EMC) is not, however, the same as the relative humidity by any means. Table 2-21 shows the relationship between relative humidity of the surrounding air and the equilibrium moisture content of wood at 75°F. Notice that even when the relative humidity is quite high, say 80%, the plywood EMC is only 16%.

Obviously, as the wood gains or loses moisture, it also swells or shrinks; this is as true of plywood as it is of any other wood product. Because the plywood is made with crossed plies, the total overall change in even a full-sized panel of plywood is quite minor. Reduced to practical terms, in normal conditions with relative humidity varying between 40 and 80% (and an EMC ranging from 7 to 16%), the total changes in a 4 foot by 8 foot panel on the average might be expected to be about 5/100 inch across the width and 9/100 along the length. Note too that certain panel constructions

Table 2-21. Relative Humidity Related to Equilibrium Moisture Content of Wood (EMC) at 75°F (courtesy of the American Plywood Association).

RH(%)	EMC(%)
10	2
20	4
30	6
40	7
50	9
60	11
70	13
80	16
90	20
100	28

are more resistant to warping than others; in general, the more plies, the less warping is likely.

Whenever high moisture conditions are encountered, of course, exterior type plywood should be employed. Numerous test programs have been run over the years on exterior plywoods made up of various species of wood, some of them long term exposures of unpainted and unprotected test panels and others in laboratory experiments. Wet-dry cycle tests, which soak the plywood and then dry it completely time after time, have frequently been run. Though a considerable amount of face-checking occurred from long-term weathering and cyclic moisture conditions, the durability, stiffness and strength of the plywood panels was affected not a bit. Such panels, even though exposed to severe conditions and completely unprotected, can be expected to last at least 25 years with no degradation in service capability.

Plywood panels are likewise unaffected by repeated cycles of freezing and thawing. Various tests have been made under different freeze-thaw conditions. The results have been that the wood is virtually unaffected by ice expansion. Repeated cycles of freezing to $-40°F$ for 24 hours, and then allowing the panels to thaw overnight, showed no evidence of any problems. The gluelines remained unchanged, the wood was perfectly all right, and there was no splitting or delamination. A good example of the effectiveness of plywood in freeze-thaw conditions is the plywood pallets that are used in cold-storage facilities. These pallets are subjected to considerable mechanical abuse. They are hoisted about by forklifts, they carry heavy loads, and they undergo endless freeze-thaw cycles. Yet these pallets, which are used by the thousands, may stay in service for years.

Water Exposure

Plywood panels of the proper type will indeed withstand repeated exposure to water. However, when they are used in direct contact with groundwater for extended periods of time in conditions under which they may be subject to insect or fungus attack, the panels should be pressure-treated with preservatives. This is the case, for instance, with planters, stock-watering or similar tanks, retaining walls, wood foundations and the like. When so treated, plywood in contact with groundwater is expectedly long-lived, just as is plain wood when properly treated. One of the most notable examples fo the usefulness of plywood in such applications is the *All Weather Wood Foundation* (AWWF), which will be discussed later in the book.

Contact with or immersion in salt water has no adverse effects upon plywood, provided the proper kind of panels are used. The salt bothers neither the wood nor the gluelines, and the moisture has no effect aside from a bit of swelling of the wood. There is a special grade of plywood (APA MARINE Exterior) that can be used for building boats (though other types serve nicely for punts, rowboats and such) and is of extremely high quality. In addition to being virtually impervious to the effects of brackish or salt water, this grade is exceptionally tough, finely made and is primarily used where bending or curving of the panels is important. Indeed, the only problem with plywood panels exposed to sea water are decay and marine borers, against which special protection must be taken.

Since water has little effect upon plywood of the proper kinds, the same can be said of condensation. In fact, the gluelines in exterior plywood serve to inhibit the passage of moisture vapor, forming a partial vapor barrier. If exterior plywood paneling is used on both sides of a building section, for instance, there will be a double vapor barrier that has a reasonable degree of effectiveness. This is not to say, however, that a true, applied vapor barrier should not be used as well. Such a barrier should always be installed on the warm side of a building section covered on the outside with plywood panels.

Note that in most instances where plywood panels are used in building construction—floors, roof decking, wall sheathing—it is recommended that a gap of at least 1/16-inch be left at all panel ends and ⅛-inch at all panel edges. This spacing allows ample room for expansion of the panels in the event of swelling from absorption of moisture. This is particularly important during the construction stages, since the panels are often exposed for lengthy periods of time. Stormy, wet weather will cause some expansion. After the panels have been covered and the construction is complete, these expansion joints will continue to act in the event of moisture pickup from high-humidity conditions. Where humidity is exceptionally high on a regular basis, as in a bathroom or farm confinement building, the expansion gaps should be even larger. If the panels are butted tightly together, expansion will no harm the integrity of the panels or diminish the structural strength of the building. But the joint lines are likely to buckle outward in an unattractive fashion.

Permeability

The *permeability* of a material is its ability to allow the passage of liquids or gases through tiny pores or openings. The rate at which this occurs is dependent upon a number of factors, including the particular liquids or vapors involved, number and size of pores, checks, spaces or other passways in the material. Our chief concern is with water vapor, and plywood differs from wood in this respect. Also, different species of woods react differently to the passage of water vapor. The *water-vapor permeance*, which indicates the value of vapor transference through a material, can therefore be assessed only in a general way. Average water vapor permeance for some different plywoods are shown in Table 2-22. Note the difference in the permeance of plain exterior type plywood, and the same after priming and painting with two finish coats. Note, too, the considerable difference in permeance between interior and exterior plywoods. This is attributable to the waterproof gluelines used in exterior type plywood panels.

Electrical Properties

Plywood, when it is dry, is an excellent electrical insulator. It cannot conduct an electrical current. However, as the moisture content of plywood rises so does its ability to carry a current. Its *dielectric* properties diminish rapidly. A soaking-wet piece of plywood can indeed carry a considerable charge. Thus, the old trick that electricians frequently use of standing upon a square of scrap plywood while working on an electrical panel box or live circuits in a damp environment is a useful one, so long as the plywood itself remains dry. As voltages and frequencies rise, however, plywood becomes less effective as an insulator. There are differences between its ac and dc resistivity.

Decay and Insect Resistance

Since plywood is made of wood, its resistance to rot and attack by insects or fungus is no different than the individual species of woods that make up the plywood panels. There are a few species of wood that have a certain inherent resistance to rot, such as cedar or redwood. However, with plywood this would only be the case if the panels were entirely made of those woods, which is seldom the case. In any event, this resistance is not a permanent one, but will diminish with the passage of time. No woods can be considered immune to termites, and very few have a natural resistance to fungus attack. Plywoods can also be damaged by such common insects as carpenter ants, bark beetles and powder-post beetles. Marine borers pose a threat in marine applications. In effect, no types of plywood panels can stand alone against fungus (which is

Table 2-22. Water Vapor Permeance—Values in Perms Representing the Average Water Vapor Transmitted Through Softwood Plywood in Grains Per Square Foot, Per Hour, Per Inch of Mercury Pressure (courtesy of the American Plywood Association).

Product	Surface Finish	Water vapor permeance in grains/sq ft /hour/in. Hg (perms)
Interior-type Plywood 1/4"	None	1.9
Interior-type Plywood 1/4"	Two coats of inside flat wall paint	1.1
Interior-type Plywood 1/4"	One natural blonde thin undercoat, plus one coat shellac	0.8
Interior-type Plywood Sheathing 5/16"	None	1.8
Exterior-type Plywood 3/8"[1]	None	0.8
Exterior-type Plywood 3/8"	One coat exterior oil primer plus two coats exterior finish paint	0.2
Exterior Medium Density Overlaid Plywood (1 Side) 3/8"	None	0.3
Exterior Medium Density Overlaid Plywood (2 Sides) 1/2"	None	0.2
Exterior High Density Overlaid Plywood (2 Sides) 1/2" & 5/8"	None	0.1
FRP[2]/Plywood 3/4"	None	<0.1

1 Range of seven species: 0.5 — 1.4
2 Fiberglass-Reinforced-Plastic-overlaid plywood

the cause of most staining, decay and mold) or attack by certain insects.

The first step, and perhaps the most important, in combating these problems is to use the best of building materials and good construction practices and procedures. This includes employing termite control measures such as poisoning, shields and clearing the grounds completely of cellulose-bearing materials, avoiding or eliminating conditions under which decay fungi can thrive, providing sufficient ventilation and installing proper screening. The second necessity is to employ materials, including plywood panels, that have been pressure-treated with suitable preservatives, wherever that step is indicated. For best results the materials should be factory-treated, and then cut edges or ends of pieces can be further treated by dipping or brushing at the job site. Only by taking all of the proper precautions and constructing in the correct manner can plywood, or any other wood product, be protected from biological attack. However, this need be done only where such attack is known to occur and constitutes a common problem.

Chemical Resistance

Ordinary plywood is quite resistant to many chemicals; naturally an exterior-type should be used for exposure to chemicals in solution. Though slight swelling can take place as a result of absorbed moisture, this factor is minimal with plywood and so reduces the possibilities of leakage or failure of joints as in tanks or containers. Solvents, petroleum-based solutions, and most acids and bases have little if any effect upon the plywood itself. The exterior gluelines are almost totally resistant.

Generalizations of the effects of chemicals upon plywood are a bit chancy, since the specific effects and the extent of degradation, if any, varies with the particular chemical and also with the strength of the solution. However, unlined or uncoated plywood is widely used in various industrial applications where constant exposure to relatively weak solutions of acids and bases are commonplace, and the service life is virtually indefinite. Of course, such containers or tanks can also be lined with a plastic or other substance and made totally immune.

In general, strong mineral acids (*nitric,* for example) are destructive of plywood. However, dilute mineral acids have little or no effect; nor do organic acids (*oxalic, acetic*). Strong caustic solutions should not be contained by plywood, but salt solutions of whatever type have no effect at all. Alkaline solutions will attack the lignite in the wood, but if cold and dilute, degradation will occur at only a slow rate. Thus, if you would like to construct a darkroom sink from plywood, as has been often done, you can expect excellent service. The biggest difficulty, in fact, with such a sink is that the corners are hard to clean. The photochemicals, however, will not affect the plywood at all.

A better choice, of course, would be High Density Overlay (HDO) plywood, which is exterior plywood overlaid with a hard, resin-impregnated fiber. This overlay is highly resistant to most acids and can be expected to give extremely long service, especially in such applications as might be found in the home. It is, however, an expensive material and one that is not easy to purchase in most locales.

Flame Spread Rating

There is no such thing as a fireproof building. However, buildings can easily be made safe from fire, provided construction is properly done and the building is correctly designed. There are two chief considerations with regard to building materials: how fast fire can spread on a room's surfaces, and the fire resistance, or protection against penetration of fire through a building section. The greater the fire resistance, the more time there is to discover a fire, evacuate the building if necessary, and put the fire out before it can spread too far.

Materials are judged by *flame spread rating*, which is a measure of how fast a flame travels across the surface of a specimen of the specific building material. Most tests are done under a special procedure designated by ASTM E-84, the *Tunnel Test*. Specimens are compared on a basis of 0 flame spread for cement-asbestos board at the low end of the scale, and 100 for red oak wood at the top end. The lower the figure, the slower the flame spreads.

Structural components, or building sections, are classed by type, depending upon the length of time after the fire starts that the section will continue to bear its full load and will not collapse nor transmit flame or high temperature. The usual designations are 1-hour, 2-hour, 3-hour and 4-hour construction; both materials and building sections may also be rated in minutes. A 75-minute building section, for example, would be acceptable construction in an application where a building code requires a 1-hour rating. Building codes require certain fire resistive types of construction in various portions of buildings, depending upon the type of occupancy. In most cases, no specific requirements are established for the building sections of wood-frame houses. Constructing to fire-resistant standards, such as 1-hour sections, affords an added measure of protection, especially for homes in rural areas far from a fire department. Table 2-23 lists some typical resistances of common building sections.

The flame spread ratings of various materials are shown in Table 2-24. Ordinary wood-frame, plywood sheathed buildings provide a substantial amount of fire safety and except for a few localized exceptions are acceptable, at least as far as building codes are concerned. Building codes for the most part establish minimum standards. The homeowner might well want to consider improving upon them, particularly as far as fire-safe construction is concerned. This adds little to the cost of building, especially when compared to the total cost of the finished structure, but does afford the occupants considerable extra security and a better opportunity for safety in the event of fire. If this extra protection is desired, one of the easiest means to accomplish it is through the use of plywood. There are four possibilities.

Types of Construction for Fire Protection

The first is to use a method of building known as *protected* construction. The procedure is simple enough and consists of

Table 2-23. Fire Resistance of Wood Walls and Partitions, Based on ASTM E119, Except Where Noted (courtesy of the American Plywood Association).

Faces	Insulation	Rating, Min.	Notes
BEARING WALLS—2x4 studs spaced 16″ o.c.			
1. 1/4″ plywood, 3/8″ plywood	3½″ Fiberglass	15	600 plf load on 8 ft. high wall
2. 3/8″ gypsum, 5/8″ T1-11 plywood	3½″ Fiberglass	30	1,250 plf load on 8 ft. high wall
3. 3/8″ plywood, both faces	Full thick mineral wool	30	(40 minutes, non-bearing rating)
4. 5/8″ Type X gypsum, 5/8″ plywood	3½″ Mineral wool	45	FRT studs and plywood
5. Same as 4	3½″ Mineral wool	60	Same; but plywood nailed to studs through 4″ gypsum strips
6. 3/8″ gypsum, both faces	None	30	
7. 1/2″ gypsum, both faces	None	45	
8. 5/8″ Type X gypsum, both faces	None	60	
9. 2 layers 5/8″ Type X gypsum, both faces	None	120	
NON-BEARING WALLS—1x3 studs spaced 16″ o.c.			
10. 1/4″ plywood, both faces	None	10	
11. 3/8″ plywood, both faces	None	20	Values obtained on small size walls
12. 1/2″ plywood, both faces	None	25	
13. 5/8″ plywood, both faces	None	35	
14. 1/4″ plywood, 3/8″ plywood	Mineral wool 2 lbs/sq. ft.	50	8 ft. high wall

merely adding a layer of fire-resistant materials such as plaster, plasterboard or acoustical tile between the plywood and the studs, joists or rafters. This slows flame passage and temperature rise, and at the same time adds reinforcement to the structural members.

The second possibility is known as *heavy timber* construction. This is particularly applicable to roof constructions where APA 2-4-1 plywood can be employed, along with certain other materials and construction specifications. This construction can also be used in floors, by laying ½-inch plywood over 3-inch planks.

A third method is known as *treated* construction. This makes use of plywood (or wood) treated with fire retardant chemicals (FRT plywood). Such plywood has a much lower flame spread, which reduces its fire hazard classification and greatly improves its performance with regard to fire resistance. As you might expect, this type of plywood is more expensive than untreated plywood, but it does do the job.

For example, roof assemblies can be made up of treated lumber and FRT plywood in such a way that they are regarded by most building codes as virtually equivalent to being noncombustible. Under the Basic Building Code, FRT wood and plywood roofs are so classed as to be permitted in so-called "fireproof" and noncombustible types of construction, as well as for ordinary wood-frame construction. There is another point worthy of consideration in conjuction with treated plywood roofs. The rating used for fire insurance purposes, and thus for insurance premiums as well, is generally better than those for unprotected roofs. This fact alone may make the installation of such a roof cost-effective over a period of time, not to mention the additional protection afforded the occupants. Note, however, that this is not necessarily the case, FRT plywoods may not be acceptable as a bona fide protective measure under some local building codes.

Fire Retardant Paints

There is one other possibility that affords a certain measure of protection, frequently very good, for otherwise untreated plywood panels and for wood as well, This is the application of fire retardant paints. These paints can be used in existing buildings as well as new ones, and they can be used either indoors or out. The paints can be applied in much the same fashion as any other paints, though of course proper selection must be made. That they do work is unquestionable; various tests have shown that flame spread ratings may be as low as 15 or 20 with only one coat of the paint, and as low as 10 with two coats.

Depending upon the specific locale, local building codes may or may not allow the use of fire retardant paints as a means of assembling specified or required fire resistive building sections.

Table 2-24. Flame Spread Ratings of a Few Common Construction Materials (courtesy of the American Plywood Association).

Material	Flame Spread Rating
Asbestos-cement board	0
Fire retardant treated PS-1 Construction plywood	25 or less
Fire retardant coated PS-1 Construction plywood	25 or less
Fire retardant treated lumber	25 or less
Red oak lumber	100
PS-1 Construction plywood	75-200

On the other hand, such sections are seldom required for residences anyway. The homeowner can apply fire retardant paint as a means of providing himself with extra protection. After all, a flame spread rating of 10 minutes is a good deal better than the usual 75-100 minutes for most construction plywood manufactured under PS-1.

Producing Fire Retardant Hardwood Plywood Panels

A somewhat different situation occurs with regard to fire retardant treatments of hardwood plywood paneling. Most of these panels that are ¼-inch or thicker meet Class III standards for fire resistivity. This is the range of 75-200 flame spread ratings, the same as was noted for softwood plywoods. Thus, the paneling as it is, in stock fashion, fulfills national building code requirements for residences (typically Class IV, flame spread rating not greater than 500), as well as HUD Minimum Property Standards. In order to meet more restrictive fire-resistivity specifications, the panels must be manufactured in a somewhat different way. These same fire retardant panels can, of course, be used in residential constructions. Though the cost is naturally somewhat higher, the amount of protection afforded against fire disasters is considerably improved.

Three basic methods are used to produce fire retardant hardwood plywood panels. One is to glue the face and back veneers to a core made of noncombustible mineral substance. The face veneer is thin, while the core is relatively thick. The panels are available in 1¼-inch 3-ply, and ¾-inch or 1-inch 5-ply panels. While the panels cannot be vee-grooved, as are many standard panels, practically any other variety of decorative appearance as far as face veneers are concerned can be manufactured.

Another method is to treat the core material of the panels—veneer, lumber or particle board—with fire retardant chemicals, much the same as is done with softwood construction plywoods. These panels generally carry the normal factory finish. If they are cut and fitted at the job site, special care must be taken to reseal the cut edges immediately with shellac, varnish, tung oil or some similar sealer. This prevents the admittance of moisture, which can deteriorate the fire-retardant salts deposited in the core.

The third possibility, less often used and less effective, is to coat the surface of the panels with what is known as *intumescent* material. This finish, which is relatively thick by comparison with an ordinary finish, turns to a layer of foam upon exposure to fire and considerably slows the flame spread rate.

These fire retardant hardwood plywood panels are of great importance in the industry with respect to paneling used in mobile homes. A fairly large percentage of industry production is consumed by panel products for this purpose. Various fire and safety codes governing the use of materials in mobile homes are constantly being upgraded and made more restrictive insofar as fire resistivity is concerned. While fire retardant paneling, as well as other building components, is of undoubted value to the homeowner as a means of upgrading home fire safety, it is extremely important to the occupants of mobile homes. Because of the relatively small size of many mobile homes and the manner in which they are built (and especially if constructed of untreated materials), the spread of flame, smoke and toxic gases tends to be quite rapid. The chances for personal injury are high in the event of fire. We strongly recommend that owners of mobile homes who contemplate renovations such as repaneling or the installation of cabinetry seriously consider using fire-retardant plywood panels in their projects.

SPECIALTY PLYWOODS

As we have mentioned previously, there are many different kinds of specialty plywoods manufactured throughout the country for one purpose or another. One example of a speciality plywood which has since turned into a stock, standard commodity in the industry is the APA 303 Series Specialty Siding. Since the list of speciality plywoods is practically endless, and most of them would be of little interest to the do-it-yourselfer anyway, there is little point in going into an in-depth consideration of them. However, there are a few specific types of that deserve mention.

HDO/MDO Plywood

The letters HDO stand for High Density Overlay, while MDO stands for Medium Density Overlay. These plywoods are composed of plywood panels which are essentially of the exterior type. They have been overlaid on one or both surfaces with a phenolic resin-impregnated fiber surface. This overlay is extremely tough and will withstand severe exposure for long periods of time, as well as being resistant to deterioration, weathering, moisture penetration and abrasion. HDO panels must have a minimum overlay thickness before pressing of 0.012 inch, and the overlay is applied to both panel surfaces. Though usually manufactured in a semiopaque natural color, other colors are available. MDO plywood panels are also overlaid with a thermosetting resin system of 0.012 inch thickness, but of somewhat less overall durability and a somewhat different surface texture. MDO has a surface that is virtually ideal for application of paints, allowing excellent paint serviceability and performance. The overlay may be on one or both panel surfaces and is available in several different forms, including texture-embossed surfaces and grooved panels.

Both of these panels have a great many attributes that make them particularly valuable in certain applications. MDO panels are widely use in both interior and exterior panels. They can be used for siding, gable ends, garage doors, fencing panels, cabinets and built-ins, shelving, partitions, wainscoating, soffits, fascias and a variety of other home construction requirements and projects. The smooth surface, great resistance to weathering, easy paintability and workability, and ease of maintenance make MDO panels a most interesting possibility for the do-it-yourselfer. Despite the higher cost of these panels, the various good qualities make them worth the extra money in many applications. Over the long run the cost-effectiveness as well may be on the positive side.

HDO and MDO panels are also widely used in industry, concrete forming, by the transportation industry (recreational vehicles, truck and trailer linings, etc.) and in agriculture and marine applications. In the latter two areas, either type of plywood is excellent for such purposes as fertilizer bins, fruit and vegetable storage chambers, spreader bodies, animal shelters, boat bulkheads hulls, transoms, hatch covers and so on. In addition, many of the highway signs that you see as you travel about the country, especially those on the turnpikes, are made of either HDO or MDO plywood. For those do-it-yourselfers who want to use the best materials in their projects and for whom versatility and ease of working, as well as final finished appearance, are very important, we suggest that MDO and HDO plywoods be investigated.

FRP Plywood

FRP plywood is another special product that consists of fiber glass-reinforced-plastic bonded to APA grade-trademark plywood. This finish can be applied to either or both sides of the plywood panel, and is put on by any one of several methods. Most FRP finishes are factory-applied, but there are some that can be applied in the field as well. The standard size of FRP panel is 4 feet by 8 feet, but much larger panels are available for special applications (Fig. 2-17). Thicknesses from ¼-inch to 1⅛-inch are available.

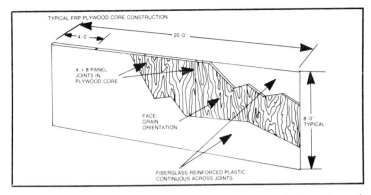

Fig. 2-17. Typical fiber glass reinforced plastic softwood plywood panel construction (courtesy of the American Plywood Association).

Various surface finishes can be had, ranging from very smooth to heavily textured, and a full range of colors can be obtained as well.

Most of the applications for this kind of plywood are commercial and industrial, such as bodies for trucks and trailers, concrete forming, walk-in coolers and railroad boxcar linings. FRP plywood can be made about twice as strong as the plywood panel without the coating. Because they are tough but lightweight, highly resistant to abrasion, impacts and similar mechanical damage, and extremely weatherable, FRP panels can be used in any application where such qualities are of value. This panel is undoubtedly the ruggedest, toughest and most generally indestructible one that you can purchase. When employed in a particular project, it could probably be expected to last longer than the builder.

Pressure-Preserved Plywood

Pressure-preserved plywood is of considerable importance to the do-it-yourselfer because of the many applications in which it can be, or should be, used. We have mentioned earlier a few of those applications, particularly the AWWF (All Weather Wood Foundation), which is an excellent method for building all-wood foundations for house additions, garages and outbuildings, as well as for an entire house. In this application, the plywood and the lumber used to build the foundation must be properly treated with the correct materials; accordingly, both the model building code organizations and various federal regulatory agencies require that each piece of lumber or plywood used for this purpose must carry the mark "AWPB-FDN" (an American Wood Perservers Bureau Standard).

Of course, there are many other applications for pressure-preserved plywood. It is used where protection is required against attack by termites and other insects, or where protection against fungi, decay and mold growth are desired. Pressure-preserved plywoods are used in applications where contact with groundwater is likely to occur. Some of the specific applications are in planters, swimming pools, garden pools or fountain bases, docks and piers, floats, retaining walls, stock-watering tanks and irrigation structures. This type of plywood can usually be bought from stock held by large lumberyards in metropolitan areas, but it may be more difficult to obtain in smaller towns or in rural areas. However, it nearly always is available, even though some effort might be required.

Numerous types of preservatives are used to protect plywood or lumber, and exactly which kind is best often depends upon the conditions under which the material will be employed. There are three main types of preservatives: *creosote, pentachlorophenol* and *salts*.

Creosote is a popular preservative that is insoluble in water, quite stable under a wide range of conditions and has low volatility. For the most part, creosote-preserved woods cannot be painted over and the odor is very strong, so they should not be used where either characteristic would be a drawback. This preservative is not harmful to pets or farm animals.

Pentachlorophenol, which can be oil, gas or other solvent-borne, is a chemical crystal that is dissolved in a vehicle and is highly preservative. There is a slight odor when the treatment is fresh, but this dissipates after a short while. The treatment is not noticeable on the plywood surface. It is noncorrosive and the panels are reasonably clean to handle, though at first they can readily pick up dust and particulate matter (easily brushed off). This preservative, usually simply called "penta," cannot be painted over if the vehicle is a heavy oil. A number of penta solutions can be used that leave a clean surface and can be painted over. Penta is excellent where severe near the ground, and for contact with fresh water.

Numerous salt preservatives are used for various purposes and can be impregnated into the plywood as either water or solvent solutions. They are clean, odorless and leave paintable surfaces. Those salts that are in common use are listed in Table 2-25.

Pressure-preservative treatments of plywood panels are very effective, primarily because of the composition of the panels. The

Table 2-25. Chemical Names and Trade Names of Various Water-Borne Salt Preservatives That Are Effective for Treating Plywood (courtesy of the American Plywood Association).

Chemical Name	Designation	Trade Name
Acid Copper Chromate	(ACC)	Celcure*
Ammoniacal Copper Arsenate	(ACA)[1]	Chemonite*
Chromated Copper Arsenate	(CCA Type A)[1] (CCA Type B)[1]	Greensalt Langwood* Boliden*CCA Koppers CCA-B Lahontuho K-33* Osmose K-33* TACO* CCA
	(CCA Type C)[1]	Chrom-Ar-Cu (CAC)* Langwood* MRC-CCA Type C Wood Preservative Wolman*CCA Wolmanac* CCA
Chromated Zinc Chloride	(CZC)	
Fluor Chrome Arsenate Phenol	(FCAP)	Osmosalts* (Osmosar*) Tanalith Wolman* Salts FCAP Wolman* Salts FMP

[1]These salts are least likely to leach.
*Trade names copyrighted by treating companies.

effectiveness is great enough, in fact, so that factory-treated plywood can be cut at the job site with no appreciable loss in preservative protection if the cut edges are not field-treated. However, such treatment certainly does no harm and may well do some good. For the do-it-yourselfer, the easiest and most effective course is to brush on (or dip the cut edges) a coating of creosote if that preservative has been used, or a coating of light-bodied penta (available at nearly all hardware stores and lumberyards) if the original preservative is either penta or a salt. Not all of the preservatives are equally effective in all applications. Use the information contained in Table 2-26 to determine which treatment might be best for your purposes, and specify that treatment. Note that the salt preservative types listed in the far right-hand column are defined in Table 2-25.

One problem that might be encountered with pressure-preservative treated plywoods is that it may not be possible to glue. Some panels may be glued only with some difficulty and with certain glues and particular techniques. Those plywoods that have been treated with salt preservatives generally can be glued in the usual manner. The same is true of some penta-treated woods where a light-bodied vehicle for the penta was used. Creosote-treated wood can not be glued. If difficulties arise in gluing preservative-treated plywoods, the best bet is to contract the treating firms to get their recommendations for specific procedures.

As to finishing pressure-preserved plywood, those panels that have been treated with salts are readily paintable. The recommendations for finishing are the same as for untreated plywood. However, if a fine, paintable surface is of great importance in the project, it is recommended that MDO plywood be used. This will give the best finished results. If the finish is to be an opaque stain rather than a paint, regular plywood panels are quite satisfactory. If the panels have a textured surface, semi-transparent stain finishes are recommended for best results. Likewise, panels treated with penta can be painted, provided that a paintable solution was used as a preservative. If this is a requirement for your project, be sure to specify a paintable penta preservative; if heavy-bodied vehicles are used for the penta preservative, they cannot be painted over.

Painting creosote-preserved plywood can be done, but it is a difficult job and not recommended. The usual procedure is to allow the creosote-treated surface to weather for at least a year. Then a primer of aluminum-base paint is applied (two coats or even more may be required). After a lengthy drying time, a topcoat is applied. Even with this extensive treatment, there is a fair likelihood that the creosote will bleed through onto the painted surface in due course, at least here and there, leaving a mottled and stained effect. In short, paint over creosote just doesn't work very well.

Block Flooring

One does not normally associate hardwood floor tiles, or *parquet* tiles as they are frequently called, with plywood. However, laminated hardwood block flooring is plywood. It is a specialty type of product that has become popular over the years and is now manufactured to such an extent that the Hardwood Plywood Manufacturers Association promulgated a new standard for the product in association with the American National Standard Institute. The standard currently in effect is ANSI 010.2-1975. This standard spells out the requirements for grading, moisture content, dimensions, construction and finish of laminated hardwood block flooring.

All laminated hardwood block flooring made under Standard ANSI 010.2 is classified by species, grade and finish of the face veneer. The two grades for the flooring face are *Prime* and *Standard*. Certain defects and characteristics are allowable/not permissible for faces, backs and inner plies, just as with other

Table 2-26. Recommended Preservative Treatments for Plywood Exposed to Various Conditions (courtesy of the American Plywood Association).

Table of Recommended Treatments

Exposure	Typical Application	Minimum Preservative Treatment[1] (Lb. per cu. ft. by assay)		
		Cresote	Pentachlorophenol[3]	Salt preservatives[2]
Contact with sea water, exposed to marine borer attack.	Pontoons, wharf bulkheads, scows, floats and flood gates, etc.	Creosote: full-cell 25.0	Not recommended	ACA 2.50 CCA Type A 2.50 CCA Type B 2.50 CCA Type C 2.50
Contact with ground, fresh water, chemicals, extreme humidities.	Permanent trench and tunnel lining, retaining walls, skirting for post and pier or pole type foundations, cribbing, snow sheds, floats, irrigation structures, tanks, troughs, linings for wet process industries, poultry dropping trays, septic tanks, some chemical storage tanks, industrial sewers, and smelter roofs.	Creosote: empty-cell 10.0	Penta: empty-cell 0.50	ACA 0.40 ACC 0.50 CCA Type A 0.40 CCA Type B 0.40 CCA Type C 0.40
	All-Weather Wood Foundation systems (4)	Not recommended	Not recommended	See Note 4, ACA or CCA (Type A, B or C). 0.60
Where plywood is subject to severe insect infestation or fungus attack in above ground uses.	Under these exposure conditions, protection may be advisable for subflooring over unexcavated areas or shallow plenums; sheathing and other uses such as fences, exterior siding, exposed structural units such as stressed-skin panels and box beams, reservoir roofs, splash boards in pole-type buildings.	Creosote: empty-cell 8.0	Penta: empty-cell 0.40	ACA 0.25 ACC 0.25 CCA Type A 0.25 CCA Type B 0.45 CCA Type C 0.25 CZC 0.45 FCAP 0.25

(1) Recommended minimums from the American Wood Preservers' Association "Book of Standards."
(2) Based on dry salts per cubic foot, oxide basis, full-cell treatment.
(3) Oil, gas or other solvent-borne.
(4) Plywood used in the All-Weather Wood Foundation in ground contact shall be APA grade-trademarked and shall be pressure treated in accordance with the treating and drying provisions of AWPB-FDN Standard.

Table 2-27. Characteristics and Defects Permitted in Various Plies of Laminated Block Flooring Under American National Standard ANSI 010.2-1975 (courtesy of the Hardwood Plywood Manufacturers Association).

Characteristics and defects	Inner Ply	Back	Prime Grade	Face	Standard Grade
Bark pockets	Yes	Yes	No	No	
Brashness, shake, doze, and decay	No	No	No	No	
Burls, tight	Yes	Yes	Few and small	Smooth ones up to 1 in. in avg diam.	
Cross breaks	Max ¾ in. in length	Yes	No	No	
Discolorations	Yes	Yes	No	Yes	
Fillings	Yes	Yes	Inconspicuous	Yes	
Gum pockets	No	No	No	No	
Gum spots	Yes	Yes	No	No	
Knotholes	Max diam. ¾ in. for 3-ply and 4-ply constructions, and ⅜ in. for ply next to the face in 5-ply construction.	Max diam. 2 in.; sum of diam. 4 in. in any 12 in. square	No	No	
Knot, pin	Yes	Yes	Occasional	Yes	
Knot, sound, tight	Yes	Yes	No	Max ¾ in. avg diam.	
Laps	No	No	No	No	
Mineral streaks	Yes	Yes	Slight	Yes	
Patches	Yes	Yes	No	Yes	
Rough grain	Yes	Yes	No	Slight	
Sapwood	Yes	yes	No	Yes	
Splits or open joints	Max 3/16 in. for ½ length of block	Max 1 in. for ¼ length of block; ½ in. for ½ length of block; ¼ in. for full length of block.	No	No	
Stain	Yes	Yes	No	Yes	
Wormholes	Yes	Yes	No	No	

plywoods (Table 2-27). The blocks can be 3, 4 or 5-ply, with the grain of each ply at right angles to the adjacent plies, except for the 4-ply blocks, in which the grain of the two center plies runs in the same direction but at right angles to the face and back plies. The face ply must be no less than ⅛-inch thick before sanding. The edges of the face ply must have a bevel of no more than 1/16-inch for finished blocks and square edges for unfinished blocks. They must be made with tongues and grooves for interlocking from block to block. The standard size of the blocks is 9 inches square plus or minus 1/64-inch in either direction. The standard thickness is 15/32-inch plus or minus 1/64-inch. The blocks must be square enough that the lengths of the diagonals do vary more than 1/32 inch.

Interestingly enough, the standard calls for the inclusion of an instruction sheet in each package of blocks that gives complete instructions for proper applications, preparation, type and application of adhesive, laying the flooring, and sanding and finishing procedures if necessary. This, of course, is of great assistance to the do-it-yourselfer who wants to lay down a new floor but is relatively unfamiliar with the details. The purchase of block flooring made under this standard will assure the consumer that, unlike many similar flooring products that are now being imported into this country (as well as some types being made right here), the product he is contemplating buying is of good and uniform quality. Each package of block flooring made under this standard must be marked ANSI 010.2. It must carry the name or identification of the manufacturer, the species of wood used in the face veneer, the grade and whether finished or unfinished, the block size and number of blocks, and the total square footage contained in the package.

Chapter 3
Tools For Plywood Work

The tools that are used for working plywood are the same as those associated with all general woodworking, wood fabrication and building construction. Both hand tools and power tools are used. They can range from very simple to highly sophisticated, depending upon the skills, interests and resources of the worker. Many of them are the ordinary ones that can be found in most homes and certainly all home workshops. Some are a bit less ordinary and well-known. Thsese too will be discussed, since many of them are quite handy for doing certain kinds of work.

In many cases it is quite feasible to build entire projects using only hand tools. The do-it-yourselfer need not feel that if he doesn't have a vast array of power equipment, he cannot work with plywood. This simply is not true. On the other hand, plywood works very quickly and easily with certain portable power tools, and some of these likewise are found in many home workshops. There are plenty of occasions when the larger, stationary power tools are most helpful, too, because they get the job done with ease and rapidity. At the same time they allow a degree of precision, especially in cutting, that is difficult to obtain with hand tools requiring acquired skills that take time and experience to develop.

The basic skills needed in using hand tools are easy to acquire. A short amount of practice and a bit of experience will enable you to use them with confidence and good results. Learning the intricacies of power tools takes a bit longer. Again, this is a pleasurable rather than a difficult chore for anyone truly interested in woodworking, and mastery of power tools does not take any inordinate amount of time. The results that can be obtained, of course, are excellent once one learns the necessary procedures and gains a reasonable amount of experience.

SELECTING TOOLS

There is no question but what tools are very expensive, and there is a tremendous array from which to choose. For those who are beginners in the field of woodworking and/or working with plywood, we suggest that you start with simple projects or constructions that can be built with only those few elementary hand tools that you might already have around the house. Plus, have a few more (if and as necessary) to make up a small, basic complement—the fundamental carpenter's toolbox, if you will. As skills, interest and experience increase, various specialized hand tools and equipment can be added to the kit, along with one or two utilitarian portable power tools.

From this point onward, the particular project interests of the do-it-yourselfer will doubtless lead him to acquire those hand tools and portable power tools that will be most practical and useful for the kinds of work he wants to do. Eventually, the slow process of accumulating tools and equipment as necessary can easily lead to a full-fledged home workshop, complete with various, more specialized portable and stationary power tools, as well as a considerable assortment of hand tools. When the tool collection becomes substantial enough, the next step usually is to arrange a home workshop. This will be investigated briefly in the next chapter.

Most of the tools that will be discussed in this chapter, both hand and power, are of the basic variety. Some of the more interesting and useful specialized tools (as well as those which are particularly practical in working with plywood) are looked at as well. There is some duplication among many of the tools in terms of the jobs or the kind of work that can be done with them. But they all have various advantages and disadvantages, including price, personal likes and dislikes, that make them worthy of consideration. Some of the tools that will be covered probably will be of little interest to some do-it-yourselfers. Those tools used for cabinetmaking, for instance, are somewhat different than the ones needed for laying a roof decking.

The fact that a great many tools will be discussed is no cause for alarm; probably you will never want or need to own all of them (although many do-it-yourselfers do). Nor would there be any necessity in having all of these tools on hand in order to complete any one project. We are necessarily covering a great deal of ground here in the interest of presenting a fairly rounded picture. The essential purpose is to acquaint you with the important tools and equipment that are available, their purposes, and their suitability for different kinds of plywood work. Once you know what is offered and which tools might be most worthwhile for you to own, you can pick and choose to suit the dictates of both your pocketbook and the particular projects or jobs at hand.

TOOL COSTS

Unless you happen to be a lover and collector of tools (as many do-it-yourselfers are), often there is no point in purchasing a specialized tool just to take care of one job if that tool can be rented. For instance, a power nailer is a great help and saves a lot of time in fastening down a large plywood roof deck. But such equipment can be rented at much less cost than the purchase price. Assess your needs for tools carefully. In the interest of economy, purchase only those that you feel will be of lasting value to you and a long-term asset to your shop.

Good tools are quite expensive, but cheap tools are likely to be even more so in the long run. There simply is nothing to be gained whatsoever by purchasing cheap, shoddy, poorly made tools at discount prices or in bargain basements. A cheap tool is just that—cheap. Tools that are not of high quality (and there are plenty of them around) will not last well, will produce an inferior quality of work no matter how hard you try, and will be most frustrating to use. They are just not worth the effort. When you purchase tools, whether hand or power, choose brand-name equipment of good reputation. Stay away from bottom-of-the-line tools, and wherever possible look for good guarantees. Make sure, too, that the tool you purchase fits your needs as exactly as possible and has sufficient power, capacity, size or whatever other parameter is important.

But note that it is also possible to go overboard in the opposite direction. It is usually not necessary, especially for home-shop use, to buy the biggest, ruggedest and best model of a particular tool at a premium price. Though nice to have, such tools and equipment are generally designed for commercial or industrial purposes where they will be used day after day and often under adverse conditions.

LAYOUT AND MEASURING TOOLS

Obviously one of the first requisites among tools and equipment needed for working with plywood is various marking and measuring devices. The layout and marking of pieces, as well as measuring, goes on continually and is particularly important because of the way many pieces are cut from a single sheet of plywood. Well-made precision equipment, as well as accuracy on the part of the builder, is necessary here. This is one area in particular where the purchase of top-quality gear is most important.

Rules

Probably the most common of all measuring devices for both woodworking and general carpentry is the *tape rule* (Fig. 3-1B). This device consists of a long, flexible length of steel tape that rolls back into a metal case. Better tapes have precision markings and can be easily read in both directions on either edge of the tape. The tape itself is coated to prevent the markings from wearing away. Most are equipped with a spring-return and a belt clip. Standard sizes available are 6, 8, 10, 12, 16 and 20 feet. The 6-foot variety is handy for around the shop, but the 10-foot tape is more utilitarian for general purposes. The 20-foot tape, while a bit heavy and bulky, is excellent for general construction work. Longer tapes are also available in this general style.

Another popular kind of measuring device, especially for cabinetwork and similar uses, is a rule variously called an extension, zig-zag or *carpenter's rule*. Most of these rules are made of wood and unfold to a 72-inch length in 8-inch sections. Cheap rules of this sort are virtually worthless. However, a top-grade carpent-

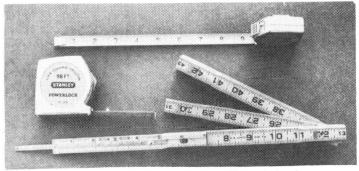

Fig. 3-1. Two useful tools. (A) Tape measure. (B) Carpenter's rule.

147

Fig. 3-2. Accurate bench rule is handy for making small layouts and measurements.

er's rule with precision markings, bound in brass and featuring a brass extension slide for making inside measurements, is a useful piece of equipment. This kind of rule is pictured in Fig. 3-1A.

There are times, however, when neither instrument will adequately do the measuring job at hand. Sometimes an ordinary *bench rule* (Fig. 3-2) is the best bet. Good rules of this type are generally made of stainless steel and carry very precise markings, available in several different modes. They are particularly useful for making short and accurate measurements, for setting dividers and numerous other tasks. These rules are usually either 12 or 18 inches long. The *folding rule* is similar, but is generally 3 feet long and either folds in the middle or into three sections.

Another possibility is to use a common, ordinary *yardstick*. If you can find a good one, made of quality hardwood and well finished, it can be a useful tool serving both for measuring and as a straightedge. Beware of the cheap, giveaway yardsticks often found in hardware stores, though; they are next to useless for either purpose. For the best buy, purchase a precisely marked, stainless steel or heavy aluminum alloy yardstick that will always remain true and precise.

Straightedges

When working with plywood, one of the most important layout tools is the *straightedge*, which is employed for laying out and marking long, straight lines. A steel yardstick or carpenter's framing square will serve the purpose for short runs. For best results, you should have either a 4-foot or a full 8-foot straightedge, and preferably both. One common trick is to save a strip of plywood or particle board, at least 4 inches wide, that has been cut parallel with a panel's "factory edge." The stock edge of the plywood or particle board, if the material is of descent quality and has undamaged edges, will be perfectly straight and will serve nicely as a full-length straightedge when carefully adjusted to the dimension marks.

One advantage of this type of straightedge is that it can also be used as a saw guide when cutting with a circular saw. To accomplish this purpose, the thickness of the straightedge should be at least ½ inch. Another trick is to fasten the steel tape from a tape measure (either one that has been broken and discarded or a new replacement tape) to the edge of the straightedge, either on the edge itself or on the top surface paralleling the edge. Thus, the straightedge can also perform the function of a full-length marking and measuring device that will cover the entire span of a panel. Note that where diagonal measurements are involved, the straightedge must be either lapped (positioned twice or more), or a straightedge of about 10 feet in length must be used.

Another possibility, more expensive but also more effective, is to purchase a device such as the "Strate-Cut" cutting guide (Fig. 3-3). This piece of equipment consists of special guide sections that are expandable from 51 inches to 102 inches and clamp directly to the plywood panel, or other workpiece. The device allows the layout of perfectly accurate lines and can also be used as a cutting guide with any jigsaw, router or portable circular saw. It is made of a sturdy aluminum alloy that will last indefinitely and is well worth

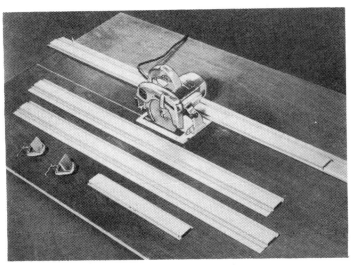

Fig. 3-3. Combination adjustable straightedge and cutting guide like Strate-Cut is ideal for working plywood (courtesy of R.A.K. Products, Inc.).

Fig. 3-4. The try square is an essential shop tool.

the investment. Various kinds of standard straightedges in numerous and styles can also be purchased.

Squares

One of the most commonly used marking and measuring devices is called a *square*. It can be used to make short measurements, square off lines, and mark right or other angles. The *try square* (Fig. 3-4) should be in every woodworker's toolbox. This square, sometimes called a *miter* square, is fixed at a right angle. Top quality and precision milling is essential to insure an exact right angle. This is important not only to layout work, but because the square is also used to set saw blades in table or radial saws to a precise 90° angle from the saw table for accurate, straight-edged cuts. Inexpensive squares, on the other hand, often are untrue. This type of square can also be obtained in a 45° configuration.

The *combination square* (Fig. 3-5) is also a useful type in that the blade slides back and forth in a slot cut in the head, so that it can be used for certain kinds of inside squaring and measurement. Most models are fitted with a level vial and a small scriber that slips into a pocket. Either 45°, or 90° angles can be tested or marked, and limiting measuring can be done as well. Combination squares are generally available only in a 12-inch size.

The kind of square that is variously known as a *carpenter's, rafter, framing* or *flat square* (Fig. 3-6), depending upon its markings, is also a useful tool, particularly for layout work on plywood and for squaring up either partially-built or completed assemblies. These squares are made of perfectly flat one-piece arms of heavy

Fig. 3-5. The combination square is versatile, accurate and features a sliding head.

metal, either steel or aluminum, and can be obtained with various different markings, rafter tables, pitch tables and the like inscribed upon the arms. The 2-foot size is perhaps the most common and is particularly useful in plywood work because it is neither too large nor too small. However, larger sizes are readily available.

For the smaller jobs, one of the handiest squares that I have used is a device called the *Squangle* (Fig. 3-7). The tool comes with a full set of instructions, which really are necessary because the Squangle can be used in so many different ways and for so many different purposes. It has a built-in level and numerous adjustments, as well as measurement markings and various kinds of angle markings. It can be used as a try square, a T-square, an adjustable-angle square and as a protractor to inscribe circles for layout work.

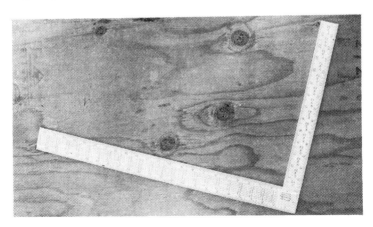

Fig. 3-6. Rafter or framing squares are available in several sizes and types, and are excellent for plywood layout work.

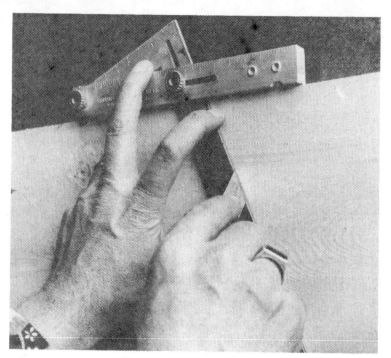

Fig. 3-7. The Squangle is one of the handiest of the small squares and can perform several functions.

Lines and Bobs

One inexpensive marking device that you will probably want to have if you will be engaging in much general construction work with plywood, or for building rough-and-ready projects or small structures where precision cutting and layout is not an essential, is a *chalk line*. The handiest kind to have is the type that is enclosed on a reel in a case (Fig. 3-8B), or "chalk box." The powdered chalk, which is available in several colors, is poured into the box and thoroughly coats the line. When the line is withdrawn from the box, it is ready to use. Once the chalk line has been snapped, the line is simply reeled back into the case and is automatically ready for marking the next line.

Some of these chalk boxes can also be used as *plumb bobs*, by running the line out to a satisfactory length and locking it in place with a special locking mechanism built into the case. The pointed cap on the case serves as the plumb bob. A plumb bob is a handy device to have around for certain tasks, especially when transfer-

Fig. 3-8. (A) Plumb bob. (B) Chalk line with chalk box.

ring a measurement from, say ceiling to floor level, or when constructing built-in cabinetry, to insure that the side panels are indeed straight up and down. For easier usage and to eliminate getting chalk dust all over your fingers, however, a standard bob with a line, either free or contained on a reel shown in Fig. 3-8A, is the best choice.

Levels

Though a plumb bob does work better on many occasions for determining a perfectly plumb or vertical line, you will need a *spirit level* for horizontal lines. In fact, many spirit levels can be used to check vertical and angled lines as well. For general purpose work and overall usefulness, the first level to add to your toolbox should be a 6-inch or 8-inch *torpedo level* (Fig. 3-9), preferably of the type with three vials. This is a handy and easy level to use. You can check horizontal, vertical or, in some cases, 45-degree lines.

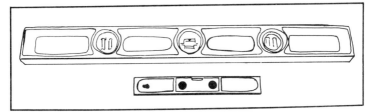

Fig. 3-9. A 2-foot carpenter's level and a small torpedo level will take care of most leveling chores.

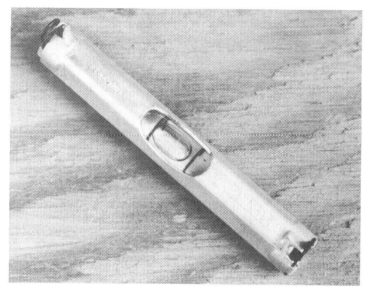

Fig. 3-10. The small line level is suspended on a taut line to check levels over long spans.

For larger work, a *carpenter's level* (Fig. 3-9) is necessary, one that can span at least 2 feet and preferably more. Though contractors use levels up to 6 feet long, for most home shop and home construction purposes, a short one (minimum 24 inches) is perfectly adequate. The purchase of a quality instrument is wise, so that you can be assured of a precision-milled frame that is true and bubbles that are and will remain accurate, even if the level is accidentally dropped.

Another kind of level that is very handy to have around, and also very inexpensive, is a *line level* (Fig. 3-10). This little gadget hooks over a length of string that can be used as a taut line. It will define levels over a fairly lengthy distance. Yet another type, known as an *inclinometer* kind of level, will automatically read the angle in degrees of whatever surface it is placed against. Conversely, it can be used to adjust a surface to whatever degree of angle is required.

Patterning Tools

There are times when it is necessary to cut one workpiece to an intricate shape or pattern in order for it to fit into or against another odd-shaped piece. This is particularly true of moldings or molded edges. For small patterns, one of the handiest devices to

Fig. 3-11. The template former is fully adjustable for patterning small shapes.

use is a *template former* (Fig. 3-11), which is a bar containing a series of adjustable pins that can be pushed back and forth to form any small pattern. Align all the pin ends in a straight row and then push the tool against the surface that must be patterned. The pins will adjust themselves to an identical configuration. Then this outline can be traced around to make a pattern, or can be traced directly on the workpiece to be cut.

Another method known as scribing can be used where the workpieces are larger, though it can be used on small pieces as well. This process makes use of a pair of *dividers* (Fig. 3-12). By adjusting the legs of the dividers a suitable distance apart, setting the workpiece to be cut against the one that must be patterned, and drawing one leg of the dividers down the workpiece as the second

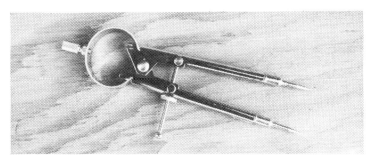

Fig. 3-12. Dividers are inexpensive and handy for all sorts of measuring and layout jobs.

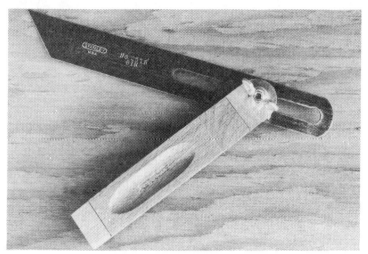

Fig. 3-13. The sliding T-bevel is indispensable for working with angles.

leg follows the irregular surface to be matched, you will end up with an irregular line on the workpiece that will match perfectly. The only trick is to hold the dividers steady and at the same angle to the surfaces being scribed at all times. Dividers can also be used for various stepping-off procedures in layout work, to make accurate measurements between two points (set the dividers and then place them against a bench rule to get the measurement), for inscribing small circles and similar tasks. In this particular case, an inexpensive set of dividers that would be of marginal quality from a draftsman's or machinist's standpoint is generally adequate for home shop work and can be obtained at most office supply stores.

Obviously, not all layout and marking chores consist of making straight lines, 90° angles and 45° angles. Working with plywood (and wood as well) invariably requires the setting or determining of all sorts of odd angles. The Squangle, mentioned earlier, works well for this purpose in some applications, but a more universal arrangement is necessary. There are two tools that can sometimes be used separately and often are used in combination. These tools should be part of your kit.

One is called a *sliding T-bevel* (Fig. 3-13). The blade of this tool slides back and forth a certain amount. It can be adjusted relative to the handle to any desired degree and then locked in place. The T-bevel can be used to establish and lay out angles, or to graduated marking angles on the T-bevel. When it becomes necessary to determine the exact angle to which you have set the T-bevel (so that you can then adjust a table saw blade to the correct cutting angle, for example), or when you must set the T-bevel to a specific angle (to lay out a workpiece, for instance), then the *protractor* comes into play (Fig. 3-14).

It is possible, in some situations, to use an ordinary half-circle-shaped plastic protractor of the kind students use at grade school, but these are not very accurate and are nonadjustable. A much better idea is to obtain an adjustable protractor with an accurate and finely marked scale containing opposite 180° graduations. With this device, you can easily establish angles to as little as ½° with no difficulty.

Another interesting and very helpful little device, if you are working a great deal with geometric shapes of various sorts and need to be able to quickly and accurately divide angles, is the *angle divider* (Fig. 3-15). It is not a necessary item, to be sure, but it is sometimes nice to have around.

Measuring and layout work requires making marks and drawing lines. The most common instrument used for this purchase is a pencil, often a dull one, but this is by no means the best way to make precise layouts or accurate dimensional markings. An inexpensive and useful tool that will do the job nicely is the *scratch awl* (Fig. 3-16). This sharp-pointed tool, much like an ice pick (which, incidentally, will also work) can be used to scribe fine and accurate but quite visible lines on a work piece, when run along the edge of a straightedge or square. It is also very handy for making small starter holes, especially in soft woods, for holding brads or starting screws. There are occasions when it makes a decent punch as well.

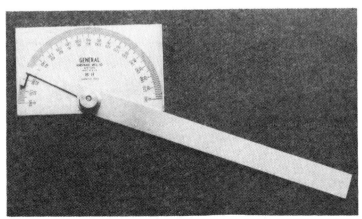

Fig. 3-14. A good protractor is the best way of determining accurate angles.

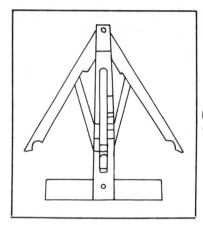

Fig. 3-15. Angle divider (courtesy of Woodcraft Supply Corp.).

Cabinetmakers and woodworkers often use either of two different tools to accomplish this same purpose. One is called a *striking knife* (Fig. 3-17), which consists of a steel blade fitted with a hardwood handle in the center. The blade at one end is in the form of a scratch awl, and at the other is an arrowhead-shaped, bevel-edged marking blade. The other tool is called a *marking knife* (Fig. 3-17), which consists of a hardwood handle fitted with a short, thick blade that is flat on one side, ground to a taper and bevel-edged for precise marking. Scratch awls are available at any hardware stores; marking and striking knives must be purchased from woodworking supply houses.

SAWS

One of the objections to using plywood that beginners sometimes voice is that all pieces for the job must be cut from a large,

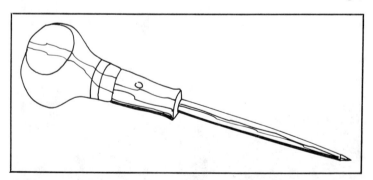

Fig. 3-16. A scratch awl is useful for marking, punching, making starter holes and other chores.

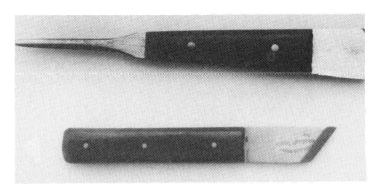

Fig. 3-17. Top, striking knife. Bottom, marking knife.

heavy and unwieldy sheet of wood. This is a lot of work. But the end results justify the initial struggle (which really isn't all that bad). If you know how to cut plywood and use the proper tools, the chore turns out to be fairly easy after all.

Handsaws

The traditional saw is the *carpenter's handsaw* (Fig. 3-18), which is just as useful today as it was a century ago—more so, in fact, because today's saws are of so much higher quality. Cutting long lengths of plywood from a panel with a handsaw is a fairly arduous job, but it also works perfectly well. Handsaws range in length from 20 to 26 inches and are available in several grades and with different kinds of handles. They are made in two different types; rip and crosscut. A *ripsaw* is used to cut with the grain of the wood, while a *crosscut* is used to cut across the grain. Obviously with plywood you will be doing both at the same time. In this situation, a ripsaw, while it will cut, does not do a particularly good job because of the wide set and spacing of the saw teeth. Incidentally, the teeth on a saw are often referred to as points. Saws are

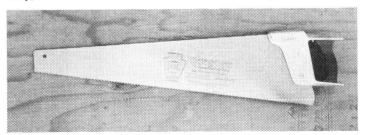

Fig. 3-18. A carpenter's fine-toothed handsaw, or panel saw, is best for cutting plywood.

Fig. 3-19. A fine-toothed backsaw is ideal for cut-off work.

classed by the number of points per inch of blade length. The best combination for cutting plywood is a 10-point crosscut at the coarsest, and a 12-point saw will allow a better cut. In general, the "finer" the blade (small teeth and many to the inch), the cleaner and smoother the cut will be, with minimal splintering.

As far as plywood work is concerned, a *backsaw* (Fig. 3-19), which is widely used in cabinetmaking and furniture building, is of moderate usefulness. While a carpenter's handsaw is flexible and spineless, so that the blade can be inserted fully through the workpiece, a backsaw has a rigid blade that usually is bound along the top edge with a heavy brass or steel spine. Thus, the blade cannot flex at all and cannot be passed through the cut. This saw, also called a *miter* or *tenon* saw, is primarily used as a cut-off saw, especially in a miter box. Very smooth and fine, uniform cuts can be made, but panel-cutting with them can be difficult though not impossible. On the other hand, when the need arises to make excellent cuts in small pieces of plywood, this is the saw to use. It need not be set in a miter box, but can be used freehand as well. A coarse backsaw has 11 points to the inch, but backsaws can be obtained in several degrees of blade coarseness. A 20-point blade will make an exceptionally fine cut and leaves smooth, true edges. A cousin to the backsaw, the *bead* or *dovetail saw*, is a smaller version with an even finer blade. It is very handy for small and intricate work.

The *keyhole saw* is an indispensable and inexpensive tool that should be in every toolbox. The long, tapering blade is particularly useful for pierce-cutting where a starter hole is drilled through the interior portion of the workpiece. Then the saw cut is made from the hole, without cutting in from the outside edge of the piece. This saw is also used to cut curved pieces and will operate even along fairly tight radii. Utility-type keyhole saws (Fig. 3-20) are perfectly

Fig. 3-20. A keyhole saw is excellent for rough-cutting curved shapes and working in tight spots.

satisfactory for all but the most demanding home shop use and are obtainable with three or more interchangeable blades—usually a coarse and a fine wood-cutting blade and a hacksaw-type blade for metal-cutting. This saw also has a junior-sized cousin called the *pad saw*. It is made in the same basic configuration, but it has a smaller, thinner and narrower blade for making very fine and tight cuts.

The *coping saw* (Fig. 3-21) is another inexpensive tool that, while used somewhat infrequently by most craftsmen, will perform cutting operations that other saws cannot. Various kinds of blades for different purposes are available for coping saws, ranging from relatively coarse to extremely fine. The blades are thin and narrow and consequently can be used for making very intricate cuts. They are of particular value and are the only handsaws that can be effectively used in cutting delicate fretwork and scrollwork from plywood.

Fig. 3-21. A coping saw will make fine, intricate cuts, either internal or external.

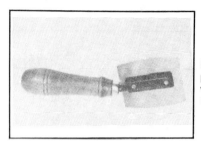

Fig. 3-22. The veneer saw is especially designed for cutting thin veneers (courtesy of Woodcraft Supply Corp.).

There is one other rather uncommon saw that is worthy of consideration, the *veneer saw* (Fig. 3-22). This is a double-bladed saw, fine-toothed on one side and somewhat coarser-toothed on the other, with each cutting edge being curved slightly upward. Its primary purpose is for cutting small pieces of thin veneer, as in inlay or marquetry work, but it has many other general purpose uses as well wherever fine surface cuts must be made. This tool is not a normal hardware store item, but it can be obtained through woodworkers' supply stores.

For those who will be crafting furniture, cabinetry, built-ins and such from plywood, a good *miter box* will be a necessity. There are many different kinds available, ranging from very inexpensive plastic or wood boxes with nonadjustable cutting slots and not furnished with saws, to very expensive, accurate, cast-frame, fully adjustable units equipped with saw guides and a top-quality backsaw. The difficulty with the inexpensive ones, aside from the nonadjustability feature, is that they quickly lose their accuracy. Most of them have only a very small workpiece holding capacity.

On the other hand, quality, precision-made miter boxes that can really do a nice job are quite expensive by comparison to most hand tools—$100 to $150 is neither an exceptional nor an unreasonable price to pay for one. Nonetheless, using a miter box of this quality is the only way to achieve perfect miter cuts regardless of the cut angle required, at least with hand-operated equipment. For the do-it-yourselfer who will be working with moldings, building frames, or doing other work that requires precisely mitered joints, we recommend the purchase of a top-grade, accurate, fully adjustable miter box and backsaw combination (Fig. 3-23) that can be securely clamped or screwed to the workbench.

Circular Saws

Perhaps the most utilitarian saw for the average do-it-yourselfer, and one that is ideal for cutting plywood, is the power

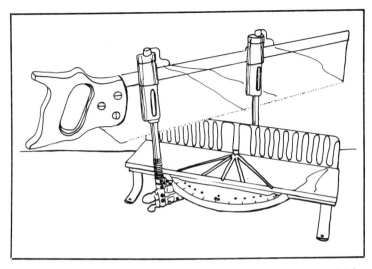

Fig. 3-23. A top-quality miter box and saw combination is essential for making precise miter cuts.

saw variously known as a *power handsaw*, *portable circular saw* or often by one particular tradename, *Skilsaw*. This kind of saw (Fig. 3-24) has the advantage of not only making the cutting jobs fast and easy, but also of having numerous interchangeable blades that can be used for various cutting purposes. This is particularly important with plywood, as we shall see later. Many different brands, sizes

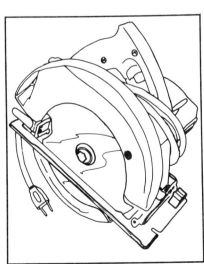

Fig. 3-24. A portable circular saw is the best bet in most home shops for roughing out plywood pieces of substantial size.

163

Fig. 3-25. Several commonly used types of circular saw blades; fine-toothed and combination blades are best for plywood work.

and quality levels of portable circular saws are available, and sometimes the choice as to which one might be the best buy for general purpose work around the home is a bit confusing. In general, neither the very inexpensive nor the very expensive saws are good choices for home use. In terms of price, something in the upper middle to lower upper range is about right. A 7¼-inch blade size is quite practical, though some workers prefer the 8-inch size. A quality brand name should be chosen, of course, and a model with standard helical gear drive is perfectly adequate. A worm-drive model is unnecessary unless you plan to use the saw for constant, heavy-duty work. A motor size of about 2 horsepower is fine, and construction should be ball and roller bearing throughout.

A wide range of different kinds of saw blades is available for portable circular saws, with the 7¼-inch size perhaps being the most common and widely stocked in hardware stores and tool supply outlets. At least three different kinds of blades should be kept on hand for working with plywood, and there are other types that may be useful as well. Some of these blades are shown in Fig. 3-25.

Not many accessories are available for portable circular saws, primarily because they are intended as sawing tools and don't do particularly well at other jobs. But you can obtain a special set of *dado blades* made especially for use with portable circular saws, which will indeed do a good job of cutting dados once the saw is properly set up. Various kinds of *saw guides* are also available, both for cutting straight lines and making accurate angle cuts across a plane surface. One device, a *miter arm arrangement*, allows the saw to be converted into a small radial arm saw. Like many such power tool accessories, it is not overly effective because it is pretty much a compromise situation.

Saber Saws

Another handy and relatively inexpensive power tool that is particularly valuable for working with plywood is the *saber saw* (Fig. 3-26), also known as a *bayonet saw* or a *hand jigsaw*. This saw is most useful for cutting both inside and outside curves, as well as for making all sorts of internal cuts within the boundaries of the workpiece without cutting in from an edge. There are many different brands and models available. Some are single-speed; others are two-speed, while a few feature variable-speed blades. The shoe, or base, of most jigsaws will tilt in one direction to any angle up to 45°, allowing bevel cuts of whatever angle is needed. A few models feature a shoe that will tilt in either direction.

Here again, the best buy for most do-it-yourselfers will be neither a cheap one nor an expensive one, but probably a saw that lies somewhere in the middle price range. The variable speed capability is a good feature to have, and a ¼-horsepower motor is generally sufficient. Construction should be of the reasonably heavy-duty type, with sleeve and ball bearings. A tilting base is essential, though the double-tilt sort is not really necessary. A

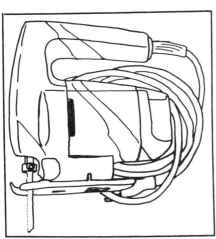

Fig. 3-26. A saber saw proves quite useful in plywood working.

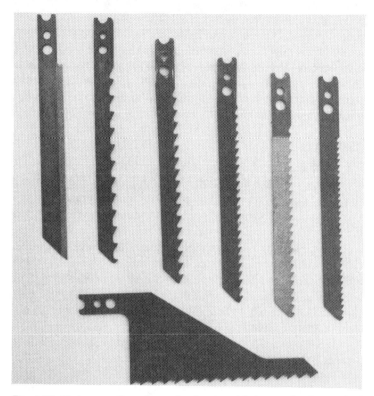

Fig. 3-27. Various configurations of saber saw blades; again, fine-toothed blades make the cleanest plywood cuts.

built-in sawdust blower is also a nice feature. It keeps the cutting line clear of sawdust and lessens the chances of wandering away from the line and making a bad cut because you cannot adequately see what you are doing.

A wide variety of blades is available for use with jigsaws, so they can be used for cutting just about any material, including metals and plastics. There is even a knife-cut blade that works nicely for slicing heavy cardboard or leather. A few of these blade configurations are shown in Fig. 3-27. Note that not all brands of jigsaw blades are interchangeable in all brands of jigsaws. When purchasing a jigsaw, choose a well-known quality brand name that is readily available in your area from a dealer who stocks a full complement of replacement blades, and one for which replacement blades are readily available at most hardware stores. That way you will not get stuck with a saw that uses only strange and relatively unavailable blades.

The *scroll saw* is somewhat similar to the jigsaw, but it is more expensive, larger and heavier, and usually of heavier-duty construction as well. The chief difference between the two is that the scroll saw is equipped with a large knob atop the head of the saw. By turning this knob, the direction of the cutting edge of the blade can be turned, instead of having to turn the entire saw as is necessary with a jigsaw. Thus, the blade direction and the angle of the scroll saw body can be independently maneuvered. This allows the accurately controlled and easily adjusted cutting lines that are necessary for intricate scrolling work. While scrolling with a jigsaw is difficult, it is not with a scroll saw. The scroll saw will also do all of the jobs that a jigsaw will. It is truly a dual-purpose machine.

Radial Saws

One of the most popular stationary power saws for home workshop use is the *radial saw* (Fig. 3-28), also known as the *radial arm saw*. This saw is rather large and bulky, but it consists of only a few main pieces: table, heavy column, and an overarm and/or arm track that extends from the column. The motor, saw blade, guards and locking and control mechanisms slide back and forth on the arm track. An almost infinite variety of cuts is possible because the motor and blade assembly is adjustable through a wide range of angles.

Cuts made in or through relatively narrow pieces of stock are made by placing the stock on the table at right angles to the arm,

Fig. 3-28. The radial arm saw has become a very popular home shop power tool and does an excellent job in plywood work.

and drawing the rotating blade across the workpiece from back to front. Long cuts are made by setting the motor and blade assembly at right angles to the track (on a built-in swivel arrangement), locking it into the proper position, and then feeding the stock along the table and into the saw blade. The angle of the blade is also adjustable for bevel cuts.

A radial arm saw is a fairly expensive piece of equipment to begin with. While the highest-price model is not a necessity for a home workshop, one of good size and with ample capacity is recommended. Radial arm saws are sized by the diameter of the saw blade; the 10-inch or 12-inch sizes are perhaps the most commonly used for home shop purposes, usually powered by a 2 or 2½ horsepower motor. Many different accessories are available for this kind of saw to substantially increase its general usefulness, so purchasing one that will accept a wide range of accessories is a wise idea even if you don't plan to buy them right away. Look for heavy-duty precision-made assemblies, a sturdy stand that does not wobble, and easy-to-use controls and locking mechanisms.

Among those accessories that work well on a radial arm saw is the dado blade. Several different types and brands are available but they all do essentially the same job—cut grooves or channels of various widths. With a radial saw this is an easy chore, since the blade is above the workpiece and cuts into its upper surface. You can always see exactly what is happening. You can also use a molding head and cutters on a radial saw for making your own moldings or for molding the edges and making decorative cuts in plywood or other stock in whatever pattern you choose. Again, molding cutters are easy to use because you can see the cuts being made and the guidelines for them. Special planer heads and blades are also available for radial saws, allowing the operator to surface-plane stock. This accessory, however, is of limited usefulness in plywood work alone, though it is handy for working solid woods. One inexpensive accessory that does work well, however, is the sanding drum, which can be quickly fitted to the saw and allows easy sanding of plywood edges.

Table Saws

The *table saw*, also known as the *bench*, *variety* or *circular saw* (Fig. 3-29), was almost always the first stationary power tool to be added to a home workshop. Today there is endless argument as to which tool is better, the table saw or the radial arm saw. The truth is that each has its own advantages, and the table saw remains one

Fig. 3-29. A table saw makes an excellent choice as the first stationary power tool for a home shop.

of the most useful and utilitarian power tools in the shop. It would be nice, of course, to have both kinds of saws. But if you can only have one, most experts would probably recommend the purchase of a good table saw. The emphasis is on the word "good," because a small, inexpensive table saw generally leads only to frustrations because it is neither large enough nor useful enough to do the number of jobs required.

A table saw consists of a heavy steel frame, usually set upon a steel stand, topped by a precision-ground cast table. Within the frame is the saw arbor; the motor is sometimes close-coupled to the arbor or may be set some distance away in the lower part of the frame or in the stand. The electric motor drives the arbor to which the saw blade is attached by means of a drive belt. The arbor can be raised or lowered by a crank mechanism so that the height of the saw blade above the table top can be adjusted. The arbor also tilts in one direction from the vertical to a 45° angle, either by means of a locking slide or by a crank arrangement. This allows the cutting of bevels to any angle from 0 to 45°.

Table saws are fitted for a variety of accessories, and most of the better saws include at least a few basic ones. For instance, one or two table extensions can be bolted to the side edges of the saw table to make a much larger work surface. In the interest of usefulness, a table saw should have at least one extension and preferably should have both. The saw should be equipped with a blade guard, preferably of the see-through type, an anti-kickback

Fig. 3-30. A carbide-tipped adjustable dado blade will stay sharper longer and is recommended for plywood work.

device to prevent small pieces from being kicked back away from the saw blade and toward the operator, and a splitter that separates the two pieces of wood as they are being cut from a single piece of stock. Under normal cutting circumstances all of this safety equipment should be in place, but for many jobs it must be removed entirely and replaced with other accesory safety devices. In some cases the blade must be left open.

The *rip fence* is a standard accessory on nearly all saws, and is locked in place parallel with the saw blade so that the workpieces being cut can be guided against and along it. Another standard accessory is the *miter gauge*, a device consisting of a precision steel bar to which is affixed an adjustable head. For straight cuts, the head is set at right angles to the bar, or 0°. The stock is placed against the head. The miter gauge is pushed forward, driving the workpiece against the saw blade. The miter gauge travels in a precision-milled groove in the table top; most saws are equipped with such a groove on each side of the blade.

There are several different brands and types of table saws on the market today at a relatively modest price that are fine for home workshop use. The features of these saws are generally rather similar, so it is a good idea for the prospective buyer to carefully compare specifications of various saws. If possible, inspect the ones that seem best suited to individual purposes. Table saws are sized according to the diameter of the saw blade. In general, an 8-inch, 9-inch or 10-inch saw is fine for home use, with a motor size varying from 1 to 2 horsepower. The larger blade size has the advantage of being able to cut through thicker stock, whether in the vertical position or in a bevel cut. The larger saws generally also have larger motors and consequently are more powerful. With regard to working with plywood, none of these factors are of much concern. Virtually any table saw is easily capable of cutting all thicknesses of plywood with no difficulty.

There are a number of other aspects to consider, however, such as easy-to-use blade tilt and elevation controls, positive and accurate rip fence locking and/or adjusting mechanisms, and precisely-ground miter gauge slots in which the gauge fits snugly and runs easily. One point to note is that 8-inch saw blades are generally more easily obtainable in a greater variety, and at considerably less expense, than are 10-inch blades. On the other hand, 8-inch blades can easily and safely be used in a 10-inch saw, provided the recommended saw blade speed is not exceeded.

There are numerous optional accessories, not generally included with a table saw, that can be added to increase the useful-

Fig. 3-31. A set of hold-down fingers is a must for working small or narrow pieces of plywood on a table saw.

Fig. 3-32. A taper jig allows sawing long tapers on a table saw.

ness of the machine. One of the first that should be added is an *adjustable dado blade* (Fig. 3-30) or a dado blade set. The former can be adjusted to cut dados from ¼-inch to 1 inch, or thereabouts, by simply adjusting the hub of the blade assembly to the desired cut-width. The latter is used by setting the several thin blades in various combinations for whatever cut-width is required. Either type works effectively, and one or the other is a virtual necessity for many plywood projects.

Another important piece of equipment that aids in working with smaller pieces and also greatly increases safety is a set of *table saw hold-downs* (Fig. 3-31). The spring-steel fingers clamp the workpiece to the saw table, yet allow it to be moved forward with relative ease. The constant pressure keeps the workpiece against the rip fence and prevents jumping and chattering, allowing for better cutting. Also, the saw operator can keep his fingers well away from the blade area.

Another device used for a somewhat similar purpose is the *universal jig*. This device can be set up to hold small workpieces as they are run past the saw blade. The jig tracks in the miter gauge slot, while the workpiece is clamped to the face of the jig. A *clamping miter gauge* is equally handy, since it can be used to clamp small pieces right into the gauge. The pieces can then be run to the saw blade.

Sometimes, especially with plywood pieces, it is necessary to make a long, taper cut. This should never be attempted freehand on a table saw, but should be done only with a *taper jig* (Fig. 3-32). The jig will adjust from 0 to about 15°. When properly positioned and adjusted on the saw table, it will allow the cutting of long tapers with no difficulty.

The table saw can also be used as a molder, by replacing the saw blade with a special *molding head* into which cutters of various types can be locked. The three-blade head (Fig. 3-33) is a good choice. With this device you can make up your own molding stock, or you can shape plywood or other stock in whatever decorative fashion you choose and the cutters will allow.

Sanding Disc or Wheel

There is one more item worthy of mention. This is the *sanding disc* or *wheel*, a cast steel or aluminum disc that can be fitted to the table saw arbor in place of the saw blade. The items come in standard 8-inch and 10-inch sizes, and replaceable sandpaper discs are readily available. The sandpaper disc is affixed to the sanding wheel, and workpieces up to a maximum of 3 to 4 inches in thickness (depending upon table saw size) can be edge-sanded with ease. Some sanding wheels consist of a relatively thin plate made so that sandpaper discs of different grits can be attached to either face. The sanding wheel is a particularly helpful device in

Fig. 3-33. With a molding head set, a table or radial arm saw can be used to make moldings and decorative cuts.

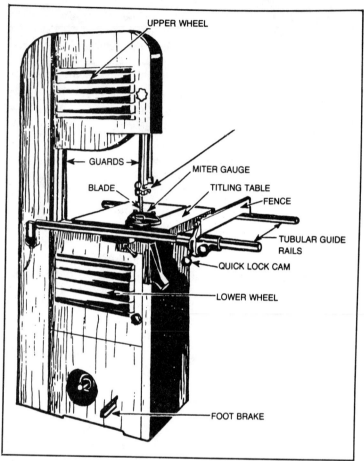

Fig. 3-34. Parts of a bandsaw.

smoothing plywood edges and in trimming off very slight amounts of material for precise and tight joint fitting.

Bandsaws

While both the table saw and the radial arm saw can be used to cut even full sheets of plywood, the *bandsaw* (Fig. 3-34) cannot. This is a rather specialized power tool that is not found in too many home workshops because of its somewhat limited usefulness, considerable cost and, at least for most models, its large size. Most bandsaws are freestanding upon a floor stand, though some models are designed for benchtop installation.

The saw consists of a narrow, fairly thin endless or loop blade that runs on either two or three rubber-tired wheels, one of which is motor driven. The arrangement is analagous to a vee-belt passing over two or three pulleys. The blade is exposed for only a certain short length just above a relatively small tabletop where the cutting takes place. The bandsaw excels at making fine, curved cuts that can be taken down to a quite small radius, depending upon the blade width. The tool is ideal for cutting such figures and patterns from plywood and can be used for other general purpose cutting jobs as well. However, the purchase of a bandsaw is difficult to justify costwise unless there is a definite use for it on a repetitive basis.

HAMMERS

Strange though it may seem, the hammer has not changed its basic form since the days of early Rome. The chief differences in modern hammers are simply slight changes in configuration for different purposes, and of course the materials from which they are made. When working with plywood, you will need at least one hammer and probably will want to have three or four different kinds, depending upon the type of work you will be doing. Hammers today are made of several different materials. All of the best ones have forged steel heads. The shanks may be of tempered steel, fiberglass or wood. The grip may be of wood, rubber, a

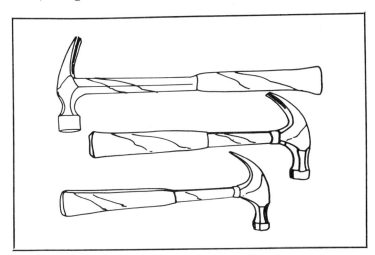

Fig. 3-35. The carpenter's 16-ounce claw hammer (center) is best for all-around utility use, but the framing hammer (top) is excellent for subflooring and similar jobs, while the 13-ounce model (bottom) is best for cabinetry and casework.

175

plastic material or leather. Just which type you purchase is a matter of personal preference and also cost, because they all work equally well. Those with rubber or plastic grips are somewhat more comfortable to use because the material absorbs shocks. The steel or fiberglass shanks are likely to last longer than wood. With the one-piece design, there is no difficulty with the hammer head loosening on the handle or flying off.

The most popular and commonly used hammer is called a *claw hammer* (Fig. 3-35) and is identified by the curved claw opposite the striking face. The 16-ounce size is standard and performs very well for virtually all purposes. However, the same head pattern in a 13-ounce size is much easier and handier to use with smaller nails. Even a 10-ounce size is very practical for driving brads and small finish nails.

For general construction work, especially in laying plywood subflooring or roof decking, the large *framing hammer* (Fig. 3-35) is hard to beat. The long length and heavy head (20 ounces or more) drives 8d or 10d nails with only one or two blows. At the other end of the scale, attaching moldings with tiny brads, driving upholstery nails into plywood framework and similar chores calls for a tack hammer (Fig. 3-35). Most of these have one end slightly magnetized, so that a tack can be picked up and started with a light tape. Then the hammer head can be reversed and the tack driven with the broader striking face.

A cabinetmaker's hammer is somewhat similar and has a 10-ounce head with a striking face on one side as usual, but the claw has been replaced with a wedge-shaped peen. Larger nails are driven with the broad striking face, while small brads can be tapped home with the peen. This is especially useful in getting into tight, cramped spots and in working with curved moldings and shapes.

Cabinetmaking, furniture building, fabrication of built-ins and similar assembly work with plywood often requires that some of the component parts that fit tightly together (as they should) must be tapped gently into place. This should never be done with a hammer because of probable damage to the wood surface. Instead, you will need a carpenter's wood mallet, a small rubber mallet (be sure to get the kind that does not leave black marks), or a soft-faced plastic mallet. These latter are available in numerous different sizes and have interchangeable/replaceable striking faces that come in different degrees of softness. Thus, the proper face can be chosen for any kind of job or material.

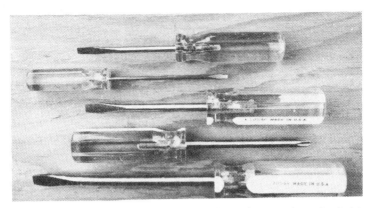

Fig. 3-36. A small set of mechanic's screwdrivers is a good way to start a screwdriver collection.

SCREWDRIVERS

Unless you will be working only in rough construction, the chances are excellent that you will need at least one set of screwdrivers. If nothing else, they will be required for mounting fixtures and builder's hardware. For general purpose work, a set of ordinary mechanic's screwdrivers (Fig. 3-36) will fill the bill quite nicely. These are most easily bought in complete sets, which vary somewhat as to the number of screwdrivers included but will generally cover most requirements. Don't buy an inexpensive set at a cut-rate price because the quality will be cut-rate as well. Expect to pay anywhere from $15 to $25 or so for a fair-sized set, including four or more blades to fit straight slots and one or more to fit Phillips screws. Look for forged alloy steel shanks and blades with quality plating applied, ground blades and comfortable, sturdy plastic grips of adequate size to comfortably wrap your hand around.

If you anticipate using screwdrivers a great deal, you might wish to investigate a ratcheting screwdriver (Fig. 3-37). This is an extremely useful tool, which usually comes in a set consisting of a ratcheting handle with various blades that insert into a handle slot. The ratcheting action can take place in either clockwise or counterclockwise directions and does indeed save a good many blisters.

Those who are interested in doing a considerable amount of cabinetmaking or furniture building with plywood and other woods should consider the purchase of a second set of screwdrivers to be used only where fine, precise work is required, expecially when driving brass or aluminum screws. High-quality cabinetmaker's screwdrivers, such as those shown in Fig. 3-38, are made espe-

Fig. 3-37. A ratcheting screwdriver (there are several kinds) makes driving screws easy.

cially for this kind of work. Most are very finely made and have the advantage of being sized to fit specific screw sizes. The turned hardwood grips are very comfortable (some are flatted slightly for even better purchase), and the shanks and blades are made of top-grade steel and are precisely ground. The shanks are flat so that an adjustable wrench can be used with the screwdriver for additional leverage and screw driving power. These screwdrivers, however, are fairly expensive and must be treated with some degree of care and respect. They are not meant to be used as pry bars, cold chisels or demolition tools.

One other type of screwdriver that often comes in handy is a spiral-ratcheting tool that features interchangeable blades. The best-known kind is the *Yankee*. In fact all brand names of this type of screwdrivers are often simply called by this term. The ratcheting mechanism can be locked or made to operate in either clockwise or counterclockwise directions. The shank is fitted with a collet chuck that accepts several different kinds of screwdriver blades, as well as a series of drill bits that are especially made for the tool. When the handle is pushed in, the blade or bit will spin in the desired direction, making screw driving or removal on a repetitive basis a relatively ease chore.

CHISELS

Wood chisels are just as important to working with plywoods as they are to working with other woods and materials. Though they are seldom needed in general construction work, such as putting an addition on your house, you will find yourself often reaching for one while building cabinets, built-ins or other similar projects. There are many different sizes, shapes, types and grades of wood chisels, ranging from quite inexpensive to as much as $25 or so. For general purpose work, a modestly priced set of *butt chisels* (Fig. 3-39) should be sufficient. They should, however, be of good to excellent quality, so that they can be kept properly sharp. Cheap chisels do not hold a good edge, and they nick with amazing ease.

Chisel sets may consist of three, four, five or more chisels. If you see no particular need to have this many, at least at the outset, buy them singly (the ultimate cost will be a bit more but not too much) and add more to the set when you need them. A good

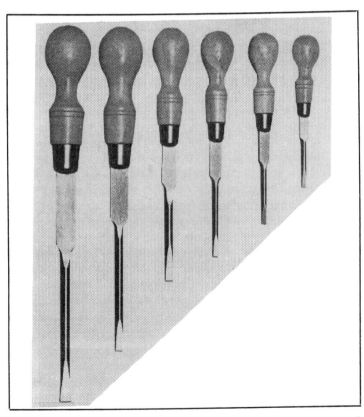

Fig. 3-38. Cabinetmaker's screwdrivers are best for cabinetry, casework and furniture-building (courtesy of Woodcraft Supply Corp.).

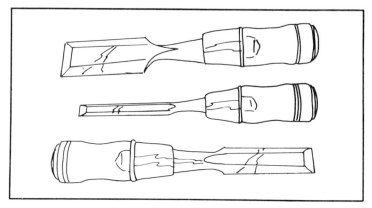

Fig. 3-39. A set of just two or three butt-type wood chisels is enough to get started with.

working combination to start out with is one ¼-inch width, and one ½-inch or ¾-inch. If you will be using the chisels for just general purpose work, rather than strictly for cabinetmaking or fine woodworking, choose the type with steel-capped, heavy-duty plastic handles. Cabinetmaker's *firmer chisels* are somewhat similar, but have a slightly different blade configuration and are meant to be used only with a light wood mallet.

PLANES

A *hand plane* is another tool that you will probably need as soon as you start working with plywood, except that it is not essential to general construction work (though often it is handy there, too). For trimming plywood ends and also for beveling edges, a *block plane* (Fig. 3-40) is about the handiest. This type of plane is usually about 6 inches long, has a low blade angle, and is small and light enough to fit easily in one hand. The fact that it is primarily designed for planing edge grains makes the block plane particularly valuable in plywood work, where end grains appear on every end and side.

Fig. 3-40. A block plane is ideal for trimming plywood edges.

Fig. 3-41. A jack plane is used for planing long edges and will surface-plane as well.

Though the block plane is also excellent for general purpose work, planing long edges is more easily done with a larger plane. Of the several types, either a *smooth plane* or a *jack plane*, both of which are *bench planes*, will work nicely (Fig. 3-41). Either type can be used for both heavy and light cuts or for smoothing edges and surfaces. The smooth plane generally runs from about 7 to 10 inches in length, while the jack plane is 14 or 15 inches long; the latter is a bit better for smoothing long edges.

Another quick and effective way to trim edges, especially of veneers, is to use an *edge trimmer*. This tool is dual-purpose in that it can also be used to trim the edges of laminated plastics, as when building a kitchen countertop with a plywood base and a laminate top. Various adjustments can be made to the blade to allow uniform and accurate beveling. It can be used for such purposes as trimming panels, drawer bottoms and fitting of lightweight joined pieces.

Fig. 3-42. A power plane is the answer for easy edge-trimming, and will also surface-plane and rabbet.

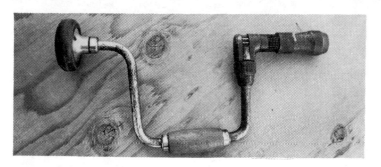

Fig. 3-43. The bitstock is the traditional boring tool and is still a useful item to have around.

The *power plane* (Fig. 3-42) has two distinct advantages. First, it takes all the work out of planing an edge or surface, especially when a substantial amount of material must be removed. The second advantage lies in the fact that the power plane will make cuts of uniform depth and smoothness, with no dipping, bowing, tilted cuts and irregular blade marks that so frequently mark the passage of a hand plane, especially when in the hands of an inexperienced user. The power plane has other advantages as well. It can plane a swath as wide as 3 ½ inches and can be set to cut bevels with uniform results. When properly adjusted and set up, the plane can make rabbet cuts for rabbet joints. Because the planer blades travel at a very high speed, they make a smooth cut even across end grains, an important factor with plywoods.

Since this is not a stationary power tool but must be hand-held and operated, the results that can be obtained when using a power plane are not quite as accurate, smooth and precise as one might reasonably expect from a stationary planer or from a jointer. On the other hand, a bit of practice and experience with a power plane will allow the user to produce excellent results in any circumstance where a hand plane might otherwise be used.

DRILLS

One of the old traditional methods of drilling holes is with a tool called a *bit brace* or *bitstock* (Fig. 3-43). The method works as well today as it did in the old days, and still remains the best way to get the job done on occasions where there is no handy electrical outlet into which a power drill can be plugged. Holes of ¼ inch in diameter to over an inch, and larger with an expansive bit, can be bored with a brace and auger bits. If you don't plan to purchase or cannot use a power drill, this is the answer. A good bitstock has a ball-bearing top grip and chuck, and a well-made center grip. The tool should be made of quality materials and have a good plated finish, including the chuck. It should also have a ratcheting feature that can be locked at center position, or set for either clockwise or counterclockwise rotation; this allows drilling in tight sport in repetitive short fractional arcs rather than in full circles.

To go along with a bitstock in nonpower situations, you will also need what is sometimes called an *eggbeater drill* or a *hand drill* (Fig. 3-44). This type of drill will easily drill holes of ¼-inch diameter and down and uses twist drill bits. A larger version, called a *breast drill*, is meant for heavy-duty use and has a chuck capacity of up to ½ inch. The Yankee (or similar) screwdriver will also drill small holes.

Electric Drills

Most folks, however, prefer to use one of the various small hand-held *electric drills* (Fig. 3-45). These drills are designated as to size by the diameter of the largest drill bit shank that the chuck will hold. There are three standard sizes: ¼-inch, ⅜-inch and ½-inch. There are many brands and models available with numerous variations between them. A typical standard drill is designed for straight-ahead drilling, since the chuck is essentially an in-line extension of the drill body. However, 90° adapters can be fitted to straight-line drills, and drills made in a 90° or 45° configuration (Fig. 3-46) are also available. Some drills are reversible, some are 2-speed, and a few feature variable-speed motors. Nearly all are fitted with Jacobs or geared chucks, which are tightened and loosened with a special key.

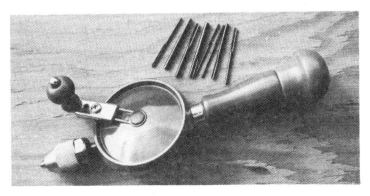

Fig. 3-44. The hand or eggbeater drill is a very useful tool for drilling small holes, even if an electric drill is available.

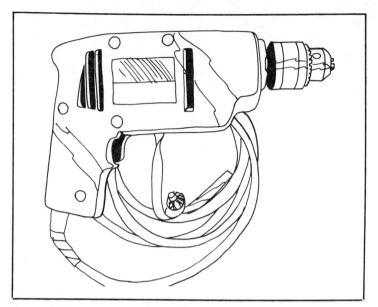

Fig. 3-45. A ⅜-inch electric drill is utilitarian and an excellent adjunct to any shop.

For general, all-around utility work and dependability, select a drill that is at least in the medium price range. The ⅜-inch size is best; choose the ¼-inch type for only occasional light-duty work and the ½-inch type for continuous, heavy-duty drilling of large holes in tough materials. The reversing and variable speed features are often worth the extra money but are not really essential. When making a decision between two similar models of the same drill size, choose the one with the greater horsepower, or current draw, and look for ball-bearing construction. Sleeve bearings and bushings do not hold up well. A drill body made of a tough, unbreakable plastic is better than metal, especially for outdoor use, because it is shockproof.

When purchasing a ⅜-inch (or ½-inch) drill of substantial power, be sure to get one with an accessory side handle. This will enable you to get a firm grip on the drill when boring large holes, so that the drill will not get away from you.

Accessories

One of the problems that sometimes arises when drilling holes with a hand-held electric drill is that it is difficult to drill absolutely straight into the material. Invariably the drill is cocked a little bit

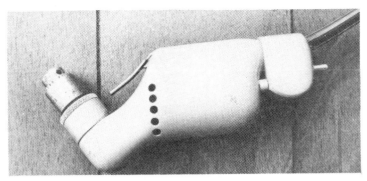

Fig. 3-46. For drilling in tight spots, an angled drill is helpful. This 45° model also converts to a power screwdriver.

one way or another, so the hole is drilled at an angle to the surface of the workpiece. Often this does not matter, but there are many occasions when it does, such as when drilling shank or pilot holes for screws that must be accurately set. There are two accessories

Fig. 3-47. A drill stand increases the usefulness of an electric drill.

Fig. 3-48. A full boxed set of Jennings type double twist auger bits, for use with a bitstock (courtesy of Woodcraft Supply Corp.).

that can take care of this problem. One is a drill stand onto which the drill is clamped. The result is a tiny drill-press arrangement (Fig. 3-47); once the drill is accurately placed in the stand, perfectly-positioned holes can be drilled at 90° (or at other predetermined angles) into the workpiece. This makes an excellent arrangement if the pieces are not too large. Another and somewhat similar accessory allows the drill to be lined up on a surface of any expanse, wherever it is needed. The drill bit can be aligned either straight or at various angles. This device also allows accurate centering and drilling of round pieces, a very difficult job to do by hand alone.

There are various other accessories available for electric drills, too. Many of them would be of no particular use in working with plywood, except perhaps in a few unusual situations. A *mandrel* with a wire brush attached, for instance, could be used to brush-texture the surface of a plywood panel. One item that must be mentioned, however, is the mandrel and sanding disc combination. The mandrel is chucked into the drill just like a drill bit, and the flexible rubber wheel will accept sanding discs of various grips.

This is sometimes a useful tool for rough-sanding or for forming and shaping plywood or other workpieces right on the job site, especially those pieces that are already incorporated into an assembly.

BITS AND CUTTERS

The type of drill bits that are used with a hand bit brace are called *auger bits*. They are available in a range of diameters from ¼-inch to 1¼-inch in either single-twist or double-twist configuration (Fig. 3-48). The single-twist type is a little less expensive than the double-twist and will bore deep holes with somewhat greater ease. The double-twist bits, however, are better for drilling plywood because they will bore cleaner, truer holes with less chipping or tearing of the wood fibers. They will cause less disturbance to the veneer plies as they pass through the plywood. Either kind of auger bit is relatively expensive. While you can buy them in complete sets, you might prefer to purchase only those sizes that you need.

For drilling holes larger than 1¼-inches, the bit to use in a bit brace is the *expansive bit* (Fig. 3-49). There are several different types available but they are basically the same. One or two adjustable cutters can be mounted at the tip of the shank and set to whatever hole diameter is desired from about 1 to 3 inches. These bits do a good job, but they must be used with care (and also a good deal of pressure). The peeling action of the cutter tends to catch in the plies of the panels. However, they will do the job.

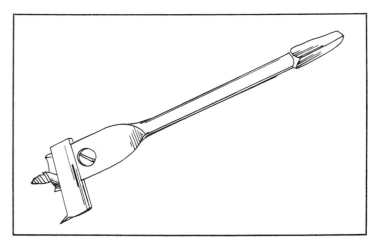

Fig. 3-49. An expansive bit will bore holes of any size from about 1 to 3 inches in diameter.

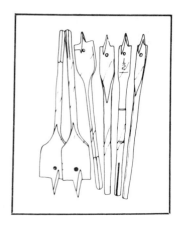

Fig. 3-50. Speed bits will cut rapidly through plywood or wood, but can leave a ragged hole.

Electric drills can be fitted with *spade* or *speed bits* (Fig. 3-50) for rapid and reasonably accurate drilling in plywood or other woods. These are available in diameters from ¼-inch to 1½-inch in increments of 1/16 inch, and may be bought in full sets or individually as you need them. They can be resharpened, are long-lived and are not particularly expensive. Spade bits must be used with care in drilling plywood, because they tend to tear the wood fibers and the plies.

Twist Drill Bits

Twist drill bits (Fig. 3-51) can be used for drilling virtually all materials, especially when one includes the several specialty variations. Carbon-steel twist drills should be used only in wood and similar soft materials. High-speed types can be used in drilling woods and other materials; they are longer-lived and hold a better cutting edge. There are three different basic types: fractional, number or numerical and letter drills. All of the specific sizes are listed in Table 3-1.

Under normal circumstances, the shank size of the bit is of the same diameter as the cutting edges. However, if it is necessary to use a 5/16-inch bit in a ¼-inch electric drill, the recourse is to use a 5/16-inch bit specially made with a ¼-inch shank. Twist drill bits can be bought individually or in sets; as far as general shop and around the home work is concerned, a set of fractional drills is the most useful. However, for drilling accurate clear-holes to pass screw shanks, for instance, then certain number drills are equally valuable. Both number drills and letter drills are more frequently used in machine and mechanical work than for woodworking.

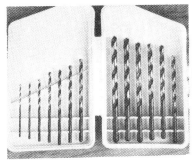

Fig. 3-51. A small set of twist drill bits will serve most wood-shop needs at the outset.

Countersink and Wood Screw Pilot Bits

For plywood work in cabinetmaking, furniture building and other projects that make extensive work of flathead wood screws as fasteners, other bits will be necessary. Where all that is necessary is to countersink the screw heads (set them flush with or below the surface of the workpiece), you will need one or more *countersink bits* (Fig. 3-52). These are available in several head sizes and

Table 3-1. Drill Size Chart.

				FRACTIONAL DRILLS		LETTER DRILLS	
Size	Dia.	Size	Dia.	Size	Dia.	Size	Dia.
1	0.2280	41	0.0960	1/64	0.0156	A	0.2340
2	0.2210	42	0.0935	1/32	0.0312	B	0.2380
3	0.2130	43	0.0890	3/64	0.0469	C	0.2420
4	0.2090	44	0.0860	1/16	0.0625	D	0.2460
5	0.2055	45	0.0820	5/64	0.0781	E	0.2500
6	0.2040	46	0.0810	3/32	0.0937	F	0.2570
7	0.2010	47	0.0785	7/64	0.1094	G	0.2610
8	0.1990	48	0.0760	1/8	0.1250	H	0.2660
9	0.1960	49	0.0730	9/64	0.1406	I	0.2720
10	0.1935	50	0.0700	5/32	0.1562	J	0.2770
11	0.1910	51	0.0670	11/64	0.1719	K	0.2810
12	0.1890	52	0.0635	3/16	0.1875	L	0.2900
13	0.1850	53	0.0595	13/64	0.2031	M	0.2950
14	0.1820	54	0.0550	7/32	0.2187	N	0.3020
15	0.1800	55	0.0520	15/64	0.2344	O	0.3160
16	0.1770	56	0.0465	1/4	0.2500	P	0.3230
17	0.1730	57	0.0430	17/64	0.2656	Q	0.3160
18	0.1695	58	0.0420	9/32	0.2812	R	0.3390
19	0.1660	59	0.0410	19/64	0.2969	S	0.3480
20	0.1610	60	0.0400	5/16	0.3125	T	0.3580
21	0.1590	61	0.0390	21/64	0.3281	U	0.3680
22	0.1570	62	0.038C	11/32	0.3437	V	0.3770
23	0.1540	63	0.0370	3/8	0.3750	W	0.3970
24	0.1520	64	0.0360	25/64	0.3906	X	0.3970
25	0.1495	65	0.0350	13/32	0.4062	Y	0.4040
26	0.1470	66	0.0330	27/64	0.4219	Z	0.4130
27	0.1440	67	0.0320	7/16	0.4375		
28	0.1405	68	0.0310	29/64	0.4531		
29	0.1360	69	0.0292	15/32	0.4687		
30	0.1285	70	0.0280	31/64	0.4844		
31	0.1200	71	0.0260	1/2	0.5000		
32	0.1160	72	0.0250				
33	0.1130	73	0.0240				
34	0.110	74	0.0225				
35	0.1100	75	0.0210				
36	0.1065	76	0.0200				
37	0.1040	77	0.0180				
38	0.1015	78	0.0160				
39	0.0995	79	0.0145				
40	0.0980	80	0.0135				

Fig. 3-52. A special countersink bit is used to cut the countersink for flathead screws.

shapes and for use with either electric drills or bit braces. A set of *wood screw pilot bits* (Fig. 3-53) is a virtual necessity for cabinetmaking and similar woodworking purposes. These special bits are sized to match the various screw diameters and tapers and perform two operations at once; the lead part of the bit drills the pilot hole, while the upper part of the bit drills the shank hole and the countersink. Some of these bit sets come equipped with stop collars, which can be adjusted on the shank of the bit to match the exact depth of hole needed.

Double-Spur Brad Point and Forstner Bits

Where precise, clean-walled holes must be drilled, there are other types of bits to use. The *double-spur brad point bit* (Fig. 3-54) with a spiral grind and polished edges will cut extremely accurate holes in plywood with no tearing or chipping. The hole will not travel or get out of round as can happen with some other kinds of bits. There is a precision type of brad point bit that is especially used for doweling, and models are also available that are particularly made for use with electric hand drills. Cutting flat-bottomed holes that do not go through the workpiece requires yet another kind of bit called a *Forstner bit* (Fig. 3-55). These bits are available

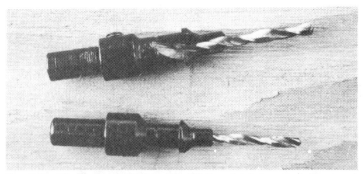

Fig. 3-53. Wood screw pilot bits will drill the pilot hole, counterbore, and countersink, and stop at the right depth.

Fig. 3-54. A brad point bit is best for precise hole-drilling in plywood (courtesy of Woodcraft Supply Corp.).

in a good range of sizes and for chucking either in an electric drill or bit brace.

Hole Saw

The best way to cut large holes (or even relatively small ones) through plywood is not with a large-sized bit, but rather with a *hole*

Fig. 3-55. A Forstner bit makes precise, flat-bottomed blind holes (courtesy of Woodcraft Supply Corp.).

saw (Fig. 3-56). For most home project purposes, the less expensive type with a mandrel and integrated pilot bit that will accept a range of different sizes of blades is quite satisfactory. Other similar and more expensive kinds for heavy-duty work are also available. The size range for hole saws is from ¾-inch diameter to 2½-inches

Fig. 3-56. The best way to cut larger holes in plywood is with a hole saw.

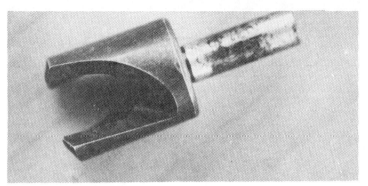

Fig. 3-57. The plug cutter makes small matching wood plugs for concealing screw heads in the workpiece.

or more. Cutting true and clean large holes is an easy chore with this type of cutter.

Plug Cutter

One other drill accessory that will be of interest to those doing cabinetwork or other projects that require hiding the countersunk heads of screws with matching wood plugs is a *plug cutter* (Fig. 3-57). You can, of course, buy ready-cut wood plugs of several different species, or you can carefully slice wafers from dowel stock to serve the same purpose. The problem, however, is that more often than not these plugs will not match the surface characteristics of the workpiece in which they wil be installed. The plugs are quite obvious. Sometimes, of course, this contrast is desired. But when it is not, a plug cutter can be used to custom-cut plugs from the same material as is used in the project, for a matched-in and unobtrusive effect. Plug cutters are available in a range of standard sizes.

CLAMPS

General construction work with plywood does not require the use of clamps, except in an occasional odd circumstance. In cabinetmaking, furniture building and numerous other general projects, though, you will find a rather steady need for them. Generally their purpose is threefold: to hold pieces of an assembly in proper position as the fabrication goes along; to hold workpieces in place on the workbench, sawhorse or in the assembly itself while work such as shaping, fitting or finishing goes on; or to hold the pieces of an assembly firmly and solidly in place while glue cures properly.

Fig. 3-58. An assortment of C-clamps is necessary in any home shop; the one on the right is a deep-throat type (courtesy of Woodcraft Supply Corp.).

In order to effectively handle all of the various clamping situations that arise, you will probably need at least three different kinds.

Perhaps the most important one with the most universal applications is the ordinary *C-clamp* (Fig. 3-58). A number of different kinds of C-clamps are manufactured in numerous different sizes. You will find that having two or three different sizes on hand is practical. Since C-clamps are frequently used in pairs, you may wish to purchase them this way. Jaw capacities range from a maximum opening of ½ inch to 12 inches or more; for general shop purposes, a pair of 2-inch clamps, a pair of 5 or 6-inch, and a pair of 8-inch clamps are usually sufficient to handle most projects. You may find occasions, however, when the standard C-clamp has

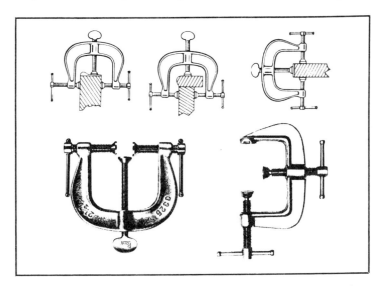

Fig. 3-59. Edging clamps will adjust and exert pressure at two or three points (courtesy of Woodcraft Supply Corp.).

Fig. 3-60. Pipe clamps are versatile and have many uses around the shop.

ample jaw capacity but insufficient frame depth to allow proper clamping. In situations such as this, a deep-throated C-clamp is called for. These clamps are made the same as the standard variety, but the back of the frame is extended substantially so that the jaws can be clamped further into the workpiece.

One of the most useful clamps for working with plywood when edging must be applied to cover the plywood bands, or for applying edge molding, are the *2-way* and *3-way edging clamps* (Fig. 3-59). Either clamp allows pressure to be applied from three directions, with adjustment from two points in the 2-way clamp and three in the 3-way. Though a more expensive and a lesser-known type, edging clamps can be a valuable addition to the tool kit, making this particular type of work much easier.

The second kind of clamp that has numerous applications in plywood work, as well as a host of additional uses, is the *pipe clamp* (Fig. 3-60). Pipe clamp jaws are sold in head piece and tail piece units packaged together. The purchaser supplies his own section of pipe of whatever length he desires and in either ½-inch or ¾-inch trade size (black iron or galvanized steel). The head piece with the screw spindle and movable jaw attaches to one end of the pipe. The tail piece slides over the other end and can be locked in place at any point along the pipe, making a clamp of whatever effective length is needed. The travel of the movable jaw is limited to only 3 or 4 inches in most cases. Since the tail piece is fully adjustable, this is no drawback at all.

Another similar type of clamp, considerably more expensive, is called the *deep engagement clamp*. This kind of clamp is used for the same general purpose as the pipe clamp, but is a bit more refined and generally a better-made piece of equipment. One advantage that this clamp has is that it can be fitted with special polyethylene jaw pads that will prevent scratching or marring of the workpiece surfaces.

A third kind of clamp that you may find need for, especially if you will be assembling items of odd shapes, is the *band* or *flexible clamp* (Fig. 3-61). This device consists of a web belt and adjustable tensioning-and -clamping head. Various standard sizes are available; small ones have a capacity of 4 to 8 feet, but they are also available in 15-foot, 20-foot and even larger sizes. The shape of the assembly makes no difference because the band is simply wrapped around the object in the best fashion to gain maximum pressure at the desired clamping points. The belt is locked in place, and pressure is applied by turning the ratcheting tensioner. Often these clamps are used in conjunction with brackets, supports and blocks so that the pressure can be directed as needed. Sometimes a bit of ingenuity must be exercised, but once properly arranged this clamp will apply uniform pressure just where it is most needed.

FORMING AND FINISHING TOOLS

There are a number of different forming and finishing tools, both hand-operated and electrically powered, that must be considered. Exactly which ones you choose depends upon the type of work that you will be doing, of course; rough construction work

Fig. 3-61. The band clamp is used on odd-shaped workpieces or articles where other kinds cannot be positioned.

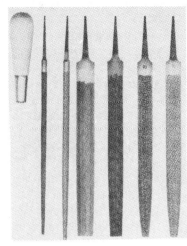

Fig. 3-62. An assortment of cabinet files and rasps, with universal handle (courtesy of Woodcraft Supply Corp.).

requires virtually none, for instance, while cabinetmaking requires several.

Hand Tools

One fast, accurate and easily controlled method of removing tiny to substantial amounts of wood from the workpiece in order to shape or form a contour is a *wood file* (Fig. 3-62). *Cabinet rasps* are even-cutting tools that leave a relatively smooth finish and are available in various lengths and degrees or coarseness. Their principal function is to remove material from the wood surface; flat, half-round or round shapes are readily available. The *cabinet file*, on the other hand, is used to smooth the wood and is designed to take off only small amounts of material per stroke. These files are available in different patterns, degrees of coarseness, shapes and lengths. Be sure also to purchase a *file brush*, which is especially made for cleaning the file teeth without doing any harm. Keep your files clean at all times for effective cutting action and to prolong the life of the teeth.

Any of the various brands of forming and contouring tools that consist of a shaped, cheese grater type of blade fitted with a handle or grip is also a good choice for this kind of work. There are several types and brands. The Stanley Surform line is perhaps the best known, but there are others as well. Square and round file patterns are available, as well as a broad, flat model with an offset handle and different sizes of plane-like types. These tools can be used for roughing out shapes, for removing substantial amounts of material and also for final contouring or forming. They do not, however,

Fig. 3-63. A rubber sanding block makes hand-sanding easier (courtesy of Woodcraft Supply Corp.).

leave a perfectly smooth finish by any means, and further work must be done if a fine finish is required.

Most often the last process to be gone through before applying a final finish to an assembly is smoothing the surfaces. This normally is done with sandpaper, either by machine or by hand. You can easily make your own hand-sanding block by wrapping the sandpaper around a scrap of wood. An easier way, however, is to fit the paper into a rubber *sanding block* (Fig. 3-63). These blocks are sized so that strips to fit them can be cut from a standard sheet of sandpaper, with no extra paper left over. The rubber block is comfortable to hold and is slightly flexible. It also allows sanding in close quarters. An alternative is to fit the sandpaper into a plastic sanding block, most of which are somewhat larger and easier to hold than the rubber ones, that is fitted with a soft and resilient bottom pad to allow smooth sanding.

Better yet, and especially useful for sanding contours, are the *sanding planes* like those shown in Fig. 3-64. These tools are lightweight and ruggedly made of hardwood with aluminum bases. The bases are covered with sponge rubber cushioned soles, and the sandpaper strips can be tightly locked in place. They are available in half-round, flat and convex models with various handle and grip

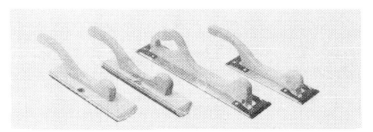

Fig. 3-64. Sanding planes work very well, especially when a lot of forming must be done (courtesy of Woodcraft Supply Corp.).

configurations. They are highly maneuverable and very easy to use with great precision. For hand-sanding, they are without a doubt the best all-around bet.

Hand-sanding is something of a chore, no matter what kind of device is used to hold the paper, especially when large surfaces must be sanded or where a relatively large amount of material must be removed. Power sanders make the job much easier, and there are several kinds to consider.

Finishing Sander

Perhaps the most popular kind for all-purpose use around the home and workshop is the *finishing sander* (Fig. 3-65), more commonly referred to as a *pad* or *orbital sander*. There are many different brands and sizes available, but the best choice for general purpose work is in most cases a medium-duty type with a 1/3 sandpaper sheet capacity and a motor of about ¼ horsepower. The small, less expensive models do not hold up well under medium-duty conditions where they are used for relatively lengthy periods of time. The large ones, while fine for production work, are both heavy and expensive. A sander with an automatic dust pickup feature, though a bit more expensive, is an excellent choice, and saves breathing and cleaning up a lot of dust. The primary purpose of this type of sander is not to remove large amounts of material, but rather to smooth the surface to the point where a finish can be applied, or to sand between finish coats.

Three different kinds of pad action are available: orbital, straight-line and multi-motion. Some sanders are dual-purpose in that they can be shifted from orbital to straight-line action by making a slight adjustment on the tool. The orbital action is fine for some purposes, such as auto body work, but is not a good choice for woodworking. Both orbital and multi-motion actions can leave strange looking whorls and whirligig patterns on the finished surface, especially in soft woods, that can easily show up in the applied finish. For sanding plywood, as well as other woods, the sanding direction should always parallel the face grain. The movement of the paper should be straight back and forth. Thus, the straight-line action is the one to choose.

For those who will be involved in a considerable amount of cabinetwork, furniture building and similar kinds of tasks, another type of pad sander will be of interest. This is a small, single-handed sander that operates at quite high speed and has a small, square pad. It is especially designed for fine finishing work and is built so

Fig. 3-65. A finishing sander makes short work of sanding chores.

that it can be used easily with one hand, even in an overhead position. It excels at getting into tight quarters and cramped places where other sanders simply will not fit. It also works well on irregular and curved surfaces, provided the curvature is not too great.

Belt Sander

There are plenty of occasions when sanding seems the best way to remove a substantial amount of material, but the pad sander will do so only very slowly, even if fitted with coarse paper. The type of machine needed for this kind of work is a *belt sander*, which is fitted with an endless belt of sandpaper that travels at fairly high speed and covers a considerable amount of area at one time. Until recently, most of the belt sanders available on the market were bulky, heavy tools geared primarily for commercial and other heavy-duty uses. They were quite expensive. None of them were particularly well suited to occasional home shop use.

Now, however, there is a new generation of small, lightweight, efficient and quite inexpensive belt sanders (Fig. 3-66) that are ideal for general purpose do-it-yourself projects. The one shown in Fig. 3-66 has ample power for most jobs and is much easier to handle than the larger, heavier types. Unlike many belt sanders, this one does an excellent job of smooth-finishing even softwoods and does a superb job on hardwoods. It is not fitted with a standard dust-pickup vacuum and bag, which is almost an essential feature of a belt sander. However, an accessory piece allows it be

Fig. 3-66. This small and quite affordable belt sander is an asset to any home shop and works well on plywood.

attached to an ordinary vacuum cleaner or shop vacuum, which accomplishes the same purpose rather neatly. One drawback is that the sanding belts are of a new size and available only in a few grits. They are not yet widely stocked by tool supply outlets. But this size will shortly be readily available everywhere in good variety as the small belt sander increases in popularity, as it is sure to do.

ROUTER

Though probably not one of the first pieces of power equipment that you might buy for your workshop, a power *router* (Fig. 3-67) is particularly useful in cabinetmaking and furniture building. The router has many other purposes and can be used in all types of projects. There are several sizes and brands available. Once again, neither the smallest nor the largest is apt to be the best choice for a home workshop. Choose a name brand, one that accepts a wide variety of accessories (and there are a good many for routers) that are readily available, and one that uses standard router bits. The first thing to look for is accurate adjustment controls that can be easily changed and whose scales can be easily read. The tool should be of good to excellent quality and not a light-duty, inexpensive one. It should be fitted throughout with ball bearings and have a motor of from ¾ to 1 horsepower. The positioning of the on-off switch should be convenient. It is a good idea, too, to purchase a cased model to keep the machine fully protected when it is not in use. There is a wide variety of different bits that can be obtained for routers. These bits can be bought individually or in sets.

As to accessories, a dovetail joint fixture is offered, which allows you to make the necessary cuts for either rabbeted or flush dovetailed joints rapidly, accurately and easily. There are butt-hinge templates for mortising-in butt hinges, and letter/numeral templates for making routed signs. With a set of trammel points, you can rout circles of practically any size. Another useful workshop accessory is a special router table. The router is mounted upside down beneath the table, so that the bit protrudes above the table top. The whole affair is designed with a rip fence, blade guard and miter gauge. It can be used as a junior-sized shaper for making moldings or shaping edges on relatively small workpieces, as well as numerous other chores.

The router is a particularly useful power tool for working plywood wherever edge shaping, mortising and similar tasks are required. Because router bits spin at such a high speed—25,000 revolutions per minute or so—since they are so sharp, the bits make a very clean, smooth-edged and accurate cut despite the cross-graining of the plywood layers. Mortising plywood with a chisel, for instance, is a bit difficult because of the cross grains. Edge shaping with hand tools is sometimes problematical for the

Fig. 3-67. A portable router can be used for many purposes; wherever possible, use carbide bits for plywood routing.

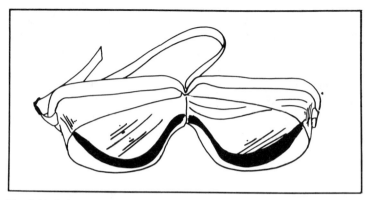

Fig. 3-68. Safety goggles are an important piece of personal protection gear.

same reason—tearing or chipping can result. But a router makes short work of these jobs.

PERSONAL PROTECTION GEAR

When you are working with power and hand tools, a certain amount of personal protection equipment may be vital to your own safety. For instance, ordinary engineered-grade plywood construction panels are not noted for their smooth and rounded edges. In fact, they are likely to be rather splintery. If you will be handling a number of sheets, as when sheathing a wall or decking a roof, a pair of leather work gloves is likely to save you some pain and some time in digging out splinters. Note that when handling panels, the wrists are just as susceptible sometimes to scrapes, abrasions and splinters as are the hands themselves. Choose a pair of gauntlet type gloves for complete protection. And you'll protect your wristwatch at the same time.

Always were some kind of eye protection when you are working overhead, whether with power tools or not, and when you are using any power tool that may throw chips, particles, sawdust or shavings. Safety *goggles* (Fig. 3-68) are a common choice because they fit tightly enough to keep even dust and sawdust away from the eyes, while at the same time offering ample protection against flying chips or splinters. Most types can be worn over eyeglasses, without discomfort and with no loss of vision. Some folks prefer to wear a *face shield*, which has the advantage of covering the entire face and giving excellent protection against flying debris, while at the same time allowing plenty of ventilation. The disadvantage to wearing a face shield is that airborne particles can drift up inside the shield quite easily and cause eye or nose irritation. Face shields protect particularly well outdoors on a windy day. Some workers also object to the added weight and to the fact that a shield does flop about somewhat.

When you are using a sander, always wear a *dust mask* (Fig. 3-69) or *respirator*. This is usually not necessary when you are only tough-sanding by hand, but for extensive hand-sanding and most certainly all power sanding the mask is an excellent idea. Some of those fine dust particles are bound to get down into your lungs, where they will do no good whatsoever. Aside from that fact, the irritation caused by sawdust in the throat and nasal passages can stay with you for hours and create most unpleasant sensations. Interestingly enough, sawdust from different kinds of woods can produce different reactions, which also will vary with different workers. A dust mask should also be worn when you are working steadily with a power saw that leaves a cloud of fine sawdust particles hanging in the air inside the workshop. Ventilation, of course, can help this situation, but is not always possible or effective. There are many times, too, during sanding or sawing operations when goggles should be worn as well as the dust mask.

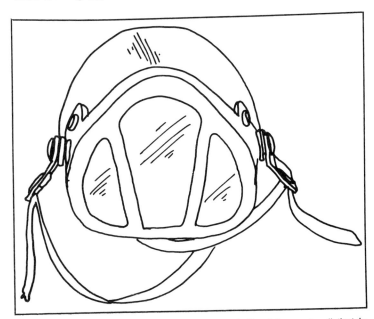

Fig. 3-69. A lightweight, inexpensive dust mask like this one is all that is needed to give your lungs and respiratory system adequate protection from sawdust.

Fig. 3-70. When using noisy power tools, wearing hearing protectors is a good idea.

Painting is another task that requires a mask. Generally this is confined to spray painting (indoors or out) with a gun, but should also extend to using spray cans. Paints and other fluids that are formulated with volatile ingredients and give off potentially harmful fumes are most dangerous. But even spray dust from water-based paints is to be reckoned with. For painting, an ordinary dust mask is a little bit better than nothing, but not much. Be sure to use one that is designed and approved for the purpose.

Recent studies have shown that the noise level, the loudness, of most home workshop power tools is well above what is considered the safe threshold for hearing. In other words, operating power tools, especially within the confines of a small workshop, can eventually damage your hearing. This is not true of all power tools; a drill press, for instance, is normally rather quiet. But a portable circular saw, a table or radial saw, a router and some other tools are quite noisy. There are two ways that you can protect yourself against possible hearing damage; both of them are simple, inexpensive and easy. One is to use a set of *ear plugs*, of the type especially designed for hearing protection rather than those used for swimming. It takes but a second to place them in your ears, and the protection afforded is good. For even better protection, use a set of hearing protectors that look much like stereo headphones (Fig. 3-70). Again, these protectors take but a second to slip on, are comfortable and will afford excellent protection. Neither type cuts out all sound; they simply minimize to a substantial extent the damaging frequencies and cut the overall sound level down to safe proportions. You can still hear many of the sounds around you and can understand someone else talking to you.

Over the years, the hard hat (and sometimes those who wear them) has been the butt of many a joke. But getting crowned on the cranium with a dropped tool, or a falling piece of lumber or plywood, is no joke at all. Nor is cracking your head against a rafter when you rise or move too suddenly in cramped quarters. If you will be doing much general construction work, such as building an addition on your house or putting up a new garage, give serious consideration to buying a hard hat. After you buy one, wear it. It's cheaper and a lot more comfortable than a concussion.

MISCELLANEOUS TOOLS

The various tools that we have just discussed are, of course, not the only ones available by any stretch of the imagination. There are a good many others, and though most of them are geared toward other types of work, there may be some that you will find valuable in working with plywood. There may be others that you will want to have for other kinds of work, either in conjunction with working with plywood or not. There are some miscellaneous items that you

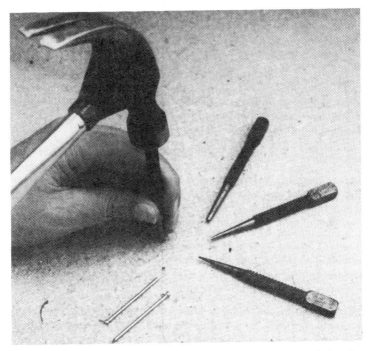

Fig. 3-71. Assorted nail sets are needed for setting nail heads below the workpiece surface.

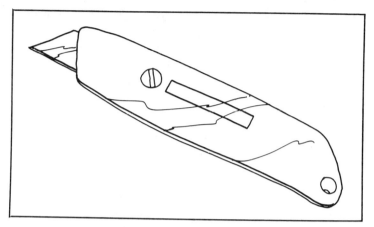

Fig. 3-72. An ordinary utility knife is inexpensive and very handy.

will need for plywood work, and a few others that could be used that need mention.

Nail sets are one indispensable item. They generally come in four sizes (Fig. 3-71), designated 1/32, 2/32, 3/32 and 4/32. This refers to the diameter in inches of the tip. A nail set is used, as the name implies, to set a nail so that its head is below the surface of the wood and can then be covered with wood filler.

Another item that generally sees a lot of service is the *utility knife* (Fig. 3-72). There are several different kinds available in any hardware store, with either fixed or retractable blades. The blades are double-ended, flexible and very sharp—almost like a razor blade—and the knife is very handy to have around.

In the clamping line, there are two additional kinds of clamps that are awfully nice to have around. One is the *cabinetmaker's hand screw*, which is primarily intended for cabinetmaking and assorted fine woodworking projects. These clamps feature heavy hardwood jaws, often felt-padded. Adjustment is made with a pair of steel screw spindles, which allow the jaws to be angled to grip nonparallel surfaces. Several sizes are available, and the well-equipped master woodcrafting shop will have at least a pair each of three or four sizes. But they are expensive and definitely not for all-purpose clamping jobs. The other kind is the *corner clamp*, invaluable for making frames of all sorts, and also for clamping right-angle edges of casework together. You can get by with one of these clamps, but two is better and a set of four is needed for convenient working. They too are expensive but practical and worthwhile for many shops, even of the smaller home variety.

Various kinds of doweled joints are considerably stronger than their unreinforced counterparts, but getting the dowel holes properly aligned so that the workpiece surfaces remain flush with one another can be a real problem. If you plan to make many doweled joints, by all means invest in a *doweling jig*. It will save you a lot of frustration. This special tool, available from woodworking supply shops, accurately positions dowel holes and then guides the drill bit through a bushed hole for perfect alignment.

In the power tool department, a *drill press* is great to have around, though certainly not essential. But it is amazing what a good drill press can accomplish. The drilling of accurate holes is its principal purpose, but when fitted with various accessories it can perform other functions quite nicely, too. A *rotary rasp* or *rotary file* clamped in the chuck makes short work of squaring internal (or external) edges, straight or curved, and can also be used for forming and some kinds of shaping. A *sanding drum* is fine for smoothing edges. With a shaper fence, adapter and special cutters, you can perform various shaping operations as well. Hollow square mortise chisels and bits, or a dovetail attachment and bits, will enable you to easily make mortise-and-tenon or dovetail joints. A surface planer attachment allows planing the surfaces of workpieces; with a *fly cutter* you can cut out perfect circles.

A *shaper* is also a lovely piece of power equipment. It is designed for making moldings or shaping the edges of workpieces to any number of shapes, both simple and intricate. Its cousin, the *jointer*, is used to square up stock, plane it to the desired thickness (a factor of little importance with plywood as a rule), cut bevels and chamfers, and make tapers of various sorts. The first tool is perhaps most valuable in cabinetmaking, while the latter is a great aid in furniture building. Both can perform all sorts of tasks for miscellaneous projects as well. However, both are quite expensive, take up quite a bit of room and are rather specialized. There are small benchtop models, but their usefulness is limited by their small size. Unless there is a definite and continuing need for shapers and jointers, they are hard to justify in a home shop unless monty and space are of little concern.

There is one very specialized power tool, which is fairly expensive, that does deserve a space in any home shop where a lot of in-shop projects are built. That tool is the combination *belt-disc sander*. This is an excellent tool for edge squaring and finishing, general sanding of any part small enough for the tool to accept, grinding chamfers, easing edges, trimming off tiny bits of material to exactly meet a dimensional guideline, and a hundred and one other little sanding chores. It can even be used for sharpening tool blades. In truth, once one has become accustomed to having a belt-disc sander around, it is very difficult to get along without one.

There are a good many other miscellaneous tools that you might need, too, such as a wrench for changing the blade in your table saw or radial arm saw, *Allen wrenches* for making adjustments to power tools, pliers, hacksaw, adjustable-jaw pliers, crescent wrenches and others. Seldom do projects consist entirely of any one kind of material. Over a period of time you will probably find yourself working with plastics, glass, sheet metal, electrical circuits, roofing materials, plastic laminates and a host of other odds and ends. These tasks will frequently require more additions to your toolbox. Whatever task crops up, use the proper tool for the job if at all possible. Buying a new tool or two, or renting one, is generally cheaper than hiring someone else to do the job or ruining part of a project by using the wrong equipment.

THE STARTING TOOLBOX

The question arises as to just what constitutes a good, workable selection of tools with which a beginner can start out. The selection should be complete enough to allow the construction of interesting and worthwhile projects, but small enough to be reasonably affordable. The following selection, based on experience, is such a kit for working with plywood as well as general woodworking:

- Pencil
- 10-foot steel tape
- Framing square
- Combination square
- Sliding T-bevel
- 16-ounce claw hammer
- 2/32 and 3/32 nail sets
- Homemade straightedge
- Torpedo level
- Scratch awl
- Back saw
- Keyhole saw
- Carpenter's crosscut handsaw
- Block plane
- Set of five assorted screwdrivers
- ¼-inch and ½-inch butt chisels

- Utility knife
- Two 6-inch C-clamps
- Four-in-hand file
- Sanding block
- Oil stone
- ⅜-inch electric drill
- Countersink bit
- Speed bits, sizes as needed
- Twist drill set, 1/16-inch to ¼-inch, by 64ths
- 7 ¼-inch portable circular saw
- Circular saw blades, one crosscut, one plywood
- Two sets of sawhorse brackets

The total cost of all these tools, few as they are, represents a substantial outlay, but they need not necessarily be purchased all at once. Spread out over several months, the process is a bit less painful. In any event, the cost compares favorably enough with expenditures for other pleasurable pastimes like a new set of golf clubs, a snowmobile or a week in the Bahamas.

The circular saw could be eliminated from the list, since it involves an investment of $100 or thereabouts. The circular saw was included because it is such a popular tool. It represents large savings in time and effort and a considerable increase in potential accuracy of cutting, always a problem for beginners. The electric drill is included as a simple matter of economics. To achieve the same range of hole-drilling capabilities (to say nothing of ease, accuracy and the additional tasks that can be performed with various accessories) with hand drilling tools would cost on the order of twice as much as the electric drill and bit sets.

You will note the absence of numerous small hand tools that are commonly used in general household and mechanical repair work—slip-joint pliers, utility pliers, vise-grips, crescent and other kinds of wrenches. Such tools are not specifically required for working with plywood, but a need for them will most likely occur in due course. Plans should be made to add these general purpose tools to the kit, too, as the necessity arises.

KEEPING TOOLS SHARP

In order for tools to perform effectively, obviously they must be kept in good operating condition. Perhaps the most important aspect of this is in maintaining a fine, sharp edge on the blades of all cutting tools. This is particularly important to woodworking in general and doubly so with plywood. Working woods, and especially hard woods, is a difficult job with dull tools and is virtually impossible with plywood. Dull blades perform poorly and result in rough, ragged and imprecise cutting because they are dangerous to boot. Extra effort is required in trying to make a decent cut, and the blade is likely to slip, catch, or bounce and cause injury to the worker. There is no substitute for sharp cutting tools, and the sharper the better. But they will not stay that way automatically. You will have to work to keep them in proper order. Exactly how this is done depends upon the tool itself.

Saws

A dull, or dulling, saw blade will announce itself quite quickly. An experienced craftsman can tell immediately whether or not the blade he is using is in top condition. Handsaws, regardless of type, will begin to stick in the kerf, bind and will not cut smoothly when they become dull. A good deal of extra muscle is required to operate the saw. The cut will be splintery, probably uneven, and there will be noticeable signs of tearing of the wood fibers. Tracking the blade along a cutting guideline may also be a problem.

Power saw blades will react in somewhat the same fashion. The blade may tend to wander from the guideline, the cutting action will be slower than it should be, and the noise pitch of the motor will change. This is caused by the fact that additional strain is imposed upon the motor because of the dull blade. It quickly becomes overloaded and "bogs down," even to the point of stalling. This is hard on power saws and can lead to eventual burn-out of the motor. On top of that, the cuts are ragged, rough and imprecise. In severe cases, the blade may bind in the cut and cause kickback, jumping the saw out of the cut (portable saws) or driving the workpiece back or up away from the blade (table saws).

Three principal factors contribute to poor cutting with either hand or power saws. The first and most important is the lack of *sharpness* of the teeth. The second is improper *set* of the teeth. The set is the slight bend that each tooth has to right or left of the centerline of the blade; this results in a kerf that is slightly wider than the thickness of the blade itself, and prevents binding of the blade in the cut. The third factor is a *buildup* of *resins* or other materials on the blade surface. This is particularly a problem with power saw blades, where friction heats the resin to form a sticky glaze on the blade surface. When especially bad, this gum becomes so hot that it will throw off clouds of bluish smoke as sawing proceeds.

Sharpening handsaws is not a difficult proposition, and the technique is easy to learn. It is a bit complex to explain. Full instructions are outside the scope of this book. However, the base process consists of first jointing the saw teeth if necessary (only if the teeth are uneven and incorrectly shaped, and done with a flat mill file). Then all of the teeth must be filed to the correct shape with a small triangular file and also filed to the proper angle if necessary. At the same time, each individual tooth is sharpened. Then the teeth must be set. This can be done with an inexpensive saw-set tool (Fig. 3-73). The blades of virtually all handsaws can be sharpened in this general way. When properly done, any saw should last for many years.

The alternative to sharpening your own handsaw blades, of course, is to take them to a saw sharpening shop as soon as you notice that they are becoming dull. This, however, is fairly expensive and over a few years will amount to the cost of a new saw. Note, though, that if you continually check you saw blades and touch up the teeth with a file (setting is not necessary every time a good blade is sharpened), the saw will cut better and more efficiently, need a complete sharpening and resetting less frequently, and will have a considerably longer service life.

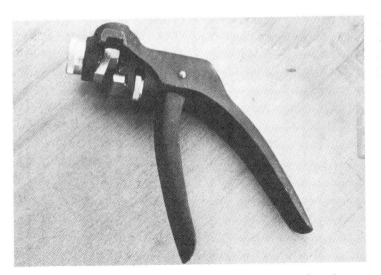

Fig. 3-73. A small saw-setting tool and files are all you need to sharpen handsaw blades.

Circular saw blades can also be sharpened by hand. The best method is with automatic saw-filing equipment, but hand sharpening is also quite possible if the blade is essentially in good condition to begin with. The process of sharpening these blades includes the following: jointing or *rounding* so that all of the teeth are of equal height and the blade is as round as possible; *gumming*, if necessary, to increase the tooth height after repeated sharpening; setting the teeth to the proper opposing angles; and filing to shape and sharpen them. As with handsaws, frequent touch-up with a file is an excellent idea and will keep your blades in better shape longer. Expensive circular saw blades are best taken to a saw sharpening shop when they become dull. As a practical matter, replacement blades for small table saws and most portable circular saws, except in the large sizes and in the case of special blades, are so inexpensive that they can actually be treated as throwaways. The price of a new blade is generally less than the cost of having one sharpened, so the easiest course is simply to throw the old blade awaq as soon as it becomes overly dull and does not respond to touch-up filing. Then buy a new one. Under no circumstances, however, should you attempt to work wood, and especially plywood, with a dull power-saw blade.

Carbide-tipped circular saw blades will retain a sharp cutting edge much longer than steel blades and are particularly useful for extensive plywood cutting, but eventually they, too, will become dull. Incidentally, carbide-tipped blades should be handled and stored with care in order not to chip or nick the carbide tips. Sharpening of these blades must be done with special equipment. While this can be done in the home workshop with the right gear, the setup is an expensive and exacting one. The usual recourse is to send the blade out for sharpening by a specialist. Carbide tips can also be replaced when they become damaged.

Most other types of saw blades are simply replaced rather than resharpened. For instance, blades for reciprocating saws, scroll saws, saber saws and the like are quite inexpensive and are generally bought by the package. They can be touched up somewhat with a small file. This is a useful procedure upon occasions if one or two teeth become slightly nicked or dulled, or if no replacement blades are ready. But simply installing a new blade when the old one becomes dull is certainly the easiest procedure. The same holds true of coping saws and hacksaws; these blades are impossible to resharpen.

Chapter 4
Basics of Working With Plywood

The basics of working with plywood are really quite simple to master and consist of a few fundamental techniques, a bit of knowledge and a healthy ration of common sense. Practice, experimentation (never be afraid to experiment) and experience automatically lead to increased skills, added knowledge and advanced projects. Many of the procedures are the same as for working plain wood; others are peculiar to plywood. But with the exception of construction work, the process generally begins with a workshop, or at least a work area.

THE PLYWOOD WORKSHOP

In many cases—perhaps in most—the home workshop neither evolves nor is planned ahead of time, but simply happens. Somewhere along the line a few tools are accumulated, in order to repair the vacuum cleaner, adjust Junior's bicycle, fix up the window shutters and repair the porch railing. The collection of tools and equipment grows. In due course some more-or-less convenient corner is used to store them in and to perform various repair and minor construction jobs. Then the need arises for a workbench. The heap of tools grows larger and larger and requires rack space. The hardware box expands into a good-sized assortment of old milk cartons, paper sacks and small boxes. The workshop has happened.

This is rather unfortunate because a good workshop has a great many more requirements than just a spare corner to work in. Anyone who becomes interested in woodcrafting or other workshop avocations, and who becomes even a little bit involved in do-it-yourselfing, might just as well resign himself (or herself) to the fact that eventually there will be a definite need for a full-fledged workshop. Having a good workshop requires some planning. There are literally thousands of different kinds of workshops, of course, and home workshops are not the subject of this book. However, there are some points that should be noted, particularly with respect to working with plywood.

Where To Put The Workshop

With the exception of those few custom-designed, spacious and fully equipped home workshops that are built right into a new house, few homeowners are lucky enough to have what might be considered an ideal spot for a workshop. However, there are a number of places where a home workshop can be built. Probably the most popular and the most readily available space is in the basement. Remodeling may be difficult, but there is likely to be more available room here than in most places in the house. A basement has the advantage of being warm in winter and cool in summer, and both plumbing and electrical facilities are usually handy, There are also usually some difficulties to be overcome; good lighting must be provided; adequate ventilation must be arranged for; dampness must often be rigorously combatted; and good access must be somehow provided. This latter difficulty is of particular consequence to a plywood workshop. The panels are large and often heavy as well, and full sheets are difficult to wrestle down the average cellar stairway. If a large bulkhead of doorway to the outside does not exist, then one should be built, of ample size to accommodate not only plywood panels but also good-sized completed projects.

Another popular spot for a workshop is in the garage. Older garages are often too small to house both a vehicle and a workshop. However, it can be managed if there is an extra 3 feet at one end of the garage, and preferably an extra 3 feet along one side as well. A switch to a compact car can result in a good bit more space. Or the vehicle can be left outdoors, and the entire garage converted to a workshop. That space of 10 × 20 feet or thereabouts can make a mighty fine workshop. Access is excellent, and there is plenty of room to maneuver. Heat and electrical capacity can generally be installed without too much difficulty, especially if the garage happens to be an integral part of the house. Lighting is not much of a chore, and ample storage space can be gained by creating a loft under the rafters of a pitched roof. Other alternatives are to build a carport and take over the existing garage as a workshop, take over an entire two-car garage for a super-shop, or add an extension to an existing garage and use that for a workshop.

A less appealing and sometimes more problematical, but yet perfectly workable, solution is to convert an unused attic to a workshop. The biggest problem here is access, because some way must be figured out to get plywood panels into the shop and finished articles back out again. In many cases, feasible arrangements can be worked out. Where they cannot, an alternative is to arrange some temporary work space, even an outdoor spot, where plywood panels can be set on sawhorses, rough layout work done, and the panels rough-cut into pieces of manageable size with a portable circular saw. Then the rest of the project can be taken care of in the attic workshop. Ventilation, heating, lighting and electrical outlets are usually not very difficult to install in an attic. One problem with attic workshops that is not common to most others is the noise factor. The rooms directly below will be constantly treated to various bumps, thumps and vibrations. Much of this problem can be alleviated, if not eliminated, by the use of acoustic tile, acoustic insulation, thick indoor/outdoor carpeting over felt padding on the floor, and vibration pads under the legs of stationary power tools and heavy-duty workbenches.

An ideal answer, even better than a garage, is an entirely separate outbuilding. If you have sufficient space on your grounds and local zoning allows it, you can put up a new building for your workshop. This is neither as difficult nor as expensive as you might think and has the great advantage of keeping noise, dust and fumes entirely away from the house. You can work to your heart's content all hours of the day and night and disturb no one. Such an outbuilding can be built from scratch. In many areas a concrete foundation or floor is not required, and you can build an insulated, wood-floored shop upon short piers. You can purchase a small kit building, or buy an existing small building and move it onto your property.

Some homeowners are fortunate enough to have a spare room in the house. Sometimes this is actually a guest room rather than a room that is totally unused, but even so it can be converted into a workshop. Though the guest room has long been part and parcel of the traditional family house arrangement, when one considers the cost of maintaining an empty room that is occupied only when Aunt Florence comes once a year for a week, it is difficult to justify. That space could be made to earn its keep if it were a full-fledged workshop, and would be of much more use to the family. Converting a spare room does have its problems. Access may be somewhat difficult in terms of plywood panels, and the available free space within the room may be a bit small. Also, noise could be a bothersome factor. Dust and/or fumes can easily escape into other parts of the house. However, the use of acoustic materials, a good ventilating system and the installation of a close-fitting weatherstripped door will help greatly. In the case of first-floor spare room, a window might even be replaced with a large window unit or even a patio door arrangement to allow good access.

Another excellent possibility is to enclose a porch. Many homes have porches that are used little if at all, and yet most of them can be readily enclosed. This is a job that can almost invariably be done under local zoning laws. Often about all that is required is a bundle of two-by-fours, a few sheets of engineered-grade plywood, a door and perhaps a window unit or two, and some exterior siding to match that on the house. If the porch is a small one, it is entirely likely that it might be constructed in such a way that enlarging it could also be possible.

You can also build an addition to the house. This is a major project to be sure, and not something that can be done in a weekend or two. However, for many homeowners it is a solution that is perfectly acceptable and practical, perhaps even ideal. In such circumstance, a complete workshop can be planned in advance and built to suit your exact needs. This allows the opportunity for custom-designing. The planning should cover not only present needs and the installation of equipment that you now have, but also future expansion and for tools and equipment that you would like to have later on.

Workshop Space

The primary requirement for workshops is space. This is also the commodity that seems to be in shortest supply in most home workshops, so whatever space is available must of necessity be carefully planned out and allocated for various purposes, many of which overlap or must be multi-purpose. Space is particularly important in a plywood workshop, both because of the nature of the work normally being done, such as cabinetwork and furniture building, and also because of the physical size of the panels themselves. While a fairly complete electronics workshop or model-making workshop can actually be set up in a closet, working with plywood requires a considerable amount of room, especially if the shop is well-outfitted. Because of these space requirements, in some home shops it is necessary to have a specific workshop area that is as well equipped and laid out as is possible under the given conditions, plus another area, either indoors or out, where some of the initial work can be done on a temporary basis.

Maneuvering Space

The first requisite is maneuvering space, so that you will have ample room in which to move about and also to move plywood panels, lumber, partially completed assemblies and completed projects around the shop area with minimum difficulty. The shop should have good access, with as wide a doorway as is practically possible. A plywood panel will easily pass through any ordinary house doorway if stood up on its side. But many doors may have an effective opening of as little as 28 inches or so, and a bit more if the door is removed from its hinges. This is really a bit slight for easy access, and 32 inches should be considered a minimum. In most instances it is not overly difficult to replace an existing narrow door with a larger one. A 36-inch door is a standard size, and wider ones are available as well. If you can arrange for an extra-wide door, double-doors or a sizable sliding door arrangement, so much the better.

The layout of the workshop should be such that material can be brought in through the principal access door without turning any sharp corners and brought directly to the initial cutting area. This might consist of a pair of sawhorses where a full panel can be laid down and cut with a handsaw or circular saw, or it may be a radial arm or table saw. Whatever the case, there must be substantial space, unhindered and uncluttered at this point. Once the panel is cut into smaller workpieces, they can then go into the shop for various other processes and assembly.

Plenty of maneuvering space is also necessary in the vicinity of lumber and plywood storage racks, so that the pieces can be easily placed and removed with plenty of room to swing the ends around or to check through the stock in search of a needed piece. By the same token, there must be plenty of space around the workbenches and also around and between the various stationary power tools and work centers.

The maneuvering space that you will need depends upon the type of equipment being used, as well as the physical size of the plywood panels. For instance, in order to cut a strip the length of the plywood panel on a radial saw, you will need from 9 to 10 feet on one side of the saw, at least 8 feet on the other side, and about a 6-foot cleared area along that entire 18-20 foot length in order to adequately handle the panel. If the panel is to be run through the saw sideways, you will need 5-6 feet on either side of the blade plus a 10-foot clear area in front of the saw. For a table saw, the situation is a bit different. In order to make full end-to-end or side-to-side cuts on a table saw, you need a field of about 9 feet to either side of the blade and about 10 feet in front of and behind the blade. Obviously a lot of wasted space results in either situation, both of which assume the saws to be immovable, because not all cuts require the same amount of maneuvering space. Therefore, the best arrangement is to make the saw movable so that it can be positioned in a clear area for the cutting chores, and then moved out of the way to free the space for other work. This reduces the amount of space required to the size of a full standard plywood panel, 4 × 8 feet, plus enough space at one end or side to be able to safely and easily handle the panel during the cutting process.

Where this much space is simply not available, there is an alternative/less attractive, to be sure, but perfectly workable. The maneuvering space at the shop doorway can be reduced to just enough room to admit passage of relatively small workpieces, equipment and articles into the shop without undue difficulty. The amount of maneuvering space around the stationary power tools and work areas can likewise be substantially reduced to no more than is deemed necessary for the contemplated projects. Benches, tool boards and stationary power tools can be aligned along the walls of the workshop, so as to leave the center of the area free and clear for assembly, workpiece processing or whatever.

Then another temporary area can be chosen where the initial layout and cutting can be done. This might be in the garage, with the family chariot moved outdoors for a while, in a carport or even out in the backyard. In fair weather, layout and cutting work can go on at any time. In inclement weather, the plywood panel can be brought indoors and set up temporarily either in the shop or in some other spot. The layout work is done, and then the panel can be taken back outdoors at some relatively opportune moment for cutting. In this situation, the cutting is done either with a handsaw or with a portable circular saw and straightedges. After the rough cuts are made, the pieces can be brought back into the shop for final trimming. Sometimes the weather does not cooperate, but with patience and frequently a heavy jacket the job finally gets done.

In most home workshops, the available space must be multi-purpose, and the easiest way to accomplish this is to make the major components of the workshop both movable and multi-purpose themselves. This is not practical, of course, for some of the equipment. But as previously mentioned, the radial arm saw and the table saw at least should be movable on locking casters. Other stationary power tools can also be made movable including the bandsaw, lathe, sander, jointer or shaper. Another good idea is to fabricate a relatively small movable bench upon which can be mounted the smaller power tools such as a bench grinder, drill press, router table, jigsaw or other device. This rolling bench can then be moved into whatever position is most convenient for the work going on. Likewise, a worktable or workbench of relatively small size can be mounted on casters and pushed around the shop to be used wherever is most convenient.

To sum up, then, when you plan your workshop be sure to allow as much maneuvering space as you possibly can. If you cannot allow ample space for full sheets of plywood, make arrangements to do the initial layout and cutting work elsewhere. Scale down accordingly the needed maneuvering space in the workshop. This procedure will allow you to put together a fairly complete workshop in a relatively small space. But even in this instance, try to arrange for as much free, open space that can be put to multiple purposes, as needed, as you possibly can. For the elements of the workshop that must be fixed, such as the principal workbench, allow even a bit more maneuvering space than you think will be necessary.

Work Surfaces

It is perfectly obvious that a workshop must have working space; yet this is another area in which many shops fall short. There are two considerations: work surfaces and free working space.

The principal work surface is the workbench; every workshop must have at least one. If there can be *only* one, the bench should be as large as is reasonably feasible. A length of 6 feet should be considered minimum for a fully workable surface, though 4-foot benches must at times be used as the primary work surface in a small shop simply because there is insufficient room for a larger one. An 8-foot bench is really no more than comfortably adequate. If it can be made longer, so much the better. Benches are sometimes made with a top surface of only 18 inches in depth, but this simply is not enough for anything but small, light work. A 24-inch depth is perhaps most common and this is just adequate. If you can afford the space in your workshop, by all means consider making a bench with a top depth of 30 inches or even 32 inches. Just those few extra inches make an amazing difference in the general working convenience of the bench. As to height, use your kitchen countertop as a guide and position the bench top at the most comfortable working level.

If there is room enough and if the principal workbench is only on the order of 6 to 8 feet long, a second workbench can be extremely handy. In this case, a 4-foot length is quite useful. Even a 3-foot bench provides enough additional work surface to be worth having. Obviously, a larger one is better.

There are other ways to gain additional surface space which in many projects can be invaluable. The extra surface area allows you to "spread out" and perhaps even run two or more projects at once, making the jobs go more easily. One possibility is a fold-down shelf with leg supports. Depending upon what wall space might be available, this auxiliary bench can swing down lengthwise on a piano hinge so that it protrudes well out into the workshop floor and can be used from both sides. The bench can be arranged to fold out like a broad shelf, perhaps extending only 18 inches or 2 feet into the room. When unneeded, it is simply swung up out of the way.

Another possibility and an excellent one for shop owners who plan to do quite a bit of fabrication of cabinets and small furnishings is a low worktable on casters. This table can be of any convenient size with a height of 12 to 18 inches. It can be made to fit conveniently under a workbench shelf, so that it can be rolled away when not needed. The purpose of a worktable of this sort is to afford a separate space up off the floor where a project can be assembled more conveniently than either on a workbench or the floor itself. In addition, the whole affair can be scooted on a workbench or the floor itself. In addition, the whole affair can be scooted around the shop floor and positioned (or gotten out of the way) wherever is most convenient The bench should be flat-topped, sturdily made and have at least two lockable casters. Other than that, there are no particular specifications and you can build in whatever fashion suits best.

Portable power tools do not have their own work surfaces, so they must be provided. Often the workbench or a worktable serves the purpose quite nicely. In other instances, it may be best to provide a special work area for them. For instance, a craftsman who does a considerable amount of work with a router might devise a special rolling router cabinet that will house the router and all of its various accessories, with a top work surface on which a pantograph can be mounted, or a router table where ordinary routing work can be done. Whatever the specific power tool, some sort of work surface area is necessary in the immediate vicinity so that you have a place to set workpieces and lay down tools and power tool accessories. Note that this need not necessarily be the surface upon which the tool is mounted or used (stand-mounted tools have little if any work surface area surrounding them anyway), but may be a small rolling worktable or cabinet with a work surface. This sort of arrangement is extremely flexible and may serve with greater facility.

A small amount of working surface is provided with most stationary power tools. Some have little or none, such as the wood-turning lathe. Others, like the drill press, have only a small surface, while a radial arm or a table saw has a rather more substantial work surface. In most cases these surfaces must be enlarged if the tool is to be fully utilized. Extension wings are available for table saws and larger, surrounding work platforms can also be added as necessary. The radial arm saw has a fairly small work area directly beneath the blade. While this is adequate for small workpieces, the radial arm saw must almost invariably be fitted with side tables that extend for a considerable distance to either side of the saw head. These work surfaces can be the tops of adjacent workbenches, which can be cleared when the need for sawing long pieces arises. Or they can be specially-built wing tables of whatever depth and length is desired or can be fitted in. Surround-type work surfaces (plywood is excellent for this purpose) can be built around the tables of bandsaws, jigsaws and similar tools. Rolling tables or cabinets can be used next to a wood-turning or metal lathe, as well as other power tools. Surround-type work surfaces can be made as permanent attachments or in such a way that they can be easily disassembled when not needed.

There is a tendency in home workshops, especially of the smaller sizes, to attempt to "get by" with inadequate and often cobbled-up work surfaces. This is not only inconvenient and sometimes frustrating but can also be dangerous. Working in a crowded, cramped, cluttered area reduces efficiency and effectiveness, takes away a good deal of the pleasure of working in a home shop, and results in finished products that are certainly of lower quality than might otherwise have been the case. When laying out your workshop, allow yourself as much work surface area as you can possibly squeeze out. This may take some doing, but you will find that the resulting freedom of movement and action is well worth the effort.

Free Work Space

Just as you need ample maneuvering space in order to be able to shift plywood panels, workpieces, rolling cabinets or worktables, power tools and partly-assembled projects around the shop, you also need room in which to work. Obviously there must be space at the benches, particularly around bench-mounted vises, so that you can work there without banging into other shop equipment or stock. Racked tools and stored items should be kept well above the bench top surface, where they are out of the way. There must be foot-room below the bench so that you do not trip over items stored there. Ample head room is important, too, so that as you are shifting long workpieces you do not inadvertently hit a light fixture or a lumber rack.

But not all work in the shop takes place directly on the bench top. Many projects are large enough that they must be built in an area of free floor space or upon a low assembly table, where they can be reached from all sides. This means that there should be an open and uncluttered space of at least 3 feet in all directions from the assembly. A proper arrangement and layout of the workshop will allow a considerable area of free working space, usually located in the center of the room or, in the case of a one-wall workshop, out several feet from the benches, where work can go on without interference from other shop elements. If possible, such working areas should also be out of, or integrated with, the principal traffic pattern of the shop. It is also helpful if the area is located so that as few steps as necessary are made from the work area to the workbench, tool racks and hardware containers.

Storage Space

The third essential space requirement for the home workshop is storage space. This is one that is often skimped or is never planned for efficient storage and to maximize the storage space available. Storage space can be broken down into two categories: active and inactive.

Active storage space is required for the tools, hardware and materials that are frequently needed. Stationary power tools should be so placed and arranged in the shop that they can be turned on and put to work with a minimum of bother. This is especially true of those power tools that are in frequent use such as the table saw, power sander or drill press. Those that are less often used or may come into play only occasionally for certain projects can be left in less accessible positions if necessary, and brought forth and set up as needed.

The usual procedure with hand tools is to rack the greater proportion of them in locations and in such an arrangement that they are instantly at hand when needed. Handsaws, screwdrivers, chisels, pliers, wrenches, layout and measuring tools, squares, levels and other tools that get constant use are usually placed in wall racks above and behind the various work surfaces. Many of the more specialized tools that see only infrequent use can be stored in drawers, cases or tool chests where they are out of the way but yet can be readily taken out as required. Delicate tools and equipment, such as precision measuring devices, dial indicators and inclinometers, should be kept enclosed in tool chests or drawers. Most of the basic hardware items should be kept in bins or cabinets, all properly labeled, for immediate access and selection. The same is true of certain supplies such as sandpaper, glue and steel wool. The trick to making the most of active storage space is a combination of logical arrangement and efficient use of every available square inch.

Likewise, there is a need for substantial amounts of inactive storage space. Unused, leftover and scrap material and stock accounts for a great deal of this space. Such odds and ends seem to rapidly and continually accumulate in any active workshop. Since few do-it-yourselfers can bring themselves to throw anything away, there needs to be a place to put these things. Lumber racks are needed and often are placed at ceiling level where they are out of the way. Small pieces of stock such as cut ends, strips and leftover doweling can similarly be racked, while small pieces can be stored in cartons or bins. Miscellaneous builders hardware and scraps of various nonwood materials, many of which will eventually find their way into one project or another, can be stored in "junk boxes" or heavy drawers made for the purpose. Certain tools, both hand tools and portable power tools, may also see quite infrequent use. Inactive storage space for these must also be developed.

Work Flow

For efficiency of operation and ease of working, a home workshop should be organized in terms of its work flow pattern. There is an infinite range of possibilities as far as specifics are concerned.

Much depends upon the shop size, the number and types of tools in the shop, and the kinds of projects that will be undertaken. Although a workshop may be set up primarily for woodworking, there may be subsidiary areas, usually of smaller scope, to handle ancillary endeavors such as sheet metal working, electronics, model-making or leatherwork.

The flow of work in most shops starts at either the door, where the materials to work with are brought in, or from the material storage area, which is also frequently located near the entrance to the shop. From this point the material goes to the layout stage, then to rough cutting (if necessary), on to fine cutting, then to fitting and assembly and finally to finishing. These steps can be arranged in sequence from the entrance of the shop to its farthest recesses (Fig. 4-1).

Also, there are always particular work areas that have different purposes. For instance, one might keep tools and equipment necessary for cutting in the general vicinity of the table saw or radial saw. Obviously, all of the finishing supplies, such as sanders, files and the like would be grouped in that area. All of the accessories and equipment needed for the table saw—clamping miter gauge, assortment of saw blades, wrenches for changing saw blades and making saw adjustments, multi-purpose jigs, rip fence and a host of other items—can be stored in the saw cabinet or in an adjacent cabinet, shelf section or other storage area. It would be rather pointless to keep all of this gear stored on the other side of the shop in the stand that holds the bench grinder. In other words, both hand and power tools and accessory equipment should be kept grouped close to the work area where they will be most frequently used.

When you lay out your workshop, try to arrange everything in logical work centers as much as you can (Fig. 4-2). This will cut down on wasted time and effort, make the shop work easier and more enjoyable, and will also help to fully utilize the space that you have available. Obviously some compromise must be made in virtually every shop, but with forethought in planning you can make the most of what you have.

Shop Utilities

The utilities systems in home workshops are more often than not discouragingly inadequate, and this is an area that you should concern yourself with right at the outset in order to avoid future problems. There are four main considerations: *lighting, electrical power, heat* (and/or *cooling*) and *ventilation*.

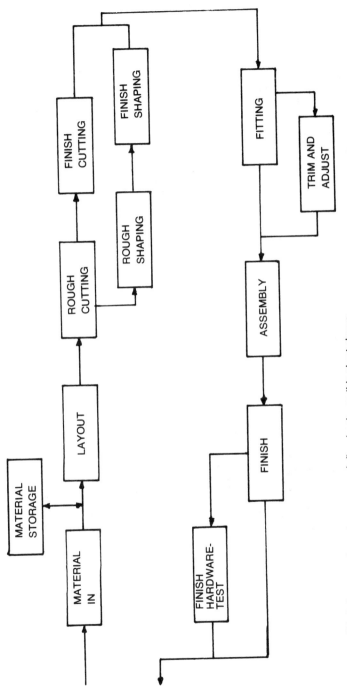

Fig. 4-1. Workshops can be set up on a work-flow basis, as this chart shows.

Lighting

Good lighting is extremely important in any workshop for three reasons. First, a dimly lit shop that is full of shadows and dark spots is just an invitation to injury. If you cannot see what you are doing, eventually an accident is bound to happen. When one is working with power tools, the consequences of shop accidents can be severe. Poor lighting equals safety hazards. The second reason is simply that if you cannot see what you are doing, you cannot work efficiently. You will spend half your time squinting at practically invisible guidelines, searching for points you have marked and shifting things around to get into a better light. Mistakes will be made and results will be poor. Frustration is likely to run high. The third reason is that working in a properly lighted shop is much more pleasant and enjoyable. After all, one of the biggest reasons that homeowners set up their own workshops is for personal enjoyment.

You need adequate background lighting which will give a satisfactory level of general illumination. To achieve this you will need plenty of ceiling fixtures spaced all about the shop area. There is no way that you can make a workshop too bright, so install as many fixtures as seems reasonable. These fixtures should preferably be industrial-type 4-foot two-bulb fluorescent fixtures with white reflectors. You can also use incandescent bulbs, and these also should have reflectors. Never use bare bulbs. The fluorescent fixtures, however, will give you far better light and consume less electrical power. At a minimum, an overall illumination level of 30 to 40 footcandles with even illumination and no dark corners is a good arrangement.

This, however, will be insufficient lighting for any sort of close work. Some of the ceiling fixtures can doubtless be located so that they shine directly on certain work surfaces, such as the workbenches or a table saw. For general bench work, you will need an illumination level of about 70 to 80 footcandles for comfortable working conditions. If an industrial fluorescent fixture is placed about 3 ½ to 4 feet above the work surface, the illumination level will be about right. However, this line of light should be continuous along the entire work surface; a single 4-foot fixture centered above an 8-foot bench will leave the ends of the bench top in dim light. If the ceiling fixtures hung for general illumination do not adequately light the work surfaces, more fixtures should be added to cover them.

Close work requires a great deal higher level of illumination.

For general close-up work, about 100 footcandles is barely adequate. For fine work, 200 footcandles should be considered a minimum. The usual procedure for attaining high illumination levels is to place incandescent fixtures right at the work site. The handiest type of fixture for shop use is a gooseneck lamp with a clamp arrangement that can be attached to the bench or worktable. Either standard or small floodlight bulbs can be used in these fixtures, and there are some fluorescent types available as well. In many cases the light should be placed quite close to the work and shine directly upon it, positioned so that no shadows fall upon the surface of the workpiece. This often means that the lamp must be continually adjusted as work progresses, so that a flexible gooseneck or a swiveling extension-arm lamp is virtually a necessity. Generally one or two of these will suffice in a workshop, and they can be moved from point to point as needed.

Power tools must be particularly well lighted in the interests of safety and efficient working conditions. A drill press, for instance, should have a light focused directly on the work. The same is true of a bench grinder, bench-mounted jigsaw or a bandsaw. A table saw needs strong light from the left rear, as well as a high level of general illumination. Particularly strong lighting is a good idea in the layout area, and should ideally come from above and all sides so that there are virtually no shadows.

Electrical Power

There seems to be a rule in many home workshops that there is never an electrical outlet where you need one. The only way to get around this situation is to install plenty of them. In order to do this, you must have sufficient electrical capacity in the house's main entrance panel.

A workshop should be served by at least two 20-ampere circuits. This is a minimum, and most shops (even small ones) require three circuits and preferably should have more. Sometimes the easiest solution is to install an electrical subpanel in the workshop, with a main power cable going directly back to the house's main service entrance panel. A six or eight-circuit subpanel is generally adequate and will afford space for that many additional circuits in the workshop. Of course, the main entrance panel must be capable of handling this added load. If you are unfamiliar with how to make a determination or how to make the installation, contract the services of an electrician.

Each power tool should have a convenient electrical outlet into which it can be plugged, without recourse to extension cords and power cables snaking all over the shop floor— a dangerous situation. Many power tools draw a considerable amount of current, so the circuitry should be carefully planned and split so that no overload is likely to occur. A good many outlets are needed at the workbenches. These can consist of standard wall outlets located behind the bench and about a foot above the work surface, or of plug-strip mounted either at that location or along the front edge of the bench itself, just below the bench top lip so that it is back out of the way and also protected. Floor outlets are an excellent idea in larger shops so that a table saw, for instance, can be plugged in directly beneath its station when the saw is rolled out into the center of the floor. When extension cords are necessary (and few shops are able to get along entirely without them), the type that retracts into a wall or ceiling-mounted reel, or long coil-cords that are especially designed for this purpose, are the best bet.

One important point should be noted here. If you elect to do your own workshop wiring, whether for receptacles, lighting fixtures or a subpanel to serve them, make absolutely sure that all of the work is done according to National Electrical Code standards. Also, be sure to check first to find out if doing your own electrical work is allowable in your area. In most instances it is, but a permit may be required. If you are the least bit unsure as to just how to proceed, have the job done by a professional. The hazard is too great, (faulty wiring is a major cause of house fires), and there is no point in taking chances. Shock hazard can also be quite high where power tools and home workshops are concerned, so play it safe. Either install or specify *ground fault circuit interrupters* (GFIs) instead of ordinary circuit breakers.

Heat/Cooling

It is axiomatic that a workshop, in order to be comfortable, must be heated or cooled to a reasonable level just as is the remainder of the house. If the workshop area was a heated portion of the building to begin with, in most instances no further work must be done. But this depends upon numerous factors, because there are instances where the heating system might well be altered. For instance, a forced hot air system in a workshop is marvelous for blowing dust around and for carrying great quantities back to the central filtering system. Some changes in ductwork or the addition of filters at a cold-air return or baffles and/or directional diffusers at a warm/air register may be in order.

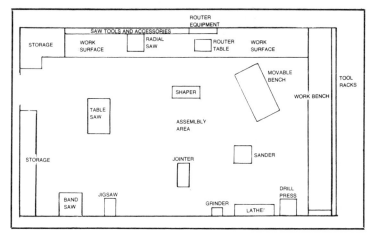

Fig. 4-2. Typical work-center arrangement for a home workshop.

In circumstances where the house's heating or cooling system must be extended into a new workshop area or where auxiliary equipment must be installed, judgment must be exercised as to what kind of equipment is installed—how and where. A wood stove throws find heat but is not a particularly safe kind of heat, especially in a woodworking shop. Probably the safest heat to use is electric resistance heating, preferably of the radiant ceiling panel variety. However, electric baseboard units will also serve well, provided they are kept vacuumed and free of sawdust and that nothing is stored or even inadvertently piled against them. Of course, extending an existing system also works very well provided that the boiler has sufficient capacity to handle the job. Auxiliary cooling can be provided by a small window or wall air conditioner.

In any event, a workshop should be heated, safely and to a reasonable degree and comfort-cooled if necessary. Many shops do not have to be heated to the same level as the house. For relatively active woodworking, a temperature of about 65° or even less is sufficient. For other kinds of work that require less physical exertion, such as electronics work or modelmaking, a higher temperature level is generally required. It is best if the heat source is relatively draft-free and, for heating units installed in the shop area itself, flameless.

Ventilation

Ventilation can be a problem in any active workshop, especially a woodworking shop. There should be a continuous flow of fresh air, and means should be found (and effort expended) to reduce the amount of particulate matter in the air as much as possible. Part of this can be achieved by keeping the shop scrupulously clean and vacuumed. Larger shops should be outfitted with sawdust-collecting systems that suck up and dispose of the sawdust practically as soon as it leaves the blade. The use of power tools fitted with accessory sawdust pickups is an excellent idea and a great help. Note, too, that many kinds of sawdust, particularly that associated with some kinds of plywood and many kinds of plastics and particle boards, can be harmful, particularly to people with respiratory problems or allergies.

There are many solvents, paints, thinners, various solutions and supplies used in workshops that are either quite odoriferous or exude fumes that may be harmful or toxic, or both. It is pointless to say that such materials should never be used within the confines of a small workshop, because no matter who says it for how long and however loudly, their use will continue. Always use such materials only when plenty of ventilation can be provided, on the order of a complete air change for the entire volume of the workshop every few moments. There just is no point in risking your health and your safety, not to mention your workshop or your house, through a buildup of toxic and/or flammable fumes or particulate matter. Installing a wall or ceiling-mounted ventilation fan is no big chore. If there is insufficient input of fresh, clean air from some other part of the house, you can install an inlet vent as well. Just opening a window or two, incidentally, may not be sufficient, especially if there is no breeze blowing outdoors. Often as not, forced ventilation is the only feasible answer.

The greatest hazard in most home shops lies with flammable paints, thinners, solvents and strippers. Any material that has a volatile vehicle can be dangerous, more especially when it is used in a small, confined space with inadequate ventilation. Spray painting, except with latex or other paints that use water as a solvent, should only be attempted in a spray booth that has a built-in ventilation system. This holds true not only for compressed-air spray painting equipment, but also for the popular spray cans. While the level of danger is not nearly as high with brush-applied paints, ventilation is nonetheless essential.

If you install a ventilation fan, be sure to get the type that has a shaded-pole motor, rather than a brush-type motor. Motor brush sparks can easily ignite paint vapors. If you contemplate setting up a spray painting booth, check into local electrical regulations. You may find that all of the equipment within the booth must be wired to explosion-proof specifications. This depends to some degree upon the types of paint that will be used, as well as the design of the booth and the kinds of electrical equipment, if any, that will be within the booth.

LAYOUT AND PLANNING

Laying out and planning a workshop requires a considerable amount of careful thought. If you do not yet have a workshop but want one, this is the time to start planning. If you do have a workshop that is something of a hodgepodge, starting all over again with a carefully thought out arrangement probably will not be amiss. In either case, the base procedure is to commit a complete design and set of specifications to paper. All of the details can then be shuffled around until you come up with an adequate and satisfactory arrangement, which then can be translated into a working shop.

The first step is to outline the workshop area on paper. Make a complete floor plan of the space, done to scale and noting the locations of doors, windows, heating units, electrical outlets and any other items that might be of importance (Fig. 4-3). Next, make an inventory of all the tools that you currently possess. The third step is to list all of the items that you plan to add to your tool collection in the immediate future—those that you definitely know that you want—and then add all of the tools that you would like to have. You should give some thought to the various kinds of projects that you want to build and work that you want to do. Particularly important are those interests that lie outside the main one for which you are building the shop. These latter interests, such as electronics, model ship building, lapidary work or whatever, may well require a small, separate corner of the shop that must be planned for accordingly.

Next, draw up small scale-sized squares and rectangles of stiff paper to represent all of the major elements of the shop: table saw, lathe, workbenches, worktables and anything else of any appreciable size. Cut them out and label them. Then you can arrange the cutouts on the floor-plan drawing and shuffle them about to your heart's content. Keep in mind the work flow, the necessary working clearances, required work space and similar factors as you make your arrangements. Once you hit upon what appears to be a logical and convenient layout for the major elements, consider how the minor elements of the workshop will fit in among them—tool storage boards, racks, lighting fixtures and vent fan. At the same time, try to build in a bit of flexibility so that as your ideas and plans change over the years, the workshop can be continuously adapted to accommodate them.

Once you have arrived at what appears to be, for the moment at least, a suitable and workable arrangement, move into the workshop area and set up a mock shop. Use cardboard cartons, chalk lines on the floor, chairs and tables, or whatever is handy to simulate the shop arrangement that you have just planned. See how the clearances actually work out, how the traffic pattern will be and how convenient the layout is. At the same time, visualize the workshop as it might be when more or less completed, with all of the various tools, supplies and materials located in the spots that you have assigned them. If your plan seems reasonable and shows promise of working out the way you envision it, get busy and start building. If not, make whatever adjustments are necessary and then proceed.

No workshop is ever a stable entity, and the plans that you draw up should not be considered immutable. Shop owners sometimes make the mistake of working everything out to the nth degree so that there is a place for everything and everything has its place. Then they become most upset when some new factor must be introduced into the scheme that had not been previously planned for. No workshop is ever complete, and no workshop should be planned so rigidly and with such fixed ideas that changes cannot be made at will. You can at best work out a reasonable arrangement, and numerous times will come along when changes will have to be made, some of which will doubtless upset the original plans. So be it. The workshop should grow along with your interests, skills and scope of activities. You might just as well resign yourself to the fact that you will never achieve the ideal workshop. You will, however, have a good place to work and a lot of fun.

HANDLING AND STORAGE OF PLYWOOD

The proper storage and handling of plywood is largely a matter of common sense; yet as often as not the appearance or usefulness of plywood panels, or sections thereof, are inadvertently lessened through carelessness or neglect. As expensive as plywood is these days, it makes sense to treat the material properly and carefully; a brief look at what is involved is therefore in order.

The appearance of construction plywood, the so-called engineered grades, is normally of little consequence. This means that a bit of rough handling will do no harm. The panels are tough

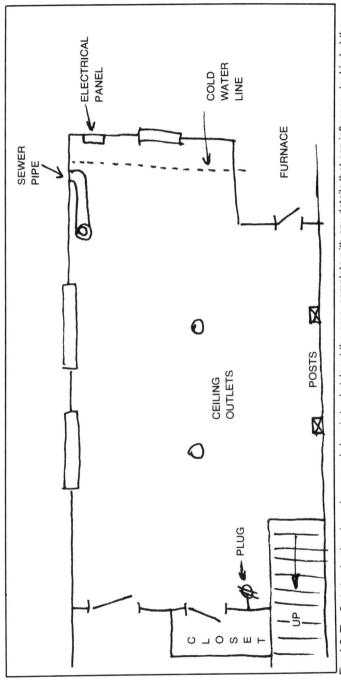

Fig. 4-3. The first step in planning a home workshop is to sketch out the area, complete with any details that can influence (or hinder) the shop arrangement.

enough so that slight mechanical damage certainly will not disturb the integrity of the plywood. The sheets can be dropped flat, skidded off trucks or trailers, scooted about on the floor of the construction site or on the ground, and little actual harm will come of it. One common occurrence that should be avoided, though, is dropping, pivoting or skidding the panels on their corners. A heavy impact or a substantial amount of abrasion on the points of the corners will break up or chew apart the squared edge. The panels should not be deliberately mistreated by dropping heavy objects on them or by handling them in such a way that severe mechanical damage might result; these plywood panels are mighty rugged but not indestructible.

The appearance grades of softwood plywood panels must be handled with a great deal more care. This is particularly true of those that have the higher grades of face veneers. When the panels are to be used in situations where the appearance is paramount, great care must be taken to prevent any damage to the face ply and frequently the back ply as well. The surfaces must be kept well protected and free of grit, tiny particles of sand and dust. Scratches and abrasion marks, as well as the relatively deep grooves left by particles of grit, are very difficult to sand out and may be obvious when the final finish is applied. When panels are stacked one atop another, be sure that the surfaces are absolutely clean and free of dust. They should not be handled roughly, walked upon or otherwise mistreated. The corners should be kept protected. When the panels are moved about, care should be taken that they do not bump into other objects. Likewise, decorative hardwood panels must be handled with extreme care, especially the prefinished ones.

It takes very little to scratch or abrade a finished face ply. These panels are often shipped face to face with a paper interleaving, and it is a good idea to leave the protective covering in place until the panel is actually put into service. Panel faces can also be temporarily protected with a layer of thin polyethylene film taped on. Don't use newspaper as the ink may come off. Wipe the panel surface down with a clean, dry cloth to remove all dust and grit before applying the film.

Ideally, all of the different kinds of plywood panels should be stored in a cool, dry place, regardless of their particular makeup. The relative humidity of the storage area does not have too much effect upon the panels, so long as it is not excessive and is comparable to the normal relative humidity range of the locale. Panels preferably should not be stored in damp areas, and certainly not where there is any possibility of contact with condensate or ground moisture. Relative humidities on the low side are best, and no harm will be done even if the percentage is as low as 10 or 15. It is a good idea, too, to keep plywood panels away from strong sunlight. This is particularly true of finished hardwood panels or any panels that will eventually be treated with a natural finish. Strong light will cause a darkening of the wood and/or may have a detrimental effect upon the finish. If part of a panel is continuously exposed to strong light and part is not, the result will be a two-toned or multi-hued effect that can only be corrected by covering the surface with an opaque finish.

Panels should also be stored away from any source of direct heat, like a baseboard heating unit or a hot-air register. The area should be clean and not subject to accumulations of dirt and dust. If necessary, stored plywood panels or sections can be covered with a sheet or tarp. If a breathable fabric is used as a cover, the plywood can be wrapped as tightly as you wish. However, if polyethylene sheeting or some other nonbreathable material is used, it should be fitted as a very loose cover so that there will be plenty of air circulation around the wood and condensation cannot form on the underside of the cover.

Horizontal and Vertical Storage

The best way to store panels if they are to be left for any length of time is flat on an even and relatively level surface. Ideally the panels should be fully supported from beneath, on a continuous surface that is flat. If laid across a couple of two-by-fours, the panels can eventually sag and droop and may "take a set," which makes them difficult to work with. If the panels are laid upon blocking, space the blocks close together to minimize this problem. Make sure that the sides and ends are particularly well supported. Thin panels can be placed atop one or more heavy panels, which are much less susceptible to warping.

Most home workshops do not have sufficient space to allow for a horizontal floor storage bunk. If the panels are stored flat, they are much more likely to be shoved into overhead racks. This is a perfectly acceptable alternative, too, but works best if the rack has plenty of cross supports and if the first sheet in the stack is of ¾-inch thickness, to avoid the possibility of warping. Where horizontal storage is not possible, the only alternative is vertical storage. This works quite well provided that the panels are stood straight up and down and rest solidly on the bottom edge. If panels are leaned against a wall, for instance, even the thicker ones will eventually bow somewhat. The best situation is to build a vertical storage rack with closely spaced slots, so that panels and panel sections can be stacked on end, supported by the rack and by each other so that they do not lean. Of course, the thicker the panel and the smaller the piece, the less of a problem warpage from vertical (or slightly tilted) storage is likely to be.

Outdoor Storage

Temporary outdoor storage of quantities of engineered grade plywood panels that will soon be used in a construction project is a different matter. The panels should be off-loaded from the delivery truck with a reasonable degree of care and stacked on a fairly level spot on top of a series of four-by-four stringers or some other support. They should be kept well up off the ground, especially if the ground is damp or muddy. If the panels are bundled in stacks, they probably will be skidded off the tilted truck bed. This should be done slowly and carefully so that damage does not occur to the bottom panels, especially at the ends that first make contact with the supports. The bundles should be dropped directly on stringers positioned under them as they come off the truck. The steel bands that hold the bundles together should be cut away as soon as the bundles are dropped.

Then the stacks should be covered with a tarpaulin or with polyethylene sheeting. A tarp can be wrapped tightly around the stack. If plastic film is used, place a small stack of two-by-fours or two-by-sixes centered lengthwise on top of the stack. Lay the plastic over the stack and pull it outward on both sides, wing-fashion, and anchor it securely (Fig. 4-4). Pad the corners with old rags so that the plastic will not puncture. This allows plenty of air circulation around the plywood; if this is not done, condensation and mold formation will probably occur.

SELECTING AND BUYING PLYWOOD

Much has already been said about the various types, kinds and grades of plywoods, and there is no need to go into detail again. Determining the kind of plywood panels that you want to purchase is a matter of selecting either interior or exterior type, face and/or back veneer grades, appropriate glueline and so on. The situation is somewhat similar with hardwood plywoods, and in most cases the type of finish, if any, as well as the face veneer wood species, the grain, figuring or patterning and similar factors are of impor-

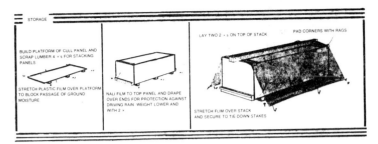

Fig. 4-4. The APA recommends this method of storing and covering plywood panels outdoors (courtesy of the American Plywood Association).

tance. The trick, and sometimes the problem, is to choose a panel that is exactly right for the application you have in mind, by outlining the correct specifications.

When you determine what kind of panels you want, the next step is to visit a lumberyard or plywood dealer and select the particular panel. It is just as well to have a second possibility in mind, in case the supplier does not have the kind you want. You can place a special order, of course, but this entails extra time and often additional cost. If possible or practical, inspect the panels you are about to purchase to make sure they are indeed what you want and that they are in factory-fresh condition. This is especially important in the case of appearance grade softwood panels, and doubly so in prefinished hardwood paneling. If you transport the panels home yourself, handle and stow them carefully so that no damage will occur. If the panels are delivered to your home, inspect each one as it comes off the truck to make sure that no damage has resulted. Remember that you are under no obligation to accept damaged panels.

The usual procedure in buying plywood is to purchase full-sized 4-foot by 8-foot panels, the standard stock size. In some cases, such as exterior siding material, the size may be 4 feet by 9 feet or 10 feet. As a rule, plywood panels cannot be purchased in other than standard stock sizes.

There are exceptions, however. Sometimes suppliers have plywood panels on hand that have been damaged in shipping or in handling at the yard. You may be able to buy this plywood in smaller pieces cut from the damaged panels, or buy the whole panels at a discount price. A few suppliers sometimes custom-cut plywood panels into special sizes for good customers and end up with a number of odd-shaped pieces lying about. These, too, can sometimes be bought. Another possibility is to contact building contractors or cabinetmakers to see if they are willing to part with any leftovers that might have accumulated from previous jobs. Also, because many do-it-yourselfers have no need for full-sized plywood panels, many lumberyards and home centers now stock half-sheets and quarter-sheets of plywood in various types and grades, especially to meet the demand for small pieces of plywood.

Buying hardwood plywood is generally a bit more involved than many of the softwood plywoods. Only the larger dealers stock any great amount of hardwood plywood panels, and then not in any great depth, because there are so many different kinds from which to choose. The usual routine in selecting such panels is to go through the samples provided by various manufacturers and make a choice from the stock items that are offered. The panels, when they arrive, will not be exactly like the sample, but they will be generally the same. Hardwood plywood panels can also be ordered to specification. This is done by going through a dealer to the manufacturer. Speciality plywoods are generally handled in much the same way, since these items are almost never kept in stock by retailers.

Softwood plywood panels are generally priced on a per-square-foot basis, while hardwood panels may be sold on that basis or by the panel. Prices do vary somewhat from dealer to dealer. If there is more than one plywood outlet in your vicinity, no harm will come from doing a bit of price shopping. You may be able to save a few dollars. If you are buying small pieces rather than full panels, be sure to compare the per-square-foot price of the small pieces against the same kind of panel in a full-sized sheet. You may find that you are paying an exorbitant price for the small pieces. In some cases that fact may be immaterial to you, but in others you might want to buy the full sheets and save the leftover portions for later projects.

WORKPIECE LAYOUT

The object in laying out workpieces on a sheet of plywood is to arrange the pieces so that there is as little scrap material left over as possible, and the fullest use is made of the panel. The best procedure is to start at one end of a full panel. Lay the pieces out so that you utilize the full width of the panel while working down its length toward the other end (Fig. 4-5). There are cases where this is not possible, because many pieces must be laid out lengthwise to the face grain of the panel. There may be more long pieces than short pieces so that fully utilizing the panel is difficult. However, squeeze as much use out of the panel as you can.

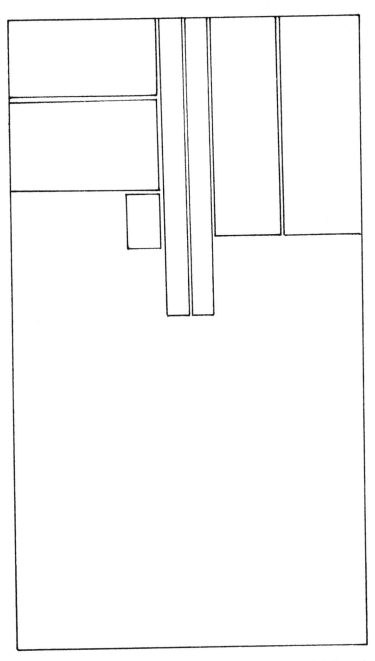

Fig. 4-5. Start the layout of small pieces to be cut from a full panel of plywood from one end or the other, for the best utilization of the panel (as a rule).

Begin by placing the panel on sawhorses or a worktable, at a convenient working height. Make sure the panel is fully supported so that it does not sag and the ends or sides do not droop. An uneven panel surface is difficult to work on, and inaccuracies in dimensions and guidelines can crop up all too easily. If the pieces are all fairly large, you can lay them out directly on the surface with a marking instrument, straightedge, carpenter's square and similar tools. However, if you have many small parts to cut, the easiest and most efficient course is to make up patterns of all of the parts from cardboard or stiff paper. Then you can shuffle the various pieces about on the panel surface until you arrive at the most advantageous arrangement that wastes the least amount of plywood. Use the patterns to trace around, and then go back over them with your marking and measuring tools to double check the dimensions and clean up the guidelines.

Whichever way you go about it, be sure to leave ample space between the parts for the saw cuts. Saw cut width (kerf) varies considerably, so take this into consideration. A fine-toothed jigsaw blade may make a cut of 1/16 inch or less in width, while a circular saw fitted with a carbide-tipped general purpose blade may make a cut as wide as ⅛ inch or more. Remember, too, that either the top or the bottom of each cut-line, depending on what kind of tool is being used, will be relatively ragged. The opposite one will be quite smooth and true. The degree of difference depends upon the specific saw blade (and also how sharp the saw blade is), but there will almost always be some difference. If slight splintering or raggedness makes a difference, allow ample space between the parts so that you can make a rough cut first. Then trim the piece to a very fine edge on a table saw, bandsaw or other tool.

Cutting Pieces From Smaller Panel Sections

Obviously it is much easier to work with smaller panel sections and cut smaller parts from them rather than to cut each part individually out of a full-sized panel. Often you can lay the pieces out so that you can rough-cut a section from the panel, and then cut several smaller parts from that section (Fig. 4-6). In this case, an even wider allowance should be made for the saw cuts, depending upon how much final trimming must be done. In most cases, it is easier to rough-cut all of the parts that are laid out on a full panel, and then finish-cut each part individually. For instance, you might cut major sections from a panel with a portable circular saw. Then cut the small parts from the sections with a jigsaw. Finally trim all

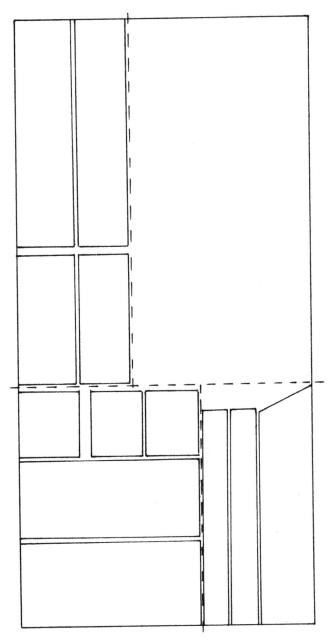

Fig. 4-6. Often a layout for a series of small pieces can be arranged so that the full panel can be cut into a few easily manageable sections, and the smaller pieces can then be cut from the sections.

the edges to their final dimensions with a high-speed extra-fine toothed carbide-tipped blade in a table saw. Though this means three cutting operations, a considerable amount of handling, and more wasted wood (which becomes largely sawdust) than would be the case with single-cutting, the finished result is excellent and the likelihood of mistakes is lessened.

Often a project will call for two or more identical parts, and sometimes these are angular in shape. When this situation occurs, try to lay them out in a reciprocal or mirro-image pattern for maximum use of the plywood (Fig. 4-7). Pieces that have irregular shapes can only be nested together as neatly as possible (Fig. 4-8), and sometimes there is no way to avoid fairly sizable waste pieces. Don't discard those pieces, however, because they surely will come in handy later on in another project. If nothing else, they can be sawed up for gussets, glue blocks or supports. Sometimes it is also necessary to orient the parts either with or across the face grain of the panel. As you do your layout work, be sure to keep in mind which is which. If one particular part is laid out with the grain, then another identical part should be laid out in the same direction and not across the grain. This is particularly important, for the sake of appearance, if the applied finish of the completed article will be natural or semi-transparent. Note, too, that in most cases (but not all) long pieces are laid out with the grain. Short and blocky pieces can be laid out either way, and angular or odd-shaped pieces can be laid out in the most convenient manner. Frequently the nature of the project and its desired final appearance will have a bearing on this layout.

Earlier it was mentioned that it is better, when only a few pieces will be cut from a full sheet, to start laying out from one end and proceed towards the other. Utilize as much of the width of the sheet as is possible, as opposed to laying out a few parts along one lengthwise half of the sheet. This is because a leftover cross-grain piece of panel—a cut-off end—is generally more useful than is a narrow, long length. This is not always so, of course, but it is true enough that the rule of thumb is a good one to follow. In any event, with a bit of care and patience, leftover plywood pieces can be joined together either end to end or edge to edge to make up a larger panel. If close attention is paid to the grain and pattern, the match can be made quite unobtrusive.

Fitting Together Pieces On One Panel

Often projects are planned and dimensioned so that all of the parts can be cut from a full plywood panel (or two panels or a half

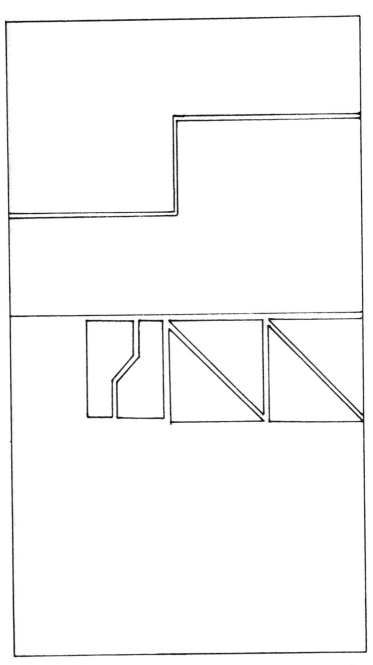

Fig. 4-7. This layout shows how reciprocal or mirror-image pieces can be arranged for maximum plywood utilization.

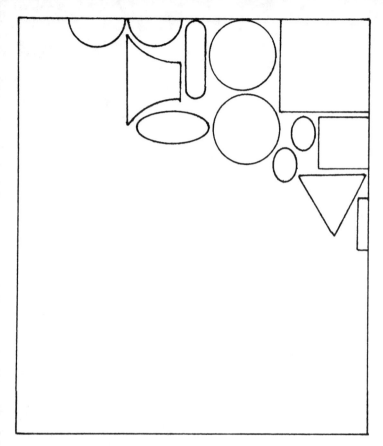

Fig. 4-8. Odd-shaped parts must be nested together however they seem to fit best; this can often be done best by juggling patterns about on the panel.

panel) with virtually no waste and nothing left over. The fit is exact. Obviously a great deal of care must be taken in such a situation to ensure that the layout and dimensioning of the parts on the panels is perfect. The cutting must also be done exactly right, because there is no room for error. Most plan sets for these projects that have been developed and published by plywood concerns, how-to-do-it periodicals and similar sources usually have been thoroughly checked, actually done and the dimensions are correct; the plans are workable. But errors do crop up, and it is wise to double check all of the dimensions and the layout before making any cuts in an expensive plywood panel. It is entirely possible that you will find an error, usually of the typographical sort.

If you are making up your own project plans that call for all of the pieces to be tightly fitted together on a single plywood panel, there are a couple of factors to keep in mind. One is that the plywood panel which you purchase for the project may very well not be exactly 4 feet wide and 8 feet long. Those dimensions can be off a bit, usually on the minus side, which in turn could have an effect on your dimensions. Keep in mind, too, that you cannot butt the part outlines directly together, as they are usually laid out on a drawing. There must be an allowance around all of the parts for the saw cuts. For a single-cut layout, this allowance must be at least 1/16-inch, which allows no room for error or for final trimming. If you want plenty of space in which to make your rough cuts, allow ½ inch between parts and make the rough cuts straight down the middle of this "aisle." Depending upon the saw that you use for rough cutting, this should leave ample room for off-vertical cuts (common with handsaws), a bit of wandering and a splintery edge, plus about 3/16 inch on either side of the rough cut for final trimming. This takes up a bit of extra plywood, to be sure, but also gives you an added measure of insurance against uncorrectable errors.

CUTTING PLYWOOD WITH A HANDSAW

Cutting plywood with a handsaw (Fig. 4-9) takes some time and some elbow grease, but it is a good way to get the job done, especially in instances where electric power is not available. This is also a good method of cutting full panels into smaller sections,

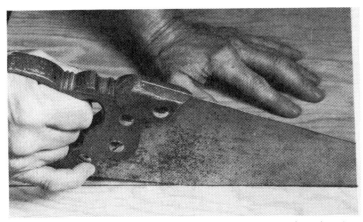

Fig. 4-9. Curved sections are easily and cleanly cut on a bandsaw (courtesy of the American Plywood Association).

with less set-up time involved than with a circular saw, after which the smaller sections can be finished up with a different kind of saw.

The panel must be well supported, especially a relatively thin one, so that it will not say or droop and the edges will not chatter as the sawing proceeds. Use a sharp, fine-toothed carpenter's handsaw, a crosscut kind of 10 or 12 points (sometimes also called a finish or panel saw). When making long cuts as across the full width of a plywood panel, frequently it is easiest to saw halfway across from each side and make the kerf meet in the middle.

Making good cuts with a handsaw takes a bit of practice, but it is not really difficult. Most of the cutting action takes place on the downstroke, and only a bit on the upstroke. The trick is to keep your forearm lined up with the cut that you are making, the wrist and hand being on extension of the forearm, and all lined up with the saw blade. The blade must also be held perfectly upright so that it does not cant to one side or the other (unless you are making a bevel cut). Keep your arm and shoulder relaxed and apply pressure as you make the downward stroke. Then release it as you bring the saw back up. The cutting action should be done in a smooth, easy motion. There is no point in trying to force the saw, either with pressure to make it cut faster or by attempting to tilt or bend the blade toward or away from the guideline. The usual procedure when sawing plain wood is to hold the saw at approximately a 45° angle to the workpiece. However, with plywood that cutting angle should be increased to at least 60° from the vertical for best results.

Since the teeth of a handsaw cut mostly on the downstroke, the best face (and the layout lines) should be on the best surface of the plywood. The cut edges of the top surface will be smooth while the bottom edges will be a bit ragged, where the wood fibers tear away from the sawing action. Some splintering may also take place, depending upon the coarseness of the blade. The end of the cut is the most critical point, especially if the pieces beign separated are supported, immovable and kept within their original relationship to one another, it is entirely likely that the parting edges will splinter, tear or break away from one another. Some repair work may be needed, or the pieces may be useless. Sometimes the pieces are best securely clamped together, across the open saw cut behind the saw. On other occasions, ful! support is all that is needed.

CUTTING PLYWOOD WITH A PORTABLE CIRCULAR SAW

Cutting plywood with a portable circular saw (Fig. 4-10) is much faster and easier. With a little practice, you will find that you

Fig. 4-10. Probably the easiest method for slicing up full panels of plywood in the home shop is with a portable circular saw (courtesy of the American Plywood Association).

can guide the saw freehand, without using a straightedge, with no difficulty in making rough-cuts. The portable circular saw is fine for making all but the smoothest and most accurate of edge cuts. For construction work, it is unexcelled. In cases where the cutting line must be very accurate, the saw can be used in combination with a straightedge or a saw guide. Since the shoe of the saw can be set to any angle from 0° to 45°, accurate and uniform bevel cuts can also be made.

The kind of blade that you use in the circular saw will make a substantial difference in the degree of smoothness of the cut. For rapid cutting where pieces will be roughed out for later trimming, and where there is plenty or margin outside of the finished piece guidelines, a general purpose carbide-tipped blade works well. However, there is a substantial amount of splintering and chipping along the edge. The advantage of using this kind of blade is that you can make hundreds of feet of cut rapidly without dulling the blade in the least. Perhaps the most popular blade for general purpose cutting is a combination blade, which leaves a relatively smooth edge. For a finer cut, use a crosscut blade. For the smoothest edge of all and the least amount of splintering, a hollow-ground plywood, cabinet or similar blade is the choice. These latter blades must be handled with care because they are thin. There is not much set to the teeth and the kerf is small. The blade will tend to waver or wander if forced. Whatever the type of blade employed, it should be very sharp for good results.

Unlike the handsaw, a portable circular saw cuts on the upstroke rather than on the downstroke. This means that splintering will take place along the top edge of the workpiece rather than along the bottom edge. Whenever possible or practical, the good surface of the piece should be face down while you are cutting, and the back or poorer surface up. Sometimes it is necessary to remember that you are cutting from the back, rather than the face, in order to make sure that the layout is correct.

Technique

When cutting with a portable circular saw, adjust the saw shoe so that the teeth will protrude through the workpiece for a distance equal to the height of the teeth or, as a rule of thumb, about ¼ inch (Fig. 4-11). Set the leading edge of the shoe on the workpiece and align the blade with the cutting guideline. Be sure to set the blade on the correct side of the guideline, so that the saw kerf is to the outside of the workpiece. If you place the saw blade to the inside of the cutting guideline, the piece will end up smaller than the measured dimension by the width of the saw kerf. It will not fit. Naturally this is not always of consequence. When you are making rough-cuts and there is plenty of leeway between the pieces being roughed out, you can saw to either side of the guideline or right down the middle of it. When you are "cutting to the mark," be sure to place the blade on the proper side of that mark.

After positioning the shoe, start the saw and allow the blade to come to full speed. Then ease the saw forward gently, checking to make sure that you are correctly lined up with the guidelines. Keep the shoe flat on the plywood, headed in the proper direction and tilted neither to left nor to right. Your wrist and arm should be firm and steady but relaxed. You can grip the saw in any way that is comfortable for you, with your arm at virtually any angle. However, the further you must stretch along the guideline, the less accurate the cut is likely to be. You will get best results by keeping your shoulder almost above the saw. Push the saw along at an even steady rate but not too fast. Never push the saw hard so that it overloads and drops speed. You will become used to the tone and pitch of the saw blade and motor, and will recognize the drop in pitch that signals that the saw is working too hard. If this seems to happen rather frequently, the probable reason is that the blade is dull or gummy.

As you make the cut, keep your eye on the leading edge of the kerf as it travels along the guideline to make sure that you stay on

Fig. 4-11. Saw blade is adjusted so that the teeth just protrude beyond the lower surface of the plywood. Note combination blade, used for rough-cutting sections from panels.

the mark. You may have to occasionally blow some of the sawdust away. If you pause during the cut, be careful not to twist the saw one way or the other as it is stationary on the workpiece. This will create gouges and uneven spots in the cut edge and may bind the blade in the cut. Make sure, too, that the two pieces being separated remain even with one another and that the kerf does not close behind the blade. If the pieces get out of line or the kerf closes, this can result in pinching and binding and may cause kickback. The teeth suddenly grab into the wood and bounce the saw right out of the cut. This can be a dangerous situation, and there is no way that you can control a kickback. It just happens too fast and there is too much force involved. The best bet is to exercise caution and be aware of, and avoid, the situation. Bowed or otherwise warped pieces of plywood or wood, and especially plain wood with large knots in it, are particularly susceptible to kickback problems.

During the cutting process, move slowly enough that the saw blade maintains its normal speed. As you reach the end of the cut, change from the downward-and-forward pushing action to a slight easing of the downward pressure. At this point there should be a bit more pressure toward the heel of the saw shoe than toward the front. At the end of the cut you should be just picking up the full weight of the saw, so that it goes through the last half inch or so of material smoothly and in perfect alignment with the guideline.

Setting Up a Panel With Long Cuts

Making long cuts in plywood panels (full-length shelf strips, for example) can be difficult if the panel is not properly set up.

There are several ways to go about this, depending upon the equipment and shop aids that are available, but here's one method that makes the job easy. Set a pair of sturdy sawhorses about 5 feet apart, and rest four planks across the horses to make a *trestle*. If you are cutting the panel in half lengthwise, separate the planks into pairs with a gap of about 2 inches down the middle. Lay the plywood panel on top of the planks with the centerline of the panel above the gap. Clamp each end of each half of the panel to the planks. Start the cut from one end and move the saw along as far as you can conveniently reach. Then stop the saw. When stopping the saw in the cut, the best way to do it is to hold the saw perfectly steady so that the teeth do not hit the edges of the wood, until the blade has completely stopped. Then climb up on the plywood, kneel beside the saw and start it up again. Make sure that the saw blade is perfectly aligned in the cut, and allow the blade to come up to full speed before proceeding. Then just crawl along beside the saw on your knees, cutting as you go.

There are many other ways to set up a panel for lengthy cutting. But the object is to give the panel full support, while allowing a half inch or so of clearance directly below the cut line where the blade can pass unobstructed. These cuts can be made freehand for rough cuts, but they should be done with the aid of a straightedge or a saw guide whenever accuracy and a clean, true edge is important. Remember that if the first cut is the final cut rather than a rough-cut, and will represent the finished edge of the piece, the rougher and more inaccurate the edge is, the more work will be involved in finishing and truing it up to a smooth edge. Remember, too, that in such cases you must make a bit of an allowance over and above the exact dimensions of the workpiece, so that you can smooth and trim the edges a bit without working your way into the workpiece itself. Otherwise, the piece will end up not quite large enough.

Making Plunge-Cuts

A portable circular saw can also be used to make plunge-cuts. Rather than cutting in from one edge and proceeding into the material, the cut is started directly in the panel itself. This is a handy procedure when fairly long and accurate cuts must be made within the borders of the workpiece. To make plunge-cuts, first secure the panel solidly to a work surface by clamping it down. Set the blade depth so that the teeth will protrude just a bit from the backside of the panel. Then align the saw and the blade with the

Fig. 4-12. This shows the beginnings of a plunge cut with a portable circular saw, with the hub of the saw being lined up with the arrow, and the blade aligned with the cutting guideline.

cutting guideline, with the saw blade resting upon the workpiece surface and the saw body tilted up and resting firmly on the nose of the shoe.

The blade should be positioned at the near end of the cutting line, but far enough along the line so that the blade will not cut back beyond the start of the line (Fig. 4-12). In other words, for a saw with a 7¼-inch blade, the center of the blade must be positioned at least 3⅝ inches from the start of the cutting guideline. Actually, its a good practice to allow at least an extra half inch, for insurance. While keeping the saw and blade properly aligned, tilt the saw forward a bit so that the blade clears the plywood surface.

Start the saw and allow the blade to come to full speed. Then gently lower the saw, pivoting it on the nose of the shoe, until the blade makes contact with the surface and cuts its way down through the workpiece. When the saw shoe is flat on the surface, then you can begin cutting forward along the guideline in the usual manner. Stop the cut just as the teeth reach the end of the guideline. Turn off the saw and allow the blade to wind down fully; then remove it from the cut. Since the blade is circular, there will be a small section of uncut wood left on the underside of the workpiece. The cut can be finished with a carpenter's handsaw or a keyhole saw.

These last cuts should be made with great care, preferably from the reverse side. Remember that with the portable circular saw, you are cutting from the "bad" surface. With a handsaw you should cut from the "good" side to avoid splintering. This will also avoid having a chunk splinter out of the good surface of the workpiece if the cut-out section should fall through.

USING A SABERSAW

A saber saw is a very useful power tool for roughing out plywood parts, though it is less so where fine, accurate cuts must be made. Usually a certain amount of edge finishing must be done after the cutting to obtain a smooth edge. Like the portable circular saw, the saber saw blade cuts on the upstroke. Splintering will take place along the upper edges of the workpiece rather than the lower. Thus, sawing with a saber saw should be done from the reverse side of the workpiece—the back or bad side—and layout work should be done on the back surface as well.

Saber saw blades are quite narrow. The saw itself is relatively lightweight, and its operation is accompanied by a great deal of vibration. The saw shoe must be kept flat on the workpiece surface, with a goodly amount of downward pressure, as well as forward motion, exerted on the saw. Even so, making a smooth and accurate freehand cut takes practice, and even with a very fine-toothed sawblade and an experienced hand the cut edge usually must receive further treatment (except in circumstances where rough edges are of no consequence).

To make a cut with a saber saw, rest the tip of the shoe on the workpiece, at the beginning edge, just as with a portable circular saw. Align the blade with the cutting guideline (and on the correct side of the line), turn the saw on, and wait a second for the blade to come up to speed. Ease the saw forward and begin the cut, making minor adjustments to align with the guideline as you do so. Keep a firm downward pressure on the saw, and move it forward slowly enough that the saw blade speed does not fall. The cutting speed will vary quite a bit with different kinds and thicknesses of materials. The cutting action is not nearly as fast as a portable circular saw, for instance, and some degree of patience must be exercised. Too much forward pressure puts undue load on the saw motor and also can easily break the saw blade. Fine-toothed blades with the least amount of set will allow the smoothest cuts.

Straight lines can be cut freehand, but they are best done with a straightedge of saw guide of some sort. If the workpiece is relatively small, the rip guide provided with the saw may be adequate. Otherwise, make some other arrangements. Curved lines must usually be done freehand, and this does take a little practice. Coordination between the eye, hand and saw in order to keep exactly to the cutting line can only be gained with experience.

For best results, whether straight or curved cuts are being made, anchor the workpiece firmly to prevent vibration and chattering, especially if the piece is relatively thin. Keep a firm grip on the saw, with your forearm about a 45° angle and your arm relaxed. As with a portable circular saw, the farther you extend your arm, the less control over the blade you are likely to have. If you must stop the saw in order to change position or brush sawdust away, turn the saw off and allow the blade to come to a full stop before letting go of it. Otherwise, it may catch and jump right of the saw cut. You can leave the saw in place or remove it entirely; picking up the cut line again after removing the saw is easy with a saber saw, and the stop/start point will be unnoticeable. If you try to turn the sabe saw too rapidly, force it into a radius smaller than it is capable of handling, or exert undue forward pressure on the blade while making a curved cut, the blade will probably snap.

The saber saw is frequently used in making inside cuts, where the cut is started not from an outside edge but at some point in the interior of the workpiece. One procedure is to drill a starter hole of about ⅜-inch diameter through the workpiece at any convenient point along the cutting guideline. The hole must be positioned so that it does not lap outside the guideline, but it can be anywhere within. The the blade of the saber saw is slipped into the hole, and the cut is started in the usual way. Back-cuts or side-cuts can be made to square up right-angle corners and trim off uncut portions. The saw is so maneuverable that this is easy to do.

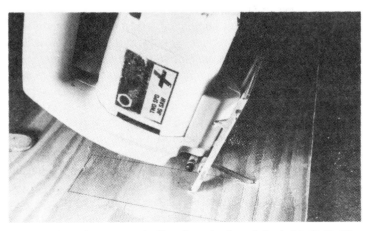

Fig. 4-13. The saber saw can be tipped over backwards to start a pierce cut.

Another method for accomplishing the same purpose is called *pierce* or *thrust cutting*. No starter hole is required. To make this kind of cut, tip the saw forward until the shoe is almost straight up and down, and the curved shoe tips and the tip of the blade make a three-point contact with the workpiece surface (Fig. 4-13). The blade and the saw body should be aligned with the cutting guideline. Tilt the saw body forward just enough so the tip of the saw blade clears the work surface. Hold the saw firmly and turn it on. When the blade reaches full speed, gently ease the saw back down and allow the tip of the blade to begin "scratching" its way into the wood. Continue the tilt, without undue pressure, until the blade finally works its way entirely through the material and you can set the saw shoe flat on the workpiece surface. Then continue the cut in the usual manner.

MAKING CUTS WITH A TABLE SAW

Cutting full panels of plywood, or large sections, on a table saw must be done with care and with the proper arrangement. Only the largest of table saws with full extensions at the sides are capable of handling a full plywood panel without additional aids. A full panel is heavy, awkward and difficult for one person alone to saw. The trick is to arrange full support for the panel, either at the sides (one or both) and at the rear of the table. This can be done in a number of ways. Adjustable roller stands are the best bet, and two or three of them are sufficient for even a large and active workshop. They are a worthwhile investment if you plan to do much cutting of full panels or large pieces on your table saw.

The stands can be adjusted to whatever height is necessary and positioned as needed to help hold up the sides of the workpieces as they go past the blade, or to catch the cut pieces as they come off the back of the table. Since the rollers work quite freely, the workpieces can be pushed through the saw with relative ease and not much danger of catching or binding. Even so, having a helper handy to assist in guiding the workpieces is not a bad idea. A note of caution is in order here, too; the worker and the helper must keep themselves well coordinated to both their motions and actions and the cutting action of the saw. Coordination is necessary to avoid having the saw blade bind or catch, and perhaps stalling the saw or having the workpiece kick off the table.

A common alternative, especially in home workshops, to using roller stands is to cobble up supports from sawhorses, scrap pieces of plywood and C-clamps. By clamping a fairly good-sized piece of plywood to a sawhorse, positioned at the correct height and perfectly level across the top edge, you can fashion supports as needed. These supports can then be placed at the sides or to the rear of the saw table, just as roller stands are. The workpiece will not slide across the top edges of the plywood supports as easily as it will across rollers, so the cutting job must be done slowly and carefully. But this system does work rather neatly. Small workpieces that will fit well upon the saw table and can easily be manipulated without fear of their binding or toppling off the table need no extra support.

As with other saws, the finer the blade mounted in a table saw, the finer the cut. The table saw is excellent for making final, finish cuts after piece have been roughed out with a portable circular saw (Fig. 4-14). By mounting a very sharp plywood or cabinetmaker's blade in the saw, you can cut and edge that is perfectly true and probably will require no further work, at least insofar as smoothness is concerned. Cuts made in this manner are ready to make up a glued joint just as they are. Assuming that the table saw is a good one and that the operator knows how to use it properly, fine cuts of excellent accuracy and trueness can be made.

Step-By-Step Procedure

The first step in making good cuts with a table saw is to make sure that the saw itself is accurate. The blade should be checked periodically to ensure that when the arbor is set at 0 degrees inclination, the blade actually is exactly at right angles ot the table.

Fig. 4-14. Making a finish cut on a table saw. The saw guard has been lifted for photographic reasons.

This can be checked with a good try square and the blade adjusted as necessary. The rip fence adjusting guide must be accurately set, and it must be reset when blades of different thicknesses and kerf widths are used. It is important, too, to make sure that the fence is at the same distance from the saw blade at both forward and back ends after it is locked in place and has not been locked at a bit of an angle. Th miter gauge should also be frequently checked, especially if it happens to get dropped on the floor, to make sure that the head is exactly at right angles to the guide bar when the pointer is set upon zero. This, too, can be checked with a try square.

A table saw cuts on the downstroke rather than the upstroke, so the good surface of the workpiece, with the layout lines inscribed thereon, should lie upward. Since the saw blade makes first contact with the workpiece along the lower edge, sometimes lining up the cutting guideline with the saw blade can be a problem. To get around this, extend the guideline down the edge of the workpiece and line the piece up at the bottom edge, rather than trying to do so at the upper edge.

When making a cut on a table saw, first double check all of the settings to make sure they are exactly right. Set the workpiece on the table against the rip fence or held firmly against the head of the miter gauge. With the saw off, run the piece forward until it just touches the blade to make sure that alignment with the cutting guideline is accurate. Make sure, too, that you are cutting on the correct side of the guideline; since most of these cuts will be finish cuts, there is no room for error and accuracy is paramount. Draw the workpiece back away from the blade and make sure the safety guard is in place. Start the saw and ease the workpiece forward inot the balde. Continue making the cut with a downward and forward pressure on the workpiece, judged so that the noise pitch of the saw blade stays approximately the same and the saw is not overloading. The blade height should be adjusted so that the teeth protrude above the workpiece top surface only ⅛ to ¼ inch. If the blade happens to be a hollow ground or other thin and narrow kerf blade, run the workpiece through the saw slowly. This will minimize blade wander and result in a truer edge.

As the end of the cut is neared, slow the rate of feed somewhat and make sure that you have a firm grip on the workpiece. The last quarter-inch or so of material should be fed through the saw quite slowly, especially with plywood. This will prevent tearing or splintering at the corners as the two pieces finally separate. When the cut is complete, ease the workpiece fully past the blade with care, making sure that it does not get a bit out of line and hit the teeth at the back edge of the blade. This will result in a gouge or nick in the cut edge or a chewed-off corner. Push the piece fully through the anti-kickback device, past the splitter and off the back edge of the table. Shut the saw off, allow it to wind down to a stop, and *then* remove the cut-off piece from the saw table.

Handling Workpieces

Many workpieces of reasonable size can be cut on a table saw just using the hands and/or miter gauge or rip fence (the two never should be used together). But small pieces must be handled with great care. Various aids should be used during the cutting process so that the operator's fingers will stay well away from the blade and the danger of kickback is minimized, if not entirely eliminated. Small plywood pieces can catch or tear on the blade and can get kicked back violently. They may shoot directly backward and hit the operator squarely in the stomach, or fly up off the table with a potential for serious injury. From another standpoint, these small

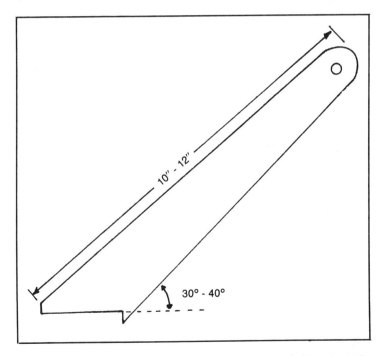

Fig. 4-15. This push stick is easy to make from scrap material and greatly increases safety when cutting small pieces.

pieces must be kept flat on the table and accurately aligned with the saw blade, or the cut will be a poor one. Wherever possible, small workpieces should be firmly clamped miter gauge or some similar device as it is moved through the saw.

Often it is possible, and quite safe and practical, to move the workpiece along with a *push stick* (Fig. 4-15); there is a myriad of designs, and you can easily make up your own. The thicker the push stick, the more rugged and easy to handle it is. The width of the stick should always be a good deal less than the width of the workpiece being cut, so that there is absolutely no danger that the push stick will contact the saw blade. Wide but stubby workpieces can be pushed through the saw with a *push block* (Fig. 4-16); these too are easily made up in the workshop. Glue a piece of fine sandpaper to the bottom of a push block or even a fairly wide push stick. This will help to hold the workpiece more securely and prevent the device from slipping. Though more time is required for setting up, a set of table saw hold-down fingers is an excellent means of keeping the workpieces under control.

The occasion often arises when working with plywood to make cuts in stock that is quite thin, such as hardwood plywood panels. This stock can be somewhat difficult to cut because it is so limber, and it tends to chatter or shift slightly on the saw table. To make clean, precise cuts in thin plywood, try this method. Cut a sheet of heavy plywood, one that is flat and smooth, such as a piece of ¾-inch A-C, to approximately the same size as the saw table. Lower the saw arbor all the way and put on a hollow-ground planer combination blade. Clamp the piece of heavy plywood to the saw table. Start the saw and allow the blade to come up to full speed. Then very slowly elevate the saw blade, while it is running, until the teeth saw through the top surface of the heavy plywood piece and protrude by an amount equal to the thickness of the thin plywood that you plane to saw, plus about ⅛ inch. Turn the saw off and position the thin stock to be cut on the plywood surface. Turn the saw on, and hold and guide the thin stock with a heavy piece as you make the cut. With a bit of practice, you can make a perfect cut every time. The plywood table can be saved for future use; make some sort of registration marks on the piece so that it can be put back on the saw table in exactly the same position each time.

WORKING WITH THE RADIAL ARM SAW

Edge-cutting with a radial arm saw has some similarities to cutting with a table saw, but there are a great many differences,

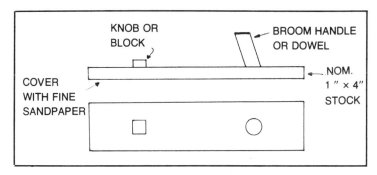

Fig. 4-16. A push block increases safety and ease of handling small pieces, and can be used with a table saw or a jointer-planer.

too. The saw blade is above the work, rather than below, and sawing to the guidelines laid out on the workpiece is simple indeed. This kind of saw also cuts on the downthrust, so the good side of the workpiece should be facing up. The saw has great versatility. Many consider it to be more utilitarian (and also safer) than a table saw. Because of its multiple-adjustment capability, the radial arm saw can be made to perform many different cutting operations as well as drilling, routing and planing. The kinds of blades used in this saw for cutting plywood, though often of larger diameter than many home shop table saws, are of the same general patterns.

Cutting relatively narrow plywood stock, up to the capacity of the particular saw (typically 12-15 inches), is done with the radial arm (and the saw blade) exactly at right angles to the rear guide fence, as indicated by zero on the miter scale. Note that this adjustment should be periodically checked and set to make sure that the blade is indeed exactly at right angles to the fence when set at the zero mark. Otherwise the cuts will be slightly out of line. The blade cutting depth is adjusted so that the teeth extend about 1/16 inch below the bottom surface of the workpiece, into a slot in the table top. The blade guard and anti-kickback fingers must be set, depending upon the thickness of the stock to be cut. With the saw positioned all the way back against the column and turned off, the workpiece is slid onto the table and the cutting guideline aligned with the blade. When everything is ready, hold the stock tight against the guide fence with the left hand. Turn the saw on and allow the blade to come up to speed. Grasp the yoke handle with the right hand and pull the blade forward across the material. *Angle cuts*, or *miter cuts*, are made in the same manner simply by adjusting the arm to whatever degree is needed. *Bevel cuts* are made by tilting the entire motor and saw assembly to the desired angle; *compound cuts* are made by doing both.

Wider pieces of plywood can be cut by placing the saw in its rip position; pieces up to a width of about 20 to 26 inches, depending upon the specific model and capacity of the saw, can be cut in this manner. The saw assembly is turned at right angles to the arm and locked in place. Then the depth of cut is set, just as for crosscutting, and the blade guard and anti-kickback fingers adjusted. The stock is fed, guided by the guide fence, along the table and into the rotation of the blade (but never with the rotation).

As with the table saw, exceptionally long or wide pieces of stock that are oversize for the saw table must be supported by other means. Roller stands or temporary props made of sawhorses and pieces of plywood can be employed. With this type of saw, though, it is common practice to build extension benches to one or both sides of the saw table. These serve nicely for handling long pieces of stock when the saw is in the rip position, and when not in use for that purpose they can be used as auxiliary worktables.

The radial arm saw is not capable of cutting the entire width of a full plywood panel—4 feet. Larger models, however, can slice a full sheet lengthwise. This is done in the out-rip position. Where full-width cuts, diagonals or other cuts beyond the capability of a radial arm saw must be made, the recourse is to rough the pieces out on a table saw or with a portable circular saw or handsaw.

BANDSAW CUTS

A bandsaw is a joy to work with. It makes the cutting of curves and certain long straight cuts very easy. However, most bandsaw cutting must be done freehand without the benefit of guides, fences and miter gauges (though these can sometimes be employed). This means that the skill of the operator is the key to fine cutting. Some practice and experience is necessary. The trick is to follow just outside the layout line, smoothly and evenly, to allow for edge-finishing later on. Bandsaw cutting generally requires both short cuts and backing out, as well as long cuts and sometimes some rather intricate patterns. Experience is needed in order to know where to make the first cuts, subsequent cuts, and which cutting pattern or sequence to follow for the most efficient and effective cutting. All of this is much more difficult to explain than it is to do. Even a neophyte should have little trouble understanding the general procedures; then all that is needed is practice.

In making bandsaw cuts, the operator stands beside the machine, facing the cutting edge of the blade and slightly to the left of it. The upper blade guide must be adjusted so that it is about ¼ inch above the top surface of the workpiece. After the saw is turned on and the blade has come to full speed, the operator gently moves the workpiece into the blade, simultaneously aligning the cutting guideline. Then the stock can be fed as fast as it will cut. The stock must be held firmly and flat on the table. It is generally held down with the left hand and fed with the right. Both arms must remain relaxed, with both downward and forward pressure steady but firm, and the forward and forward pressure steady but firm, and the forward motion through the blade must be smooth. Any sudden motions, twists or jerks will result in a gouged workpiece edge and perhaps a broken blade.

Short cuts and clearing cuts should always be made first; make the long cuts last. Rather than backing out along a cut, it is always best to cut out through waste material and then make a fresh attack. Cutting complicated patterns, long curves and the like should not be done all in one cut but rather by being broken up into segments, with as much waste material being cut out of the way as possible. On many occasions it is desirable to make one or more rough cuts to separate pieces of stock or to clear away large areas of waste material before going back and making the final finish and detail cuts. Sharp curves can be cut by making a series of tangent cuts, whittling away at the sharp curve until it has been completed.

Bandsaw blades are made in a continuous loop of thin steel; unlike circular saw blades, they do not come in a great variety of tooth patterns. The blades are available in different widths from ⅛-inch to 1½-inch. All of the standard blades for wood-cutting applications have teeth set alternately in opposite directions. The size of the teeth is related to the width of the blade—the narrower the blade, the finer the teeth. On most occasions, the saw blade is chosen on the basis of the minimum radius that can be cut by a given blade width. Where extremely tight radii must be cut, for instance, one might select a ⅛-inch blade, which has very fine teeth, that will cut to a minimum radius of ¼ inch. A ½-inch blade, on the other hand, is coarser and will cut to a minimum radius of only 1½ inches. Various lengths of blades are also available to fit different kinds of saws. In some of these lengths and brands, special blades having 15 teeth to the inch are available for cutting plywood. This allows a much smoother cut than the normal 6-tooth per inch kind of blade. For fast rough-cutting where a bit of splintering or a ragged edge is of less consequence than the time involved in making a smooth cut, either skip-tooth or hook-tooth blades with only 3 or 4 teeth to the inch can be used.

SQUARING

Squaring, or "squaring-up" as it is often called, can be one of the most troublesome and problematic details in working with plywood. There are several aspects of squaring-up to be considered.

The first is in insuring that the 90° angles laid out on a workpiece are actually perfect right angles, so that all of the sides of a square or rectangular piece, or any rectilinear shape, are in square. If they are not, the piece cannot be properly fitted into an assembly. The various angles and joints will end up out of line. Sometimes this is a matter of checking each right angle with a true carpenter's or try square to make sure that the layout line is correct, or that the cutting line did not deviate from the layout line during the cutting process. If the angles are off, then further trimming must be done. This can be a difficult proposition that often results in the loss of enough material that the workpiece may prove worthless, and a new piece has to be made up.

At other time, the necessity is to start off with a piece that is squared on two sides, so that when run through a table saw or radial saw the cut line will be square to the first two sides. Squaring up an odd-shaped or perhaps rough-cut and out-of-square piece of stock to convert it into a usable workpiece is not a difficult chore. With a straightedge, find one edge of the piece that is perfectly straight. If there are none, use a straightedge to establish a straight line and cut one edge with a handsaw or a portable circular saw and saw guide. Set the straight edge of the piece against the rip fence of a table saw or the rear guide of a radial arm saw. Cut a second, parallel edge at whatever point is appropriate. To finish squaring the piece off, just trim each end square to either of the two edges you have just cut.

Another occasion where squaring up is necessary is when you have completed rough-cutting a piece, or a series of pieces from a plywood panel, leaving slight margins that now must be trimmed off to the finish cutting guideline. This required some careful finish cutting, since the rough-cut edges may not be perfectly square with one another. On small pieces, the out-of-squareness may amount to so tiny a fraction of an inch as to be virtually unnoticeable and of no consequence. When cutting long or large pieces from a plywood panel, though a deviation of only a quarter-degree or so can result in such a difference in dimensions from one end to another of a workpiece as to unworkable. Therefore, accuracy is extremely important, particularly in such precise jobs as cabinetmaking.

Another circumstance applies to the cut edges themselves. The face of each cut edge, unless deliberately cut on a bevel, must be at exact right angles to the face and back of the panel. The edges must be square, in the exact meaning of the term. If they are not, the joints of the finished article will not go together well and probably will be out of skew. Squaring these edges, as well as trimming to the finish cutting guideline of the workpiece, are both done at the same time with either hand or power tools. Power tools almost invariably will give a better finished edge with much greater ease than will hand tools, but each type has its plane.

Sawing

The easiest way to make finish cuts and square the edges of the workpiece is with a power saw. To achieve a cut that needs no further work (decorative finishing excepted) and is ready for making a joint, the best bet is either a table saw or a radial arm saw. A portable circular saw will also work fairly well, provided it is used with a saw guide. The first step for all three types of saws is to make sure that the saw blade is set to an exact right angle to the saw table, or in the case of the portable circular saw to the saw shoe or base. The first two types of saws can be checked with a good try square, and the blade adjusted as necessary. After making an adjustment, make a short cut on a piece of scrap wood. Check the cut with the square to make sure that the adjustment is indeed correct. The portable circular saw does not have an adjustment. Make a cut on a scrap piece of wood and check the cut with a square. If the cut is tilted slightly but the saw shoe is locked at zero bevel, the shoe step is incorrectly positioned. The only recourse is to slightly bend the stop in one direction or the other (depending upon the direction of tilt) and make repeated trial cuts, checking each time with a try square until you achieve a cut that is perfectly vertical.

Once this is done, install the finest cutting blade that you have—a hollow-ground planer combination blade, so-called cabinet blade, or one of the several types of plywood blades is best—in the saw. Adjust the various controls and safety devices and make all of the necessary trim cuts right to the scribed layout lines. Be sure the cuts are on the correct sides of the lines. If you have done everything right, the result should be a workpiece with perfectly squared edges that require no further work, and with all dimensions right on the button.

Planing

Another method of trimming workpieces to the proper size and at the same time squaring up the edges is by *planing*. Using a hand plane for this purpose requires some skill. Since the tool is hand-held and there are no guides, it can easily get out of line. For planing short pieces, a *black plane* is best; for longer pieces, use a *bench plane*. A *smooth* or *jack plane* is the choice in most cases, but a *fore* or *jointer plane* is better yet for long edges (around 4 feet or more). Whatever the specific kind of plane, the blade must be extremely sharp for trimming plywood edges, or the result will be very poor indeed.

When planing a plywood edge, rest the forward part of the plane shoe squarely on the plywood edge, tilted neither to left nor right and with the blade not yet contacting the wood. A block plane can be held with one hand, but it is advisable to apply the forward-and-down cutting pressure with the right hand while holding the nose of the plane down firmly with the left (Fig. 4-17). Larger planes must be held with both hands. Start the plane forward with a long, even stroke, keeping it flat and aligned with the plywood edge at the same time. A considerable amount of downward pressure is needed along with the forward motion. Don't allow the plane to swing off to the right or left. The plane blade must be perfectly aligned in the slot in the shoe of the plane, and not cocked one way or another. The amount that the blade protrudes beyond the shoe determines the depth of cut, or the amount of wood that you will shave off. Several light cuts are generally easier to make and result in a more accurate edge than one heavy cut. A few trial runs on a piece of scrap stock will quickly help you to determine a good cut depth for the particular stock that you are working with.

As you move the plane forward, try to keep a relatively even pressure on the tool. Be sure not to tilt it to one side or the other. Do not, however, plane all the way across the edge and off the far end. Stop at the halfway point, or at any other point at least a plane-length or two (or more) away from the end. Reverse directions and plane back along the edge in the opposite way, meeting the first cut smoothly so that no transition shows. With plywood, you must always plane into an edge and never off an edge. Planing off an edge results in chipping and tearing away of portions of the plies at the corner of the workpiece, causing damage that is both difficult and bothersome to repair.

Planing a plywood edge with a power plane is much easier as far as physical effort is concerned, but it also requires a delicate

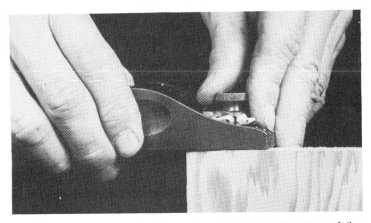

Fig. 4-17. Planing plywood edges with a block plane (courtesy of the American Plywood Association).

touch on the machine for good results. Power plane blades cut at very high speeds, and you can make a mistake before you know it. Also, the tool is heavy and a bit awkward. It must be guided carefully. The first step in using a power plane is to set the cutting blades to the required depth of cut. If you are roughing off fairly large amounts of material, a deep cut is fine. The tool will make heavy cuts with little effort. For finish trim cuts, a very light, shallow cut is advisable because this gives you greater control and makes it easier to accurately follow the guidelines. Then the guide must be set to an exact right angle to the plane shoe. It's a good idea to make a trial cut on a scrap of wood and then check the edge for squareness with a try square.

Set the front portion of the plane shoe on the edge of the workpiece. Snug the guide tight against the surface. Make sure that the plane is being held level with plenty of downward pressure on the tip, and is tilted neither to left nor to right. Turn the plane on, allow the blades to come to full speed, and run the plane along the edge. At the same time, keep a firm and steady downward pressure on the tool. As with the hand plane, do not plane off the end of the piece. Rather, reverse the plane (or the workpiece) and plane back in the opposite direction. With a power plane it is usually best to make a single cut in both directions. Make a second cut in both directions rather than making several passes in one direction and then making several more in the other. Generally a smoother edge will result with less effort.

A power plane throws a mighty cloud of shavings, chips and dust. A shower of heavier pieces may be thrown as much as 10 feet from the chute, depending upon the design of the plane. Using this tool outdoors in an area where a few shavings will make no difference is a good idea when possible. Otherwise, position yourself so that the debris does not plaster your workbenches, bury a partly completed project or sail into the living room.

Filing

A perfectly acceptable way to smooth down plywood edges, trim them to the dimensional marks and square them up all at the same time is with a wood *rasp*. This takes a bit of time and elbow grease, but the results are fine. The process is easily controlled and you can achieve good accuracy. A file is particularly valuable for smoothing out minor irregularities, humps and bumps, splintered corners and the like. If you have a fair amount of material to remove before you reach the guidelines of the workpiece, use a flat rasp which is designed for cutting away material. As you get close to the line, switch to a cabinet file, which is designed for smoothing and will remove only small amount of material per cut.

Filing wood is quite simple once you get the hang of it. For edges, a flat file or the flat side of a half-round file is used. Be sure to fit the file with a handle, lest the tang jams into your wrist (both painful and dangerous). Hold the file handle with one hand and either press down on the top of the file blade with the fingers of the other hand, or hold the tip of the file firmly. Place the file flat on the edge of the workpiece, not in line with the edge but cocked at a slight angle, so that the handle overhangs one side of the workpiece and the tip the other (Fig. 4-18). Never file across the edge, from front to back, as splintering of the surface ply on the far side is sure to result; keep the angle of the file to the workpiece surface acute. Run the file along the edge with steady and moderate downward and forward pressure in a long, smooth sweep. At the end of the cut, lift the file clear of the edge. Do not drag it back along the edge surface. All of the cutting is done on the forward stroke and dragging a backstroke merely roughens the surface. If the file tilts and catches, it may also leave a gouge on the corner of the edge.

The speed with which the file teeth will clog depends upon the material being filed, but this is not much of a problem with plywood. Even so, inspect the teeth frequently to make sure that they are staying clear. If they clog, the cutting action will slow markedly. A smart slapping of the file against your pantleg or the

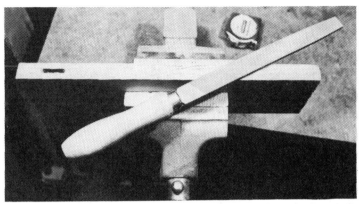

Fig. 4-18. Filing is a good way to smooth and square plywood edges; note angle of file to wood.

palm of your hand may clean out the dust. When it does not, resort to the file card. As with planes, files should always be used from the end of an edge inward and never outward across a corner. Splintering is bound to occur. As you file, be sure to continuously check the edge for squareness. Until you get close to the guideline, an occasional "eyeballing" will generally suffice, usually you can readily see a tilt of the edge surface merely by sighting along it. As you reach the guideline where squareness becomes critical, pause often to double check with a try square.

Filing an edge with one of the cheese-grater or open files or plane-type smoothers is done in very much the same way. Most of these are relatively coarse, however, and may not cut as smooth an edge as you need. One problem that arises when using these tools on plywood edges is that they can catch in the tough end-grain of the cross plies, and at the same time they tear gouges or splinters out of the opposing plies. The tool may file large quantities away from the soft wood portions of the lengthwise plies, but only small amounts from the much harder end-grain plies, leaving a ridged effect. When either of these circumstances occur, there is a natural tendency for the file to tilt or twist slightly, which can result in rounded, splintered or gouged corners along one or both sides of the edge. Since there are numerous different cutting patterns used with this type of tool, the best bet is to try those that you have to see how they work. Once again, use cheese-grater files only from the end of an edge inward and never the other way around.

There is another method of squaring and smoothing edges with a file that is sometimes used, and it is a rather effective one.

By chucking a rotary file or a rotary rasp in a drill press, or in a portable electric drill secured in a drill stand, you can achieve smooth, square edges of plywood workpieces by running the edges against the rapidly revolving rasp. This is most effectively done when the workpiece can be guided along a fixed guide or fence and is also an excellent way to square inside edges.

Sanding

One excellent way—perhaps the best for a modestly-equipped home workshop—to achieve perfectly-squared and extra-smooth edges taken down precisely to the dimensional guidelines of the workpiece is by sanding. This can be done by hand, but for a number of reasons power sanding is a much better bet.

Hand-sanding to trim and true up workpiece edges must be carefully done to avoid rounding of the corners, crowning of the edge, or sanding the edge off to a tilt, bevel, or a wavy configuration. A rubber sanding block that is relatively hard can be used to hold the paper. A sanding block or plane that has a cushioned sole, however, is not firm enough. Rounded corners are sure to result because the workpiece is narrow. The soft sandpaper surface will tend to curl down over the edge corners. The best bet is a perfectly flat piece of scrap wood of the same length as a sheet of sandpaper. Wrap the paper around the block and sand with long, even, smooth strokes, keeping the block at right angles to the workpiece surfaces. Where a good deal of material must be removed and/or the edge is rough and ragged, you can start with a fairly coarse grit of paper. Progress to finer paper as you approach the dimensional guideline. If not much stock must be removed and you are close to the line at the outset, start off with a fine grit. In both cases, check your progress frequently by eye and with a try square.

Power sanding is the best way to get the job done, for reasons of both accuracy and control, as well as the elimination of the problems that go along with hand-sanding. A pad sander will not do the job, for the same reason that the cushioned-soled sanding block will not—the corners along both sides of the edge will be rounded over. The best tool for the purpose is a stationary, bench-mounted disc sander, belt sander or combination disc-belt sander. Another possibility, reasonably workable, is a portable belt sander locked into a specially made finishing stand, which in effect converts the tool to a small bench-mounted sander. Other alternatives include a sanding wheel mounted in a table saw, a sanding drum mounted in a radial arm saw or a drill press, a sanding disc mounted on a wood-turning lathe, or a special lathe sanding table that can be fitted to a wood-turning lathe. All of these devices are used in approximately the same way and provide about the same results, although obviously some are easier to set up and work with than others. A large stationary belt-disc sander, for instance, has the greatest sanding surface area, the greatest versatility, and is always ready to go at the flick of a switch.

The key to achieving fine edges with these sanding tools is two-fold. First, the table, guide or fence must be set exactly at right angles to the sanding surface (or at any other appropriate angle if a bevel is to be sanded). Second, the sanding surface must be of an appropriate grit for the job at hand. In most cases this is a relatively fine grit (or may be quite fine) to produce a flat, smooth edge that needs no further attention. The effectiveness of the remainder of the process depends upon the skill of the operator. The workpiece must be correctly lined up, fed across the sanding surface carefully and smoothly, and the correct amount of material must be removed. Just a little practice will enable the average worker to quickly gain confidence in turning out perfectly formed, smooth edges.

Jointing

There is another method for squaring up and trimming edges of plywood—probably the best one of all—that you can employ to make perfect edges every time with practically no effort at all. Use a jointer or jointer-planer. This stationary power tool is fairly expensive and not found in too many home shops. It is an ideal piece of equipment for anyone who will be doing a considerable amount of cabinetwork or any other projects that require substantial edge finishing. The tool is capable of performing a great many precise operations, including squaring one or all four edges of a workpiece, surfacing, rabbeting, beveling, chamfering and cutting long, short or stop tapers.

The jointer does a particularly fine job of squaring, trimming and truing plywood edges. Basically, the job is done simply by adjusting the cutter edge to the desired height above the table (depending upon how much stock is to be taken off the edge of the workpiece), turning the machine on and feeding the piece through the machine. A bit of time is needed to gain the skill that enables the operator to turn out a perfect edge every time, but this really is no problem. Some practice on scraps will turn the trick. As with the other methods of working plywood edges, one of the most critical parts of the operation is avoiding splintering at the workpiece corners.

CUTTING SCROLLS AND CURVES

Brief mention has already been made earlier of the equipment that will cut curves in plywood. Plywood is often used in making curved pieces and scrollwork. The large size of the stock lends itself to making long, unbroken curves. The techniques involved deserve a bit of further attention.

Laying Out Curved Lines

There are a number of procedures that can be used in laying out curved lines. One simple method where the radius is fixed along the curved line is to employ circular objects that can be traced around in a series of interconnected curves or, wherever necessary, joined by straight lines. These objects can be anything from tin cans to pot lids to dinner plates or whatever else is handy. Circles and arcs can be easily laid out with a draftsman's compass. Where free-flowing curved lines of no fixed radii are required, relatively small patterns can be laid out with the use of a set of draftsman's templates known as *French curves*. This can be done directly on the workpiece but more often is first traced out on heavy paper. Thus, after the pattern is adjusted and corrected as necessary, it can be cut out and used to trace around.

Large and/or long curved patterns are often drawn to scale in small size and then transferred to a full-size sheet of paper that is marked off in the form of a graph. The line measurements from the small drawing are transposed, scaled out and spotted at the graph-line intersections. Then these points are connected by a free-flowing line that represents the necessary curve. If the paper is relatively stiff, it can be cut to form a pattern or template. If not, it must be transferred again to a piece of stiff paper or heavy cardboard to make up such a template. This can then be traced around directly on the workpiece surface.

Once the layout is made, making the cuts depends upon the kind and complexity of curves involved as well as the tools available in the shop. Smaller workpieces can be effectively cut with a coping saw, a keyhole saw or a compass saw. The trick in making cuts with any of these saws is to proceed slowly, keep a sharp eye on the cutting line, and try to maintain the saw blade position exactly at right angles to the surfaces of the workpiece so that the edge will be in square as much as possible. It is necessary to do all of the cutting outside of the cutting guideline—on the scrap side of the line. A good margin should be left so that the edge can be trimmed, squared up and smoothed without encroaching upon the workpiece material. Exactly how much should be left is up to the worker; the more material you leave, the more you will have to remove during the finishing operation. If you allow too little margin and then inadvertently tilt your cut to a bevel or gouge into the workpiece area, you will have repairs to make.

Sawing Operations

Curved cuts and fairly open scrollwork can be cut with a saber saw with very little difficulty and on pieces of practically any size. The cutting can be done fairly close to the cutting guideline, leaving only a minimal amount to be taken off during the finishing process. The saw blade can be kept straight up and down and the cut direction is controllable with relative ease. Remember that with this kind of saw, you should be cutting with the backside of the workpiece facing up. Use as fine a blade as seems to work well. Make sure that thin plywood stock is securely anchored to a work surface to prevent chattering. Make your cuts into angled corners to true them up. Do not cut off the end of the plywood; rather, before you come to an edge, stop the saw and reverse the cut, starting in from the edge to meet the previous cut. This will prevent splintering and chipping of the workpiece corners. Much the same situation occurs when using a *scroll saw*. The principal difference is that the scroll saw is capable of doing more intricate work and is easier to control.

A stationary jigsaw does a marvelous job of cutting intricate small shapes. When the cuts are made with a fine-tooth blade, the amount of edge finishing left to do is minimal. Cutting with this saw requires a gentle touch and a bit of practice to acquire some skill. The cut cannot be forced. A bandsaw is likewise an excellent stationary power tool to use for cutting curves and also leaves a good edge that requires little further work. Cutting can be done quite close to the cutting guideline. Where several identical pieces must be cut, they can be stacked and run through the blade all at once. This assures fine uniformity. With the right sort of jigs, perfect circles can be cut quite handily. As with most sawing operations in plywood, care must be taken when sawing off the end of the workpiece that splintering of the corners does not occur. Sometimes it is not possible to conveniently reverse the cut, but this should be done wherever feasible.

Though under most circumstances a broad, straight saw blade is by no means well-adapted to cutting curves, it is indeed possible to cut them with a portable circular saw. The curved lines must be long and gentle. The workpiece—often a full panel of plywood—must be fully supported so that the panel stays perfectly flat. Very gentle curves can be cut with the standard size of blade for the particular saw being used. However, better results can be obtained by using a small-bladed saw (5½-inch) or by putting a smaller blade in a larger saw (a 6½-inch blade in a 7¼-inch saw). Then set the saw blade so that the teeth will just go through the material. Guide the saw firmly but gently and slowly along the curved cutting guideline, somewhat outside the line. Under no circumstances should the saw be forced into the cut. Note that when making this kind of cut, the kerf will be quite wide so a substantial allowance must be made between the pieces being cut from a panel.

Squaring and trimming of the edges of curved or scrolled pieces is done in much the same fashion as for straight cuts. Most of the same tools and equipment can be used, but depending upon the curvature of the lines some will work better than others. A stationary belt sander, for instance, is excellent for smoothing curved parts and can also be used to shape curves in some cases. This is done by sanding at the end of the sander where the belt goes over the drum or wheel. Likewise, rotary files, rasps and sanding drums do a fine job. Whereas the work must be done freehand on a belt sander, as rule, with a drum or rotary rasp at least there is a table upon which the work can be firmly positioned.

Another bench-mounted power tool that does an excellent job of sanding curved workpieces is the small belt sander of the type often used by modelmakers and hobbyists. It uses a belt only 1 inch wide. Even better is the kind of small bandsaw on which the saw blade can be removed and replaced with a special sanding belt only ½ inch wide. Either of these tools will allow easy sanding and edge-squaring of all but the smallest and most intricate curved and scrollwork parts.

EDGE FINISHING

The edge trimming and squaring operations that we have just been discussing are a form of *edge finishing*. Generally, however, the term edge finishing is taken to mean some sort of processing other than the normal trimming and squaring necessary for making joints or clean edges. Rather, it connotes some sort of change in the form of the edge from the normal rectilinear pattern, or the addition of an edging material to accomplish the same purpose, or simply to hide the contrasting ply bands that are characteristic of plywood edges.

Easing

The simplest of all operations on a plywood edge is *easing*. We will assume that the edge has been properly squared and that the surface of the edge is perfectly smooth, or as much so as need be for its particular purpose. This process has left corners that are probably very sharp all along both sides of the edge. These corners will easily nick or gouge and, depending upon the wood species, may splinter easily as well. In addition, a sharp corner will not take paint—there is nothing there for it to cling to. To take care of these various problems, the corners should be eased by gently rounding them just a bit (Fig. 4-19).

If the wood is soft, fine sandpaper and light finger pressure is all that is needed. Sand straight back and forth, but at varying angles to the corner to achieve the rounded shape. The sanding action is really quite slight and takes off only a small amount of material. A heavily-eased corner, for instance, would be curved to a radius of perhaps 1/64 inch.

If the wood is hard, you can use a small, fine-toothed cabinet file first and follow up with fine sandpaper. Incidentally, an old piece of sandpaper that has been well-used for other purposes and is now soft, flexible and rather dull is often ideal for this purpose. The easing process is never used on edges that will be formed into

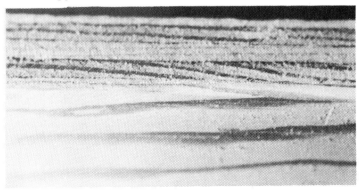

Fig. 4-19. Edge corner on the left is square and sharp, just as it left the saw. To the right, the edge has been eased or slightly rounded.

a joint, or when the edge will be covered with a band or molding. It is usually done only to exposed edges that will later be coated with a finish.

Chamfering With Hand Tools

A *chamfer* is made by cutting away on an angle part of the edge of the workpiece. In effect, the edge corner is lopped off. The cut, however, does not extend from the top surface to the bottom surface of the workpiece—only for a portion of the thickness of the stock (Fig. 4-20). Generally the corners that are made by chamfering—one along the top surface of the workpiece and another along the surface of edge—are left angular and are not rounded off. Sometimes it is desirable to ease them just a bit, particularly if the wood fibers right at the apex of the corner have a tendency toward splintering or feathering. The width of the chamfer is variable and can be made however wide the worker desires and the thickness of the stock allows. Sometimes with plywood, depending upon the particular type of panel, it is well to calculate the chamfer so that its lower corner, the one along the surface of the panel edge, is coincidental with the glueline between two plies. This reduces the incidence of splintering or chipping.

Chamfering can be done with hand tools, though a bit of practice is needed in order to get a perfect line. The easiest way is with a hand plane, and the block plane is the one to use. The blade must be extra-sharp and set to a very fine, shallow cut. Whereas the plane is pointed straight ahead for straight and square planing of edges, for chamfering it must be held in a paring position (similar to filing an edge), with the body of the plane both tilted to the angle of the chamfer and also cocked about 15 or 20° off to the outside of the edge. To avoid corner damage, always plane in from the corners and never outward across them.

You will also note that planing in one direction along the workpiece edge may result in a much better cut than planing in the opposite direction. This depends upon the angle of the wood grain to the plane blade. Planing into a grain angle will result in splintering and tearing of the wood, while planing with the grain angle will result in a smooth cut. There may be some feathering that must be trimmed off with fine sandpaper. When this situation arises, you will have no difficulty in determining which direction is the best for getting a good cut.

A chamfer can also be cut by using a cabinet file or one of the finer varieties of cheese-grater files. The difficulty with this pro-

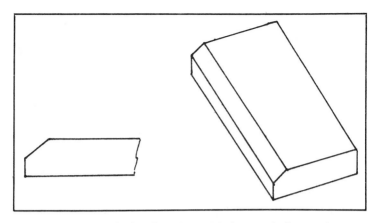

Fig. 4-20. A chamfered edge. Part of the original squared edge remains.

cess is maintaining sharp corners on the chamfer and true, straight chamfer lines. However, it can be done with a little practice and is a particularly useful process for small workpieces, woods that have a touchy grain pattern, and where only a small chamfer is required.

Chamfering With Power Tools

The easiest way to make chamfers is with power tools. With a table saw, for instance, all you have to do is put on a fine-toothed, hollow-ground blade. Then tilt the arbor to the desired degree. Line the workpiece up on the saw table so that the saw only cuts a corner off the edge. Run the piece through, guided by the rip fence (Fig. 4-21). Careful smoothing with a hard-faced sanding block, or against a stationary sander wheel or drum, will finish the job up nicely. You can perform the same task with a radial arm saw, though often the workpiece must be securely jigged and clamped on the table. Relatively small parts can be quickly chamfered on a stationery belt-disc sander by simply setting and guiding the piece past the sandpaper at the proper angle, making as many repeated passes as are necessary. One difficulty with this process, however, is that making matching chamfers on two or more pieces is difficult, unless some sort of jig is made up that will assure uniformity.

A portable router also does an excellent job of chamfering, especially when it is more convenient to take the tool to the workpiece than the other way around. By setting the router to the correct initial chamfer cut and then either leaving it that way or making a note of the setting, you can make endless chamfers on dozens of pieces that will all be perfectly identical. Incidentally,

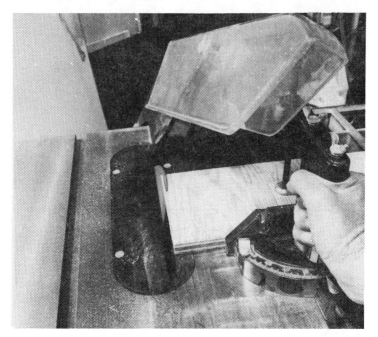

Fig. 4-21. If the stock is relatively thick, a chamfer can be cut with a table saw, as shown here. The saw guard has been lifted for photographic reasons.

the router does a particularly good job of chamfering plywood with minimal splintering or grain tearing (Fig. 4-22). If much of this kind of work will be done, we suggest that you use a carbide-tipped router bit since plywood can be hard on regular steel bits. One possible drawback to chamfering with a router is the angle of the chamfer is determined by the shape of the router bit. It is a fixed angle and only a few varieties of bits are available. On the other hand, chamfering on a table saw allows a wide variety of chamfer angles.

To chamfer with a router, chuck a V-groove or 45° chamfering bit in the machine. Set the bit depth to match the width of chamfer that you desire, and make a trial cut on a piece of scrap wood. Make whatever adjustments are necessary until you get just the cut you want. Then move to the workpiece and make the finish cuts, using whatever guides are necessary to maintain a true edge. Be extremely cautious at the corners of the workpiece, as splintering can very easily take place here. Some workers prefer to begin the cut ½ inch or so in from each corner, and then trim the corners by hand. In some cases splintering is not a problem. This can be determined by making a cut or two on a piece of scrap of the same material as the workpiece.

There are two other stationary power tools that can also be used to make perfect, uniform chamfers. Few home workshops are fortunate enough to have one or the other of these tools, let alone both, but they deserve mention. One is a jointer, or jointer-planer, and the other is a shaper. Chamfering on a jointer is done simply by positioning the adjustable fence, setting the cutter head to the proper height and passing the workpiece through the machine. Much the same can be accomplished with a power plane, only in this case the machine is passed over the workpiece and must be guided by hand. Chamfering with a shaper involves selecting the proper shaper bits, installing them and adjusting the cut and the fence, and running the workpiece past the blades.

Beveling

A *bevel* is similar to a chamfer in that the edge of the workpiece is cut off on an angle, but in this case the cut goes entirely through the stock from top surface to bottom surface, leaving the entire edge cut at an angle (Fig. 4-23). Almost invariably this job is performed with a saw. While exposed workpiece edges are sometimes purposely bevel-cut, as often as not this is part of the process of forming a joint between two pieces.

Bevels can be cut with a handsaw, but it is a very difficult job to maintain a true angle and a plane edge surface for the full length of a cut. The only practical way to bevel is with a power saw. With a portable circular saw, this merely involves tilting the shoe of the saw to the desired angle. The blade must then be run out a bit

Fig. 4-22. Using a router to make a chamfer; the angle and size of chamfer are determined by the kind of bit used.

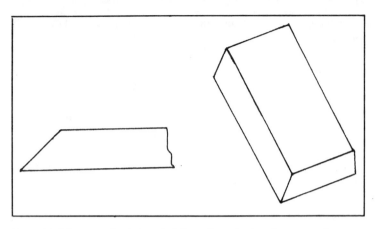

Fig. 4-23. When an edge is beveled, the entire edge is cut on an angle.

further than is necessary on a straight cut because of the diagonal involved. Note, too, that the adjustment on a portable circular saw is not all that exact. If you need a precise angle for your bevel cut, set the saw and make a trial cut on a piece of scrap. Measure the angle that results with a protractor, and make further adjustments to the saw shoe as necessary until you achieve the correct angle. Since the saw tilt can only be adjusted in one direction, you must also double check to be sure that you are positioning the cut correctly as you start. It's too easy to end up with an outside bevel when what you really wanted is an inside bevel.

Making bevel cuts on a table saw or a radial arm saw is a simple chore. On good equipment, you can be sure that the angle you set the saw arbor to will be precise. All you have to do is dial in the correct angle and go to work. Again, be sure to set the blade height properly and to make sure that you are cutting the bevel in the right direction.

Rounding

Sometimes it is desirable to round a plywood edge, such as on a table top. The two standard rounds are quarter-round, where the radius of the curve along the edge is equal to the thickness of the stock, and half-round, where the radius of the curve is equal to half the thickness of the stock (Fig. 4-24). In the former case, the top corner of the edge is rounded off and the bottom corner remains square, and in the latter case both top and bottom corners are fully rounded. Any other degree of rounding that is desired can be chosen as well.

There are a number of methods of rounding. When done with hand tools, the first step is to cut a series of accurate chamfers—the number and width of the chamfers depends upon the degree of rounding desired—with a file or hand plane. Then the corners of the chamfers are carefully rounded with a file and/or sanding block to achieve the final rounded configuration.

Rounding may also be done with a combination of hand tools and power tools. This process consists of making a series of chamfers with a router, jointer, circular saw or belt sander. Then these chambers may be cut across with more, smaller chamfers that take the corners off the original ones. Eventually, the final curvature is made with cabinet files and/or sandpaper.

The easiest way to round an edge is with either a router or a shaper. Routing rounded edges is done by chucking a bead-and-quarter round bit or an edge-rounding bit into the machine and proceeding in the same manner as when cutting a chamfer. These bits will not cut a half-round edge, but by cutting a quarter-round edge first on one edge corner and then on the other you can accomplish a fully half-round edge. Generally, the meeting point of the two cuts is a bit obvious or possibly a little ragged, so further touching up with sandpaper is necessary.

Making rounding cuts with a shaper is the simplest way to go about the job, since the entire cut can be made in one pass. The configuration of the cut can be anything that is desired. Quarter-round and half-round edges can be formed, as well as various degrees of edge convexity or whatever rounding shape is desired. If commercial cutters are not readily available in the configuration

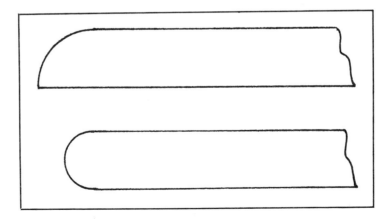

Fig. 4-24. Top, quarter-round edge. Bottom, half-round edge.

needed, they can even be custom-ground to suit. Then all that is necessary is to mount the cutters in the shaper spindle, make the various adjustments to the machine and pass the workpiece by the cutters. A good shaper will do an excellent job at rounding plywood edges. As always, great care must be taken at the corners of the workpiece to avoid splintering and chipping. If a blade happens to catch just right, it can actually tear a sizable chunk of ply right out of the piece.

There is an alternative to a shaper, not quite as good but perfectly workable, that is ideal for occasional use in a small home shop that has only limited equipment. If you have either a table saw or a radial arm saw, you can obtain a molding head that can be mounted in place of the saw blade and used in combination with an appropriate safety guard. The molding head is fitted with appropriate cutter bits, allowing you to use the saw for rounding edges in much the same fashion as a shaper is used.

Shaping

The process of rounding the edges of plywood (or any other stock for that matter) is actually a *shaping* process. But the term is generally taken to mean forming or molding the edge of a workpiece to some configuration other than simple rounding. Shaping can involve cuts ranging from quite simple to very intricate. Edge shaping is most easily accomplished on a stationary shaper. Perhaps the next best tool, though not quite as utilitarian, is the portable router. This is the machine to use when the tool must be brought to the workpiece rather than vice versa. A router could be used, for instance, to shape the edge of a table after the assembly has been completed and the table top installed. Obviously this could not be done with a shaper.

Edge shaping can be done with a molding head installed on a table saw or radial arm saw mandrel. It also can be done to a limited extent by chucking the same three-lip cutters that are used on a stationary shaper in a drill press, provided that spindle speeds of at least 5,000 rpm can be achieved. A special shaper adapter must be used, and never the same chuck that is used for drilling. Jigs, guards and guides must also be arranged.

There are many kinds of cuts that can be made with stock shaper or molding head cutter blades. The greatest variety is available for the stationary shaper, including such patterns as the simple *ogee*, ogee and bead table edge, door lip, various kinds of flutes, beads and radii, sash patterns, and cove and bead combinations. Some of these cutter patterns are decorative only, while others have more practical applications such as the glue joint, tongue and groove, and panel cupboard door patterns, all of which are used to quickly and easily cut certain kinds of joint configurations. Special shaper cutters can also be made to suit particular purposes. Regardless of the cutter patterns or the tools used to do the work, shaping plywood edges is done in much the same manner as rounding the edges. Once the machinery is properly set up and the workpiece jigged, clamped or guided as need be, the edge is simply passed through the cutter blades and the job is done. Remember to take extra care at corners to avoid the possibility of splintering and chipping.

DRILLING

Drilling a hole through plywood would seem to be a perfectly straightforward and simple exercise. It is if you know the tricks of the trade.

The first step in drilling holes in plywood or any other material is locating the exact center of the hole-to-be at the proper location on the workpiece. Obvious as that may sound, it is astounding how many times it doesn't happen. The next step is to make a starter hole. This is a good idea regardless of the type of plywood you are drilling or the type of bit that you are using. By making a small indentation with a sharp scratch awl at the centerpoint of the hole, you are assured of two things. The bit point will be accurately placed when you start to drill, and the bit will center itself in the starter hole sufficiently that it will not "walk" across the workpiece, marring the surface. This happens more often with a hardwood, but it can happen as well with soft wood. From this point onward, how you proceed depends upon the kind of hole you are drilling and the kind of tools you are using.

Using Speed and Twist Drill Bits

Perhaps the most commonly used combination in the home workshop for drilling holes is an electric drill and either a speed bit or twist drill bit. A speed bit, as the name implies, will cut through a plywood panel very rapidly. It will also leave a very ragged edge to the hole, particularly as it passes from ply to ply. As it bursts through the backside of the panel, chips and chunks will fly in all directions. A great deal of splintering will take place on the back surface. Speed bits should not be used in plywood for holes whose sides must be clean and accurately cut (such as for dowels); nor should they in any case be rammed through the wood with maximum power and push. Better results are obtained by drilling slowly at a fairly moderate bit speed, keeping the drill squarely aligned to the hole. Tilting in any direction will result in a hole that is far from round, especially in plywood. Speed bits are next to useless for drilling accurate holes in plywood on an angle. They walk around, rip and tear, catch and generally leave a mess.

To avoid splintering on the backside of the hole, clamp a piece of scrap wood over the back surface of the workpiece. This will prevent most of the trouble, though some splintering will probably take place anyway, depending upon the type of plywood and the size of the bit (and its sharpness). As you near the last ply of the panel, ease off on the pressure so that the cutting pace shows markedly. Another method, usually both easier and more effective, is to watch the drilling process carefully. As soon as the point of the bit begins to break through the backside of the material, stop drilling and remove the bit. Turn the workpiece around and use the small point-hole as a starter hole. Drill back through in the opposite direction. Seldom will the two bores align perfectly, but in the relatively rough kind of hole that this type of bit drills anyway this is usually of small consequence. In any event, a rattail wood rasp will clean up the sides of the holes.

Drilling holes in plywood with twist drill bits is less problematical in many respects. The hole sides will be fairly smooth, depending upon the wood species, and will usually be drilled reasonably true. Splintering on the backside can still be a problem, especially with the larger sizes of bits. The only recourse to prevent splintering is to clamp a piece of scrap wood to the backside of the workpiece; by the time the point of a twist drill bit comes through the wood, the hole is drilled.

It is also possible to drill at an angle with this kind of bit. The best way is to use a special jig that will hold the drill and bit steady in the desired position. The bit should be fed into the material slowly until it gets a good "bite," and then drilling can proceed at a normal pace. Extra caution must be exercised with small bits so that they do not deflect from the starting point, or possibly break. If the drilling is being done without the aid of a guide, the easiest method is to first drill straight into the material for just a short distance. Then, with the drill still running, begin to tilt the drill toward the desired hole angle. As the drill sinks further into solid material, pick up on the drilling speed. The edges of the hole at the surface will be a bit ragged, but otherwise the procedure works well. The larger the bit diameter, though, the more difficult it is to do a decent job.

Boring With Bitstock and Auger Bits

A good method of boring holes in plywood, especially if you have only a few to do, is with the traditional *bitstock* and *auger bits*. The double-twist bit generally does the best job on plywood, since there is less opportunity for large chunks of ply to lift away and tear. The hole sides will be relatively smooth, and fairly precise holes can be bored. Again, the major problem is with splintering on the back surface of the workpiece. To prevent this, use either of the methods suggested for speed bits. It is also quite easy to get a hole out of alignment when using a bitstock, probably because of the circular sweeping motion involved in operating the tool. If you are not sure if the stock is properly square to the workpiece so that you are drilling a perpendicular hole, use a small try square periodically to check the bit position.

You can drill holes up to 3 inches in diameter with an expansive bit chucked in a bitstock, but this must be done with caution in plywood. The cutting blade "goes hard" and has a tendency to catch in the gluelines, tear the grain or both. The larger the hole, the more noticeable this is. Sometimes the bit must be "rocked," by tilting the bitstock back and forth in various directions in order to make the blade cut properly. Of course, the blade must be as sharp as possible. Tearing and splintering on the backside of the workpiece is a definite likelihood; as the bit point comes through the workpiece, reverse directions and bore in from the opposite side.

Using a Hole Saw

The best way to cut sizable holes through plywood is with a hole saw. The specific type that you use doesn't matter much. If you expect to be drilling numerous holes, be sure to buy good quality blades. If hole-sawing will be a routine matter, consider investing in carbide-tipped blades. Plywood is touch on ordinary saw blades, and you will get much better blade life with carbide.

A small starter hole should be made with a punch or scratch awl before you begin sawing the hole, just as for drilling ordinary holes. This will center the pilot bit of the saw and get you off to a proper start. A hole saw can be chucked in a hand-held electric drill—the ⅜-inch size (or ½-inch) drill is best for this purpose—or in a drill press. When using a electric drill, be sure to get the pilot bit started on a perpendicular line. As the teeth of the hole saw approach the wood, make whatever minor adjustments are needed so that all of the teeth contact the surface at about the same time. This will happen automatically, of course, if the hole saw is being used in a drill press. The rim of the hole on the surface of the workpiece will be clean and smooth; however, the one at the backside will be a bit rough and ragged, though usually not much splintering will occur. If this is of not consequence, simply bore right through the stock. If the rim of the hole must be trim on the reverse side, clamp a piece of scrap stock to the backside of the workpiece. Or, as soon as the pilot bit breaks through the reverse surface, remove the saw and bore in from the other side. This must be done with care in order for the two holes to properly meet. Often as not they will miss by a bit, and the ring that is left must be trued up with a file or other tool.

If you are hole-sawing with a drill press, a piece of scrap stock must always be placed under the workpiece. Otherwise you will saw into the table. Both pieces should be securely clamped, too, especially if small. Boring holes through extra-thick stock, or multiple piece for identically aligned holes can be done by sawing from one side to the full depth of the hole saw; then bore in from the opposte side to finish the cut. The starting point for the pilot bit on the backside must be exactly opposite that on the front surface.

Hole saws can also be used to cut large holes at an angle to the workpiece surface. The angle cannot be too sharp, or the blade will merely skid away. A good starter hole must be provided. The job is a tricky one and requires some practice to be able to drill clean and accurate holes. It is most easily done in a drill press, or with a device used in conjunction with an electric drill that will hold the drill and hole saw steady at the required angle.

Circle or Fly Cutter

There is another device that can be used for cutting large round holes, called a *circle cutter* or *fly cutter*. Several sizes are available, and it is possible to cut clean holes in plywood up to 6 ½ or 7 inches in diameter. This device is intended to be used in a drill press and must be operated at slow speeds. The bit must also be kept extremely sharp. The particular danger in using this kind of tool for cutting plywood is that the cutting bit, which rides on a bar and travels circularly around the pilot bit, will catch in the crossgrains or piles, causing splintering and grain tearing. The circle cutter should never be used in a hand-held electric drill because it is very difficult, if not impossible, to control. You will be unable to make a cut, and the whole arrangement can be extremely dangerous.

Drilling Blind Holes in Plywood

There are times when it is necessary to drill *blind holes* in plywood. A blind hole is one that does not pass all the way through the material, but stops at some point usually predetermined. The purpose might be for joint dowels or setting rails. Sometimes these holes must be made flat-bottomed which compounds the difficulties.

Drilling to a predetermined depth where there is plenty of stock left beyond the bottom of the hole, such as into a plywood edge, is not much of a problem. Any kind of bit can be used. As usual, the bit must be correctly aligned to drill a perpendicular hole. You can proceed by drilling briefly and then measuring the hole depth, but this takes time and is imprecise. A better solution is to measure the required distance back from the tip of the drill bit. Make a fine mark all the way around the circumference of the bit. This can be done with an ordinary pencil, but the line does not show up very well. A permanent-ink marking pen will do the job, and the line can be easily scrubbed off afterward. If the drill is of sufficient diameter, one simple trick is to put a small rubber band around the shank, set to the right position. Then all you have to do is drill until you reach the band on the bit. You don't have to bother with any of this monkey business if you are drilling with a drill press or with an electric drill in a drill stand, where the travel of the quill can be adjusted to stop at any point you desire. This setup will make holes of absolutely uniform depth every time.

Drilling blind holes into the surface of relatively thin stock is a bit more tricky, because you have to be very careful to not burst through the reverse side. Shallow holes can sometimes be drilled with any ordinary bit. A hole ¼ inch deep in ¾-inch plywood, for instance, would work fine. But if you want to drill a hole so that there is only ¼ inch of stock left between the bottom of the hole and the reverse surface, using an ordinary bit is chancy. What you need is a Forstner bit; models are available in numerous sizes and for either electric or hand drills. This bit will bore out a flat-bottomed hole, clean-sided and accurate, with no danger of punching through the reverse side of the stock.

Depending upon the kind of plywood being drilled and the bit being used, the holes often are not overly precise. They have somewhat ragged walls and may also tend to wander a bit. If precise and accurate holes are required, with clean walls, no out-of-roundness and no travel, use a brad point drill bit. These bits are especially made for the purpose and will usually cut as clean a hole as you could hope for. Best results are obtained by using the bit in a drill press at optimum spindle speed or with an electric drill held in a stand. These bits, however, can also be used with excellent results (as can some other types) in a wood-turning lathe.

Countersinking and Counterboring

One very common procedure in working with plywood wherever fastening must be done by flathead wood screws, and for a few other purposes as well, is *countersinking*. *Counterboring* is often necessary as well. Countersinking involves drilling a bevel-edged hole into the surface of the workpiece, into which the head of a screw can be recessed. Counterboring, on the other hand, may involve drilling a two-sized hole, often with a countersink at the midpoint.

Countersinking is an easy process—so simple that it is also easy to foul up. Countersinking is done with a special countersink bit. There are several sizes available for use with either electric or hand drills. The job is done simply by setting the bit into the opening of a hole already drilled and chamfering the hole sides. The problem is that the bit cuts quite fast, especially in an electric drill, and it is easy to get the countersink both too deep and of too large a diameter at the surface. Another difficulty is that unless the countersink is held perpendicular to the workpiece surface, the chamfer will be on an angle and the screw will not seat properly. The best way to countersink most kinds of plywoods is with an electric drill operating at a relatively high speed, preferably held in a stand or jig that will keep the drill perpendicular to the workpiece. Then ease the countersink bit into the material slowly, pausing frequently to check for proper depth and surface diameter.

Where the object is to countersink for a certain size of flathead screw, two somewhat different types of bits can be used to accomplish the job easily and neatly. One method is to use a special tapered drill, to which is fitted an adjustable *countersink* and a *stop collar*. Once the stop collar and the countersink are properly adjusted on the bit, the pilot hole for the screw and the countersink for the screw head can be drilled in one operation. The stop collar will stop the drill bit at precisely the correct depth. An alternative method is to use a countersink pilot bit. This too is a bit fitted with a countersink, and some have stop collars as well. This bit will drill, in one operation, a pilot hole for the screw threads, a counterbore for shank clearance of the screw, and a countersink for the screw head.

This is all well and good for screws of particular sizes, but often there is a need to counterbore for bolt heads or other purposes. The degree of accuracy required for the holes determines to some degree the kinds of bits that will be used. But for ordinary purposes, auger bits, speed bits and twist drill bits will do an adequate job. In counterboring, the larger hole closest to the workpiece surface is drilled first, followed by the second, smaller hole. Suppose you want to run a ¼-inch carriage bolt through the edge of a narrow strip of plywood, and recess the head into the wood. The first step would be to make the centering marks and starter hole. Then bore a hole of diameter just slightly larger than the diameter of the carriage bolt head, to a depth of just slightly more than the thickness of the head. Using the center hole left by the point of the large bit, center a ¼-inch bit and bore the rest of the way through the piece for the bolt shank.

Proper Alignment of Holes

One of the biggest problems in drilling holes in plywood or other materials, especially for beginners, is in proper alignment of the holes. In most cases a hole should be drilled perpendicularly to the workpiece surface (or sometimes at some specific predetermined angle). With practice this can be done by hand with a reasonable degree of accuracy. Where precision is required, though, or even just as a matter of course to ensure better results, the best method is always to fix the drill bit in the correct position relative to the workpiece. Then make sure that neither one can move out of alignment with the other. This generally is a matter of clamping or firmly holding the workpiece and using a drill press or an electric drill in a drill stand, rigging up some sort of jig or clamp for the drilling tool, or making frequent checks of the position of the drill bit with a try square.

Correct alignment is often extremely important when two pieces of stock must be joined surface-to-edge with fasteners, particularly if the workpiece edges are thin. Drilling into an edge of a plywood sheet poses two potential difficulties. One is that, depending upon the thickness of the panel, there is not much wood left on either side of the hole. If it is drilled at an angle, the hole may break through one surface or the other. The other difficulty is that when drilling into an edge, the drill tends to catch in the cross plies and also in the gluelines. It may wander or try to cant off at an angle. The smaller and more flexible the bit is, the more troublesome this can be. Wherever possible, it is best to calculate the hole size to fit within a single ply and/or to not drill into either the face or back ply. There is a danger, too, when the hole edges are too close to the surfaces of the plywood. When the screw or other fastener is inserted in the hole, it will force the laminations apart and cause bulges on the surfaces. Pilot and shank holes should always be correctly sized so that the screw fit is not so tight as to force the plies apart.

Aligning a pair of holes in the edges of two separate workpieces can be an even more difficult chore. This is necessary, for instance, when a joint is to be doweled for extra strength. If the holes do not line up perfectly, when the dowels are inserted one piece will be out of alignment with the other. Often as not, the most careful measurements, followed by the most careful hole drilling, will still not result in a perfect match. The starting point for correct alignment in such a situation lies in a set of small devices called *dowel centers*. These are available in sets containing standard sizes, typically ¼-inch, 5/16-inch, ⅜-inch and ½-inch. The second piece of equipment needed is called a *doweling jig*, a special tool that clamps to the edge of the workpiece and aligns the drill bit perfectly. A third item, not totally essential but certainly recommended, is a special *precision drill bit* made for boring dowel holes, which cuts rapidly, freely and accurately.

The procedure is to first determine the location of the first hole in workpiece number one, make a centering mark, and fit the doweling jig into place. Then drill the first hole. Remove the jig and press one of the dowel centers of appropriate size into the hole. Set both workpieces on a perfectly flat work surface, and align them carefully without touching the mating edges together. Clamp workpiece number two in place. Then push workpiece number one, and particularly the dowel center, against workpiece number two, while keeping the two pieces in perfect alignment. The dowel center point will mark the centering point of the hole to be drilled in workpiece number two; fix the doweling jig and drill the hole. If you have done the job correctly and carefully, the two pieces will align perfectly with the dowel in place.

MORTISING

Mortising is an operation that is frequently performed on plywood for a variety of purposes. One of the most common is in setting hinge leaves flush with the workpiece surface, inletting ring pulls and similar processes. Also, mortising is down for certain kind of joints and may also be done for decorative purposes, a process that is generally called *routing*.

This kind of work can be done by hand with ordinary wood chisels. This is the usual method used in the home workshop for recessing hinge leaves and other builder's hardware for a flush fit. The wood chisel must be as sharp as possible and used with care so as not to gouge out more wood than is necessary. The procedure is to make an accurate outline of the item to be inletted; if installation is along an edge, the depth of the cut can be marked as well. Then a series of chisel cuts spaced just a short distance apart, and at right angles to the direction of the face grain, are made within the borders of the outline. The chisel is used to make single connected cuts along the outline. The pieces between the cuts are then chipped out and the bottom of the mortise smoothed. This requires a light touch and some practice—a good eye helps, too—in making smooth, even mortises.

There are two things in particular to watch out for when mortising plywood with a wood chisel. One is that the grain does not tear or splinters lift out of the surface when cutting across the grain. The chisel must be sharp. The second is to not tear out chunks of wood outside of the cut area when removing stock with the grain. The tendency is for the shearing or splitting action to continue right on past the guidelines, and possibly beyond the cuts as well. There is a third potential problems, too, and that is when you run into the gluelines that separate the plies. These must be attacked with care if you expect to get a clean, smooth cut. The basics of chisel mortising are simple enough, and you can get the idea in a minute or two. But this is an instance in particular where only practice and experience will build the skill necesary to make fine, clean mortises. The same basic procedure, incidentally, is used for any kind of mortising in plywood or other woods.

Hinge and similar mortises can be made with consummate ease if you use a portable router. All that is necessary is to set the router bit to the proper depth—always make a trial cut and measure the depth to be sure that it is accurate—and follow the outline of the area to be cut away. The most precise method of doing this is by arranging guides for the router or setting templates made for the purpose. Operating the router freehand takes practice, and making smooth-sided cuts is a bit difficult at first. The router cuts very rapidly, and the machine can get away from you with surprising ease. Careful control and close attention are necessary.

The router, of course, can be used for other purposes as well. We have already discussed edge-shaping with a router; the same principle can be applied to cutting decorative patterns of all sorts in panel surfaces. Many different kinds of bits are available to allow you to straight-groove, vee-groove, vein, bead and dovetail. The router can be used in making certain kinds of joints. Rabbets are easily made, and a special fixture is available for routing dovetail joints. Decorative cuts can be done freehand or with templates or patterns. Different kinds of guides allow crafting circles and ellipses. You can also make signs, freehand carvings or *pantograph* carvings. Special templates, guides and bits are made particularly for routing out decorative doors and panels.

When using a router on plywood, always guide the router so that the bit cuts into the grain rather than away from it. When cutting straight edges, move the router from left to right. When making circular cuts, move the router in counterclockwise direction. As you move along the cutting guideline, watch carefully for any evidence of splintering from the top surface of the workpiece. If this occurs, try reversing directions by turning the workpiece. If the difficulty persists, stick a band of masking tape on the surface, right along the cutting guideline. Don't feed the router into the stock too rapidly, as this overloads the motor and causes excessive wear on the bits. On the other hand, too slow a feed will result in a poor cut.

For deep cuts, you will probably find it best to make a shallow cut first and then a second cut to finish up. Try to make the first cut depth equal to or less than the thickness of a single ply; then go into the next ply on the second cut. When starting a cut, hold the router above the workpiece, turn the machine on, and allow it to come to full speed. Never start the machine while the bit is resting on or against the wood as this will overload the machine. The sudden starting torque will throw the machine sideways and may cause damage. Once the machine is up to speed, lower the bit into the wood, well inside the cutting guidelines, and start the cut. When starting and stopping the router while it is still resting on the workpiece, make sure that the bit is completely clear of the wood. Always wear goggles while working with a router. You might want a dust mask as well, and hearing protectors are a good idea if you will be using the tool for more than a moment or two at a time.

Chapter 5
Plywood Joining,
Fastening And Finishing

The various tools and techniques used for cutting and shaping plywood (and other) parts are, of course, only part of the story. There is still a lot of work to be done in order to assemble a pile of cut pieces to complete a project. The parts must be properly joined and then secured in one fashion or another. Though proper dimensioning and cutting of the parts is extremely important to the quality and appearance, not to mention the effectiveness, of a completed project, the joining and fastening of the parts is even more so. The finishing process, irrespective of its specifics, will truly make or break a project no matter how carefully it has been fabricated. There is much to consider in all three areas.

JOINTS

There are many kinds of *joints* used in general woodworking when putting the pieces of an assembly together. Many of them are unsuitable for plywood itself, though they may come into play in projects that consist both of plywood and solid wood. Here we will consider only those that are particularly adaptable to plywood work. Since there are a number that may well be unfamiliar and perhaps confusing, the place to start is with a few definitions. Other terms will be explained during the discussion of making joints in plywood.

■ **Butt.** The end of a piece of material, across its narrow dimension, that is pushed up against another piece of stock. With plywood, the end of a piece is usually at right angles to the surface grain, but may upon occasions be parallel to the surface gain.

■ **Edge.** The side of a piece of stock, along its long dimension. With plywood the edge usually runs parallel with the surface grain, but this is not necessarily true.

■ **Splice.** A joining of two pieces of stock, usually end to end.

■ **Lap.** A piece of stock that extends over another piece of stock, crossing or overlapping in virtually any fashion.

■ **Miter.** An angle cut, or the oblique surface formed as the result of making an angle cut.

■ **Dowel.** A length of round wood stock, usually short, and fashioned of hardwood. It is often used to strengthen joints.

■ **Spline.** A thin strip of wood, metal, hardboard or plastic, tightly fitted into matching grooves in two mating pieces of stock, for added strength.

■ **Rabbet.** An L-shaped channel or groove cut out of the edge or end of a piece of stock.

■ **Dado.** A rectangular groove cut into and across the surface of a piece of stock.

■ **Mortise.** A hole, groove or slot in a piece of stock into which another piece fits. Also, the term is often used to signify the slot or other recess made into the surface of a piece of stock, into which another piece of material or builder's hardware is recessed or flush-mounted.

■ **Tenon.** A piece of stock that inserts into a mortise, with particular respect to a mortise-and-tenon joint.

Butt Joints

Butt joints, or *plain joints*, are perhaps the most widely used joints in woodworking and are particularly important in plywood work. Most butt joints are formed across the narrow ends of the workpieces and may be made flatwise or edgewise. The commonest of all of these joints is the *flat butt* (Fig. 5-1A), where an end is joined to an edge. Obviously, this is a weak joint. With the addition of a set of dowels to make a *dowel flat butt* or a *frame butt joint* (Fig. 5-1B), a considerable amount of strength can be gained. Similarly, a *corner butt joint* can be made without any added reinforcement. By inserting dowels to make a dowel corner butt joint (Fig. 5-1C), the assembly is greatly strengthened.

A similar joint where the pieces are joined at some point other than the end of one workpiece is the *middle-rail butt joint*, which can also be either an *edge joint* or a *flat joint* (end to edge or end to surface). This joint should also be made with dowels (Fig. 5-1D) for the extra strength provided. Note that the doweled joints can be made with screws or nails instead of dowels, in many cases. However, a joint made in this fashion is not as strong as a dowel joint and also has the drawback that the heads of the fasteners must be hidden.

When making doweled joints of this sort in plywood (and they are quite effective), great care must be taken to align the dowel holes perfectly with the use of dowel centers and a doweling jig. The holes must be carefully drilled. This is particularly true of decorative plywoods that have thin face veneers. Trying to correct a mismatched joint by sanding the surfaces to an even plane may

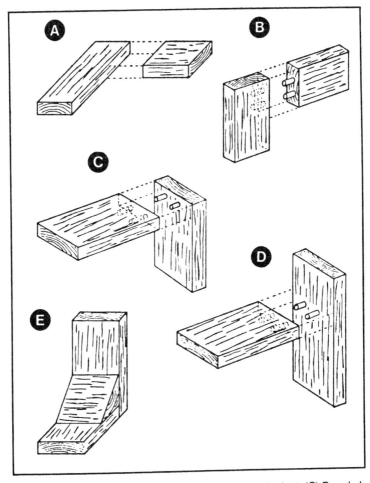

Fig. 5-1. Butt joints. (A) Flat butt. (B) Doweled or frame flat butt. (C) Doweled corner butt. (D) Doweled middle rail butt. (E) Glue block corner butt.

result in either an unsightly bump on the surface or removal of a portion of the face veneer. Also, extra care must be taken with multi-ply plywoods to insure that the holes do not wander as the drill bit passes through the gluelines. In all cases butt joints must be made with edges and/or ends that are perfectly smooth and perfectly square; for best results, the work should be done with power tools that will allow a precise edge or end surface.

One of the most commonly used joints, and perhaps the most practical of all of the butt joints, is the *glue block corner butt* (Fig. 5-1E). The joint is shown here with narrow pieces of stock and a triangular glue block. However, the same principle can be applied to—joints of any length and size. The glue block can be square or rectangular, as well as triangular. When you go about the various sawing operations in your shop, be sure to save all of the small strips of leftover material that are cut away from workpieces, whatever their size or shape. Many of these scraps will serve nicely as glue blocks. This saves the bother of cutting glue blocks from new stock and is also a worthwhile economy. On the other hand, don't skimp on the size of a glue block just because you have a piece of scrap on hand. The block should be plenty large enough to allow a sizable gluing surface against each of the two workpieces being joined.

Edge Joints

Edge joints are easy to make and are very important in plywood work. They allow the joining of two or more pieces of plywood to make one large piece. They are sometimes used in various applications during project fabrications. The key to achieving a good edge joint lies in perfect squaring, shaping or milling of the edge configuration so that the mating edges fit together perfectly with no gaps or sloppiness. Because of the nature of plywood, edge joints made in this material can be considerably stronger than those made in solid wood.

The simplest type of edge joint is the *butt* or *plain edge* (Fig. 5-2A), where the edges of the workpieces are sawed square and smooth and fitted together with glue. The edges are best cut on a fine-toothed carbide-tipped saw blade; they will need no further attention. However, if other types of blades are used, the edges should be dressed on a jointer for perfect smoothness. If the proper glue and clamping procedures are used, a fairly strong joint will result, but reinforcement helps. This can be done by making a *dowel edge joint* (Fig. 5-2B), where dowel pins are inserted in the

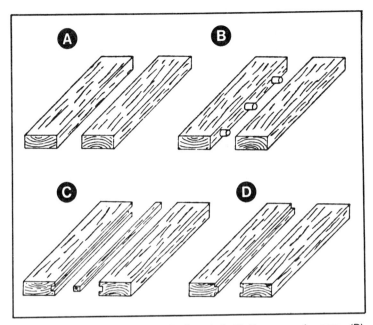

Fig. 5-2. Edge joints. (A) Plain. (B) Doweled. (C) Tongue and groove. (D) Rabbet.

edges every few inches. This kind of joint, though, carries with it the problems of perfectly aligning and drilling the dowel holes. All in all, making a *spline edge joint* is easier and also stronger. This joint is made by cutting a groove down the centerline of the edge surface of each workpiece to receive a spline. With 5-ply plywood, for instance, a little precision cutting will excise a portion of the core, leaving the other four plies untouched. Then a spline of hardboard (or plywood) can be fashioned and the two pieces mated.

Another possibility is the *tongue and groove edge joint* (Fig. 5-2C). Exactly how this joint is cut depends upon the kind of plywood being used. Generally, with ¾-inch stock, a ¼-inch groove is cut down the centerline of one edge surface, while the top and bottom surfaces are cut away to leave a matching tongue on the opposite workpiece. With some kinds of plywood, the tongue and the groove can be calculated to match the plies of the panel.

The *rabbet edge joint* (Fig. 5-2D) is also a popular one and easy to make. The usual procedure is to make the depth of the rabbet equal to one-half the thickness of the workpiece, and the width of the rabbet equal to the thickness of the workpiece. The rabbets are best cut on a radial arm or table saw fitted with a dado blade, but there are other ways to do the job. You can use a power plane, a portable router, a jointer or a shaper with equally good results.

Often when making edge joints with plywood, the grain pattern is of little consequence if the surface will be out of sight or covered with an opaque finish. Where appearance is important, the surface grain of the adjoining pieces should be matched insofar as is possible. If properly done and if the grain and pattern align reasonably well, the joint will be virtually unnoticeable. Alignment should be done before the workpieces are cut to size, so that there is plenty of stock that can be shifted back and forth to make the best match. Once you get the arrangement that seems most satisfactory, make light registration marks on both pieces so that after cutting they can be quickly and positively realigned when making the finished joint.

Rabbet Joints

The rabbet edge joint has already been discussed, but there are other kinds of rabbet joints that are easy to make and particularly effective for use with plywood. they are often employed in all kinds of casework, as well as other kinds of projects. These joints work well for edge and end fastening of plywood parts, for connecting sides and fronts of drawers or boxes, recessing cabinet back-panels and similar chores.

The *double rabbet* (Fig. 5-3A) is a relatively strong joint that is made in just about the same fashion as the rabbet edge joint—matching L-shaped grooves are cut across the ends or edges of the workpieces. Care must be taken to cut the rabbet into the correct surface of each workpiece, depending upon its position in the finished assembly. Again, the usual procedure with both the edge rabbet (Fig. 5-3B) and the double rabbet (which are essentially the same) is to make the depth of the cut equal to half the thickness of the material, and the width of the cut equal to the full thickness of material.

Another kind of rabbet joint, though not as strong as either the double or edge rabbet, is often used with plywood. This is the *simple rabbet* (Fig. 5-3C). This consists of a single rabbet cut on only one workpiece to accommodate the full thickness of the second workpiece that will be joined to it. With solid wood, the depth of this rabbet is generally one-half the thickness of the stock. This system can be used with plywood, too. However, where strength is not a primary factor, the rabbet can be cut to a depth that leaves only the face ply (or even a bit less) of the workpiece. This thin

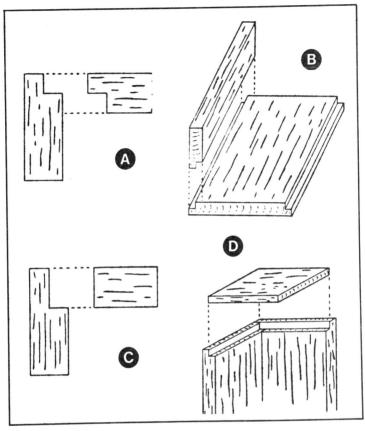

Fig. 5-3. Rabbet joints. (A) Double. (B) Edge. (C) Simple. (D) Back panel.

strip of wood then serves primarily to hide the banding or edge grain of the adjoining piece, while showing only a thin strip of end-grain at the corner. This type of joint is best made at right angles to the grain of the surface ply. If made with the grain of the surface ply, the small strip of wood that remains will have almost no strength and can easily shear away; strong joint reinforcment is needed.

For installing cabinet backs, the *back-panel rabbet* (Fig. 5-3D) is frequently employed. This consists of a simple rabbet cut all the way around the back edges of the cabinet or box, to the full width of the cabinet back panel's thickness. Once the cabinet or case is assembled (this can be done with simple or double rabbets, too), the back panel is merely dropped into the channel and secured. One common procedure is to cut the rabbet in the case sides extra-wide, perhaps an inch or more. After the back panel is installed, there is a lip of ½ inch or more protruding all the way around the back of the cabinet. This material can then be scribed and trimmed as necessary to fit the cabinet tightly against a somewhat irregular wall surface. If the back of the cabinet is flush all the way across the cabinet must be shimmed when installed or perhaps it may not fit well in any case.

Dado Joints

Dado joints are simple to make and are particularly useful in cabinetry, shelving and similar projects. They impart a great deal of strength and rigidity to the finished assembly and are essential wherever edge support is needed and extra supports such as rails or glue blocks are undesirable. Dados are generally cut with a dado blade on a radial arm or table saw, but can also be done well with a portable router. In a pinch, even a portable circular saw can be used.

The *plain dado* (Fig. 5-4A) is the one you will use mostly. It consists simply of a plain, flat-bottomed groove cut across the workpiece. The groove is just a hair wider than the thickness of the stock that will fit into it. While the depth of the groove is generally about half the thickness of the stock into which it is cut, this is variable; often a shallower groove is desirable.

One difficulty with a plain dado is that the joint is exposed at the edges. Frequently this is of no consequence, but in the case of a bookshelf that has no additional trim or face frame, the appearance may be objectionable. The problem can be neatly gotten around by using a *stop dado* (Fig. 5-4B). The dado is cut in from the back edge of the workpiece and stops an inch or so from the front edge. Then the adjoining piece is notched to exactly fit the dado. The result is an appearance identical to a simple middle rail butt joint. When the dado is stopped, the round dado blade will leave a curved end to the cut, rather than a squared end. You can get around this problem by cutting the end of the dado square with a wood chisel, or by extending the notch in the adjoining piece to a point just beyond the start of the curve. Also, you can match the curve in the notch removed from the joining workpiece for a rounded stop dado.

The *corner dado* (Fig. 5-4C) is a useful joint when installing a transverse workpiece into an open support frame. An example would be a lower shelf set between the four legs of an end table. The cuts must be made carefully and at the correct angle to receive the transverse piece; there should be as much bearing surface in

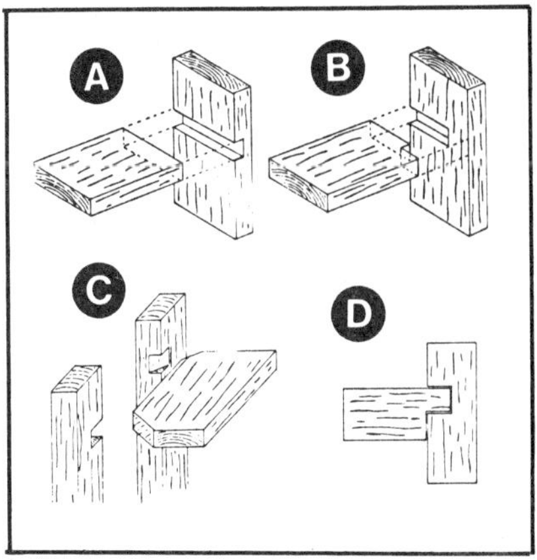

Fig. 5-4. Dado joints. (A) Plain. (B) Stop. (C) Corner. (D) Dado rabbet.

the joint faces as possible, but the cuts should not be made so deep as to weaken the stock. The easiest way to make the cuts is by jigging the workpiece at the correct angle—usually 45°—on a radial arm saw table. Then adjust the height of the saw and make a single pass with a dado blade set to the correct cut width.

There are several other kind of dado joints that might be used in plywood (and are often used in general woodworking). Because of the way plywood is made, the dado joints are not quite as effective as they might otherwise be and generally are not worth the extra effort to construct. For instance, one might make a *dado-rabbet joint* (Fig. 5-4D) by cutting a narrow dado in one workpiece and a rabbet in the other. However, this is not as sturdy a construction as a plain dado and requires two separate cutting operations instead of one.

Miter Joints

You will use various *miter joints* time after time, especially if you do much in the way of cabinetwork. Miter joints may be *flat miter*, where the stock lies flat, or *edge miter*, where the pieces stand up on edge. They also may be *simple*, where only one angle is joined; or they may be *compound*, where the angles must be joined in two planes. A simple flat miter is used, for instance, in making a picture frame. A compound edge miter would be used in making a hexagonal planter that tapers from top to bottom like a flowerpot. Thought miter joints look perfectly simple to make, they must be cut and fitted with great care if they are to present a decent appearance and form a good joint. A miter joint is essentially a weak one and in most applications requires some reinforcement.

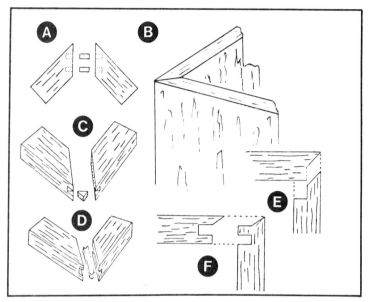

Fig. 5-5. Miter joints. (A) Doweled flat. (B) Simple edge. (C) Keyed flat. (D) Splined flat. (E) Offset. (F) Lock.

The *dowel flat miter* (Fig. 5-5A) is one commonly used construction. The same joint without the dowels is a simple miter which is used when fitting moldings together where no strength is required—just a clean, precise corner. This joint cannot normally be used as an edge miter because there is insufficient material thickness to allow the insertion of the dowels. However, the *simple edge miter* (Fig. 5-5B) is often used in making cabinets, boxes and similar items, with the joint held together simply with glue and nails or screws.

The *keyed flat miter* (Fig. 5-5C) is not as strong as the doweled flat miter, but it is relatively easy to make and produces a fairly effective joint, depending upon what strength characteristics are needed. The *splined flat miter* (Fig. 5-5D) is a much more rugged construction and is an excellent choice for cabinet panel-door frames. This joint is made by first cutting the pieces to 45° angles (or other, as required) and checking for a perfect fit. Then a groove is dadoed along the centerline of the miter cut on each piece to a suitable width and depth. A matching spline piece is cut. Glue is applied and the pieces are clamped together. The spline is slid into place. The spline can be left square-ended and trimmed with a coping saw after the joint has cured.

The *offset miter* (Fig. 5-5E) is an excellent edge miter to use in cabinets, boxes and the like where good strength is a factor. This joint is a strong one and a particularly useful one for plywood. It cuts easily, presents a substantial surface area for gluing (because of its shape), helps hold long pieces straight and tight, and readily accepts nails or screws from both outside workpiece surfaces, making an interlocking corner joint. The joint requires that several precise cuts be made (use a radial arm saw, table saw or a shaper), but the added work is worth the effort in the finished result.

To cut an offset miter (also sometimes called a miter-with-rabbet joint), first make a rabbet cut along the edge of the first workpiece into the inside surface. The cut should be equal in width to the thickness of the workpiece, and the depth should be one-half the thickness of the workpiece. Then bevel-cut the edge on a 45° angle, with the bevel slanting inward from the outside surface. Make a rabbet cut along the inside surface edge of the second workpiece, one-half as wide and one-half as deep as the thickness of the stock. Cut a bevel along the edge to 45°, slanting inward from the surface. If the cuts were accurately made (Fig. 5-6), the joint will be a perfect fit.

The *lock miter* (Fig. 5-5F) is an excellent joint to use where maximum strength is required, such as in attaching drawer sides to front or back. This joint can be made on a table or radial saw, but the cuts must be so precisely made that it is at best a difficult job. Also, the effectiveness of this kind of joint depends to some degree upon the type of plywood being used. The construction of the plywood may make accurate cutting somewhat difficult. The small components of the joint may have a tendency to tear, chip or even break off, especially in relatively thin plywood. Once the proper kind of glue is applied, the joint will actually become stronger than the wood itself. Meantime, the joint must be kept in one piece during the sawing and fitting process. By far the best method to use in

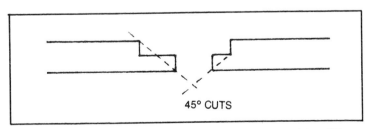

Fig. 5-6. Methods of making offset miter with two rabbets and two 45° saw cuts.

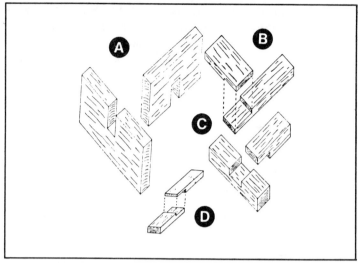

Fig. 5-7. Lap joints. (A) Edge cross. (B) End. (C) Half. (D) T-lap.

making a locked miter is to fit a set of lock-miter cutters on a shaper or router and let the machine make a perfect match of the mating edges.

Lap Joints

For the most part, *lap joints* are not used much in working with plywood, but there are occasions when no other joint will serve quite as well. Perhaps the most important one of the entire group is the *edge cross lap* (Fig. 5-7A), which is commonly used in fitting together the component pieces of different kinds of furniture. This makes a strong joint, as well as an attractive one, and it can be left unfastened and unglued so that the article can be disassembled.

The trick to making a good looking joint of this sort is to make sure that the notches in the workpieces are accurately sized to the thickness of the stock for a close, tight fit. If the depth of the cuts is relatively shallow, they can easily be made with a dado blade fitted in a radial arm or table saw. But very deep cuts can also be easily made for joining two sizable pieces, as when making a table base or some similar item. One easy method is to drill a hole of the same diameter as the thickness of the stock, so that the edge of the hole lines up exactly with the end of the slot. Draw two parallel guidelines from the edges of the hole to the edge of the workpiece. Make the long cuts with a table saw (then finish with a hand saw) or a jigsaw. Then square off the rounded corners of the slot with a jigsaw, and dress the cut edges with a file. A little careful work will insure a perfect fit.

There are times when an *end lap joint* (Fig. 5-7B) is useful for making a corner or similar joint. Just cut out half the thickness, in matching sections, from each workpiece, one from the bottom side and the other from the top side. Then mate the two pieces together. A *half lap* (Fig. 5-7C) is made in the same fashion, but it is used to join two pieces end to end. This is similar in principle to an edge rabbet joint, except the degree of overlap is considerably greater. An edge lap joint could also be made using the same principles. In frame building or casework, a *T-lap joint* (Fig. 5-7D) might also be used, where a rabbet cut is made in one workpiece and a dado in the other. As with the other lap joints, usually half of the material is removed from one piece and half from the other so that they meet in the middle. Care must be taken that just the right amount of material is removed from each piece, or the surfaces will not be flush.

Mortise and Tenon Joints

Mortise and tenon joints are widely used in woodworking of all kinds when the material is solid wood, and especially where the pieces are of substantial thickness or overall size. They require care and patience in the making and provide exceptionally strong and rigid joints. These joints can be made with special mortising tools, but they can also be made with hand tools once one develops the skill and degree of craftsmanship necessary. They can be made either rounded or squared. These joints are not widely used in plywood work, principally because of the way in which the plywood itself is made, its relative thinness, and the nature of the projects

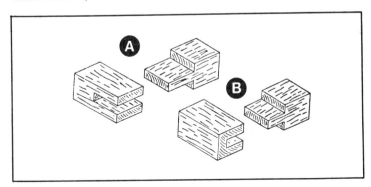

Fig. 5-8. Mortise joints. (A) Open. (B) Stub.

that are made with plywood. However, there are two mortise and tenon joints in particular that might be successfully used with plywood: the *open* and the *stub* (Figs. 5-8A and 5-8B).

The open *joint* is primarily used for making corners, as on a frame. The stub joint can be used for corners, end-to-edge joints, or even end-to-end joints in the manner of a splice. Both types can be readily fashioned on a radial arm or table saw an appropriate groove for the mortise. Saw away the front and back faces on the second workpiece to form a matching tenon. As with all joints, the trick is to get the dimensions accurate and the joint faces smooth and even. Often mortise and tenon joints are fitted with pegs for extra strength, with the peg faces left exposed for a decorative appearance.

Joinery Techniques

We have already mentioned several times that good joinery depends upon careful and accurate cutting. The dimensions must be just right. The saw blades, cutters or bits must be extremely sharp for working with plywood. The joint surfaces must be correctly squared and smoothed. These facts cannot be overemphasized. The better the work that goes into the joints, the solider and ruggeder the finished joint will be. Power machinery will generally allow you to make better joints more quickly and more easily than will hand tools. On the other hand, if you have the time and patience, you can turn out many different kinds of joints without ever flipping a switch.

As a rule of thumb, choose a joint type that can be made as simply as possible, is the most practical for the equipment you have on hand, and one that will do the desired job with sufficient strength and rigidity for the demands that will be place upon it. Consider also that the type of stock we are speaking of, plywood, has different machining characteristics than does solid wood. Those characteristics vary somewhat with the specific type of plywood. A bit of experimentation will soon show you just which joints are the easiest for you to make with the equipment that you have and the kind of plywood that you are using.

Consistency is another good approach to joinery. Once you find the two or three or more joints that work particularly well for you, stick with them. Use the same kinds of joints for most of your work, unless conditions specifically dictate the use of something different. Also, maintain consistency of joint strength and type throughout each indivicual project. There isn't much point in coupling two or three rugged, complicated joints with one or two simple and relatively weak ones in the same project. Then, too, consistency of joint selection makes the joinery itself easier, since you can set your power tools up to make all of the joints for a project in a series of steps. First, tool up to do all of the rabbet cuts of one size, then all those of another size. Make up all of the dado cuts of one size and those of another. Making a series of identical cuts is a lot simpler and faster than having to readjust your equipment for cut after cut after cut.

There is one more point that must be noted, having to do with the tolerance of fit of the component parts of a joint. In many cases, where the faces of the joint are relatively open, this presents no problem. But there is sometimes a danger of making up the joint so that the pieces fit too tightly together. The parts must slip neatly but freely together when they are dry. If they must be forced together, something surely will crack, splinter or break. This is particularly true of joints whose sections are so thin as to be easily damaged, a not uncommon circumstance with plywood. Joints with openly mated faces must fit together so that they mate nicely with just a hairline crack between them. Joints with interlocking faces must fit together just a tad loosely. In both circumstances, this allows sufficient space for the glue, which must be compressed between the faces. If the fit is loose, the joint will be weak, even though certain glues do have a substantial degree of gap-filling capability that helps in such circumstances.

Where blind, pegged or doweled joints are to be glued, enough extra space should be allowed so that any excess glue that finds its way into the bottom of the cavity has a place to rest. The glue should not split the workpiece or delaminate the plies of the plywood under the high pressure that develops when the pieces are forced together and clamped. If dowels with spiral glue grooves are used, this obviates the problem. Likewise, just a bit of clearance should be allowed in splined joints, so that the spline will slip easily into place without forcing workpiece apart.

There is also a special note with regard to doweled joints. Dowels, or dowel stock, should have a moisture content of less than 5% if possible. A dry dowel fitted into a correctly sized hole will absorb moisture from the glue and swell, locking firmly into place. If the dowel has a high moisture content, it eventually will shrink within the hole, and possibly break the glue joints and become relatively ineffective. All you have to do is place your dowels or dowel stock in a warm, dry spot for a period of time before you use them. One effective method is to place them on a screen over a heater shortly before use, or pop them in a low-temperature oven for a short while.

GLUING PLYWOOD

Plywood is a natural for glued-up assemblies. The use of glue joints is important to the overall strength, rigidity and quality level of the finished article. Since plywood already has a low moisture content and the wood species used in plywood can be easily glued, the process is neither difficult nor tricky. If you use the proper techniques, you will achieve a good glue bond every time. Plywood, of course, is a glued wood product to begin with and contains at least two layers of glue and frequently more; these are called *primary gluelines*. The glue layers that are made between workpieces during assembly in the shop or on the job site are called *secondary gluelines*.

In plywood work, there is a good rule of thumb that you can follow. Any time you join two pieces of plywood, plywood to ordinary wood, plywood to some other material or vice versa, apply glue to the joint if at all possible. This is in spite of the fact that you might also plan to use nails, screws, bolts or whatever in the assembly. The only time glue should not be used is when the article or particular construction is designed so that it can later be taken apart or disassembled.

The subject of glues can be a most confusing one, especially to anyone unfamiliar with glue properties and applications. There are dozens of different glues on hardware and lumberyard stock shelves all over the country, each of which seems to claim to be the best. But the brand or trade name of the glues that you use are relatively unimportant, so long as you purchase products of good quality made be reputable manufacturers like Borden, Franklin, National Casein Goodyear, etc. What is important is the type of glue that you use for particular applications. As far as gluing plywood to plywood, or plywood to solid wood, is concerned, you can get along very nicely with just three different types of glue. In addition, there are several others that *can* be used and some that are recommended for certain special applications.

Aliphatic Resin

This glue is the most important one for probably 95% of the gluing operations around the workshop and home. It is a high-strength chemical glue, ready to use, and of a creamy consistency and usually identifiable by its cream-buff color. It is available at any hardware store or lumberyard in several sizes of applicator-type squeeze bottles, and also in larger quantities for refilling squeeze bottles. The glue has a long shelf life; some separation of glue and vehicle may occur, but remixing takes care of that. This glue is particularly good for edge-gluing and is a fine choice for furniture, cabinetry and casework of all kinds. It has good gap-filling properties, is relatively heat resistant and sands and works very easily. It is also non-staining. *Aliphatic resin* lacks moisture resistance, however, so it can only be used for interior work where the joints well not be exposed to high moisture levels.

This glue should be mixed before using to make sure there is no separation, and then it can be applied directly from the bottle. It can be used at any temperature from 45° and up, and it should be spread on the joint faces in a moderate coat. The workpiece surfaces should be mated immediately after spreading the glue, as it sets up fairly rapidly, especially in warm-dry conditions. The joint should be clamped immediately, and the normal clamping/curing time is about 1½ hours.

Plastic Resin

Though considerably more demanding and more difficult to work with, *plastic resin glue* is an important one because it is highly water-resistant. Note, however, that it is not waterproof. It comes in powdered form and must be mixed for each application; effective pot life is only about 3 to 4 hours, so mix only what you will need during that period of time. This glue can be used for all general purpose applications. Since it has poor gap-filling properties, though, it should be used with closely fitted components and tight joints. If there are gaps or the joint is loose, the glue will not take up the space. The glueline will be quite brittle and the joint a weak one. If the joint is properly fitted, though, the glueline will be very strong.

Plastic resin glue is applied to the joint surfaces in thin coats to both surfaces, and must be used at an ambient temperature of at least 70° F or higher. After application, the joint faces should be immediately mated and clamped. Clamping pressure is high, and the clamping/curing time is about 16 hours at 70° F. The higher the temperature, the faster the glue will cure. But there is a danger in removing the clamps too soon.

Resorcinol

This is the glue that must be used on articles that will be exposed to extreme dampness, exceptionally high humidity levels or weather. Patio furniture, children's sandboxes, boats and items of this sort should always be joined with resorcinol glue. This resin type of glue has very high strength and comes in powder form. It is the only general purpose wooodworking glue that is fully waterproof. It can be used for all kinds of gluing of plywoods and other woods (whether exposed to moisture or not), and it is strong enough to be used for structural bonding. It is not particularly effective, however, in bonding wood to other materials.

Resorcinol has good gap-filling properties and works well on relatively poor or ill-fitting joints. It must be used in ambient temperatures of at least 70° or higher, but it is easy to mix and to work with. Application is made by coating each surface with a thin layer of glue, mating the parts and clamping for about 16 hours. The glue must be mixed for each job. Since it has a useful pot-life of about 6 hours (sometimes more), though, an entire small project often can be glued up with one batch.

Casin Glue

There are a number of other woodworking glues that can be used with good results and in fact are preferred by some woodworkers to the types mentioned. Just to keep the record straight and to give you some fair basis for comparison, we will note them briefly.

Casein glue is a natural glue derived from milk and available only in powdered form. This is a strong and reasonably water-resistant glue with excellent gap-filling properties that can be used in any temperature above freezing, but the warmer it is the better. It must be mixed with water for each job. The pot life is fairly short, so mix only enough at one time to take care of the immediate gluing tasks at hand. This is a very good all-purpose glue for woodworking, and it is the glue to use for oily woods such as *teak* or *yew*. On the other hand, it will leave an undesirable stain on acid woods such as redwood or oak. Gluing is done by covering each joint surface with a thin coating, mating the parts immediately and clamping for about 2 hours with hardwoods and 3 with softwoods.

Hide Glue

Hide glue is the old traditional glue for cabinetmaking and furniture building, as well as general purposes. It is a natural glue made from animal hides and bones (fish glue is much the same, made from fish parts). Among particular craftsmen, this glue is the first choice for all interior woodworking purposes, especially where particularly strong joints in fine cabinetry or furniture is a requirement. The glue is very strong, will not become brittle and has very good gap-filling properties. However, it is neither waterproof nor particularly water-resistant.

Hide glue is available in solid form, which must be heated and mixed before using, or in liquid, ready-to-use form. Often it is applied after heating. It can be used in temperatures of 70° F and up, or applied hot in somewhat lower temperatures. Both joint surfaces should be covered with a thin coating, which is allowed to become tacky. Then the mating surfaces are joined and clamped for about 2 to 3 hours.

Polyvinyl White Liquid Resin

Polyvinyl white liquid resin, usually known as white glue is readily available in a wide variety of brand names in any store that sells glue. It comes ready-to-use in liquid form, packaged in various sizes of applicator-spout squeeze bottles and can be applied at any normal temperature. It is quick-setting, easy to use and is a reasonably good choice for general purpose shop and household use. It will work relatively well for bonding plywoods and woods. Though it can be used in casework, cabinetry and furniture building, it is definitely not the best choice. It is only half as strong, for instance, as aliphatic resin. Because of its characteristics, the glue is particularly good for small jobs where either a good fit or tight clamping would be difficult. It sets up quite fast in low-humidity situations, so speed is essential here. Spread the glue on both surfaces to be joined, and immediately mate and clamp the parts for about 1½ hours. White glue should never be used for exterior purposes or where high-moisture conditions prevail.

Urea Resin

Urea resin is primarily a thermosetting resin glue made from chemicals, more often used in high frequency and steam heated pressure gluing applications. This glue must be applied at temperatures of 70° or warmer and comes in powder form to be mixed with water. Pot life is 4 hours or less, and gap-filling properties are poor. It must be used on woods of low moisture content (10%) and can be effectively used on plywoods. It is moisture resistant and handled in much the same way that plastic resin glue is. Tight joint fits are a must, and the clamping/curing time is about 16 hours.

Contact Cement

The glues that we have just discussed all can be used for general woodworking purposes, but they certainly will not do the job in every possible application that arises. For some jobs, there are certain kinds of glues that are especially designed for those purposes.

Contact cement is made from a base of synthetic rubber. It is the glue to use in bonding laminated plastics to plywood (in making countertops, for instance). This glue will also bond leather, canvas or heavy fabrics, plastics and some other materials, including veneers, to a smooth base. Some contact cements are extremely explosive and give off toxic fumes. For home shop use, seriously consider using the nonflammable type. Though the bond is not quite as strong, it is far safer for the average home mechanic to work with. Application is somewhat involved to explain, but it is not difficult. The chief drawback that once the parts are mated, they cannot be shifted even a little bit. The bond is instantaneous and no clamping is required. Work should be done at a temperature of 70° F or better.

Panel and Construction Adhesives

Panel adhesive is a special glue that is packed in standard containers for use in caulking guns and used to secure wall paneling in place, as well as for similar applications. There are several different brands and formulations, most of which work quite well. Your local hardware supplier can recommend a particular one that will fulfill your needs. These adhesives reduce the number of nails required in putting up wall paneling and afford a better installation. In some cases, small parts of decorative paneling work on cabinetry can be secured quickly and easily with this adhesive and no fasteners at all.

Construction adhesive, somewhat similar to panel adhesive, is also packaged in tubes for caulking gun application. It is used in general building and construction work, particularly in gluing plywood subfloor of considerable strength and stiffness. A particular type of construction adhesive is required in constructing in All Weather Food Foundation (AWWF) system, for bonding the polyethylene sheeting to the treated plywood sheathing.

Hot Melt Glues

Hot melt glues are popular but generally unsatisfactory for most plywood gluing applications. They should primarily be used on only small parts with light loading factors, and where joint stress is not a problem. These glues are not overly strong. They cool and set quickly.

Silicone Caulk

Silicone caulk, also known as silicone sealant, is a tough, rubbery substance that can be used as a bonding agent. It adheres reasonably well to dry plywood but will not stick at all to even a lightly damp surface. Once cured, however (which take only a few hours), silicone is completely waterproof and has the advantage of forming a flexible joint that will never crack or dry out. This is an excellent material to use when setting or bedding plastic or glass into a wood frame, as when building a shadow box or display case. Keep a tube of this material around the shop. It is available in clear, several colors and a special paintable variety. You will find that it comes in handy for a variety of purposes.

Epoxy glue

Epoxy glue is a synthetic known also as *epoxy resin* or *epoxy cement*. This glue is a bit difficult and awkward to use, since it must be mixed for each job immediately prior to application. It is also an expensive glue and not one that is particularly suited to making wood-to-wood joints. But it can be successfully used for joining small parts and also for bonding other materials to wood such as ceramic, metal, glass, plastics or foils. Epoxy does have certain advantages. It can be used at any temperature. Curing can be speeded by heat. The material once hardened is fully machinable and can be sanded, shaped, filed, drilled and even tapped. It does make a good filler for plywood and can be smoothed and painted very nicely.

Mastic

Mastic is a thick, tacky fast-bonding paste-like adhesive generally packed in either small tubes or caulking gun cartridges. These adhesives have medium strength, good gap-filling properties and generally are either moisture resistant or waterproof. They do not dry hard and should not be used wherever a joint edge is exposed. Mastic adhesives are sometimes used to apply wall paneling to a backing such as hardboard, plywood or wallboard. Though of little use in wood-to-wood joining, mastics can be used to join other materials to wood or other surfaces, such as mounting mirrors or setting mirror tile, securing metal wall tiles and similar jobs. Mastics are frequently used in setting ceramic tile on plywood countertops and wood or other kinds of tile on plywood subflooring.

Super Glue

Super glue is very expensive and generally packaged in tiny tubes or bottles for special purpose applications. These glues, of which there are many brands, will stick anything to anything, including your fingers. They must be used with great care and are designed for very small and precise gluing operations. The glue bond has tremendous strength and is waterproof and impervious to practically anything. Shelf life is short. By and large, super glues are of no value in working with plywoods. But there are times when nothing else will serve as well; if you break a chip out of a piece of laminated plastic while applying it to a plywood countertop, save the chip. After the job is done, apply a tiny dot of super glue to the chip and stick it back in place. The repair, if carefully done, will be barely noticeable.

Techniques

There are a few points to keep in mind regarding the general technique of gluing joints. The first, and perhaps the most important, is to choose the correct kind of glue for the conditions that the joint will be exposed to, the type of joint and the materials being used. If the joint happens to be a poorly fitted one, choose a glue that has good gap-filling properties, as well as the other characteristics suitable for the job. When applying the glue to the joint surfaces, do so in smooth, even coats of a thickness recommended for that type of glue, and on either one or both joint faces, whichever the manufacturer recommends. After clamping the joint, be sure to immediately wipe away all excess glue. It is much more difficult to remove later and may cause difficulties with the final finish, especially if that happens to be a clear or semitransparent one. Applying too much glue does nothing for the joint (and in the case of a doweled joint may burst the workpiece) and is just a waste that causes extra work after the joint is made.

Frequently more glue will be required when joining end grains than when joining edge grains. With plywood you will often be doing both at the same time, so the trick is to spread neither too much nor too little. Eventually you will develop a sense of rightness, gained only from experience, as to just what constitutes the right amount of glue.

When you set out to make a glue joint, make sure that all of the tools and parts that you will need are right close at hand. The faces of the glue joint should be completely free of dust or small particles, and oil is a particular enemy of most glues. The best bet is to position the parts, apply the glue, mate the parts, clamp them and drive whatever fasteners are called for in as short a period of time as possible, within reason. As noted earlier, some glues set up much faster than others. Use the slow-setting types for intricate assemblies where a fair amount of positioning and adjusting time is needed. Curing time with many glues is variable depending upon temperature, humidity and the moisture content of the wood. If you are unsure as to the particular length of time needed in any given instance for a joint to cure, leave the clamps on for a fairly lengthly extra period of time. It is best to err on the safe side. For best results, the moisture content of the wood should be less than 10%. Never attempt to glue damp wood or wood that is "right off the stump" and still green. With plywood, of course, the moisture content will be very low unless the panel has been subjected to water in some fashion.

When you make edge joints in plywood, it is best to allow the joints to cure for at least a couple of days before attempting any surfacing operations that include the joint area, be that sanding or any other kind of finishing. The reason is that the glue may cause the wood in the immediate area of the joint to swell up a bit. If you surface the material, you will peel this swelling away. When the joint finally dries completely, an obvious indentation will remain. It is also a good idea to allow the stock being used in the project, especially if for interior use, to remain stored under the same conditions in which the finished project will finally be used. This will give reasonable assurance that few if any further changes will take place in the material after it is cut and assembled.

Clamping sometimes poses some questions and also some problems. Perhaps the most common question is, "How much pressure?" One of the most common problems is having the workpieces out of alignment as the clamps are tightened. The glue acts as a lubricant and the pieces slip and slide. Regarding the question, most glues require only a moderate pressure. Assuming that you have applied a coating of approximately correct thickness to the joint faces, you will have achieved just about the right moderate pressure when a thins bead of glue squeezes out along the edge of the joint. It is possible to apply too much pressure, which results in what is called a *starved joint* where all of the glue is forced out. But the greater danger is in applying insufficient pressure, which creates a weak joint, too. However, after assembling a few joints, you will begin to get a feel for the correct pressures.

As to the alignment problem, this can always be tricky. It pays to keep a close eye on the joint edges as you apply pressure, and tap them back into line as necessary. Often the trouble lies in the fact that the clamps are not properly aligned. Any clamp will exert maximum pressure in the exact direction of its alignment, so it must be lined up evenly with the workpiece, with equal pressure applied at both jaw faces. It is a good idea, too, to check the workpieces with a square immediately after clamping to make doubly sure that they are actually properly aligned. Once they cure, if they are out of alignment or mismatched, there is no recourse but to start over.

Unless you are using clamps with padded jaws, always place a piece of scrap wood or hardboard between the jaw faces and the workpiece to prevent damage to the workpiece surface. This is sometimes a juggling act, as you have to wiggle four pieces into place between the clamp jaws. With a bit of perseverance, though, it can be done. Often it is easier to lay heavy clamps on a work surface or even clamp the clamps to something, and then insert a lightweight assembly into the clamp jaws, rather than trying to juggle the assembly around in midair. Sometimes a combination of ingenuity and common sense is needed.

It is important to fully understand that a glue joint can be taken apart before the glue sets, but not afterward. This seemingly simple fact is ignored by many woodworkers. They will fight and struggle with a joint that does not want to align properly, until the glue finally sets up into a joint that is neither precise nor strong. Then they give up and try to correct the error during the finishing process. This seldom works out satisfactorily. The simple solution, and one that should always be followed, is to quickly yank the joint apart when you first see that it is not going to go together properly. You've lost a bit of time and some extra work will be needed, but at least you will save the workpieces. Wipe the glue off to cure and dress the faces down for a proper fit again. Then try it once more.

FASTENERS

Though some small articles can be turned out with glue as the only fastener holding them together, some type of mechanical fasteners are most often used in practically all projects. There are many different kinds of fasteners, many of them specialized and only a few with more or less universal applications. A thorough knowledge of what is available and how the fasteners are used, and in which applications they might serve best, is invaluable to any do-it-yourselfer regardless of the particular task at hand. If you use the right fastener in the right place, it makes the job easy and the result effective. Using the wrong fastener invariably leads to difficulties.

Four Important Nails

Ordinary nails are sold by the pound and manufactured in standard head and shank diameters and overall lengths. They are designated by the penny system which is symbolized by the letter "d". An 8-penny nail, for instance, is symbolized by 8d and is 2½ inches long. Table 5-1 shows the size breakdown and the approximate numbers of nails per pound for the four most important types. Here are the characteristics of these four types (there are many other specialized types).

■ **Common.** It has the largest head diameter and the largest shank diameter in any given size, along with the greatest head thickness. The nail is used primarily in heavy work and has the greatest strength.

■ **Box.** This one has a slightly thinner head and slightly smaller shank and head diameter than the common nail, in any given size. It is not quite as strong as the common nail, but it is perhaps the most widely used for general purpose work including construction and building.

■ **Casing.** This nail is used for finish work where the head will be countersunk. It has a small head with tapered sides; the head has essentially a flat surface for setting either flush or coun-

Table 5-1. The Various Nail Size/Type Combinations That Are Readily Available. Figures Indicate the Approximate Numbers of Nails Per Pound.

Nail	2	3	4	5	6	7	8	9	10	12	16	20
Common	847	543	294		167		101		66	61	47	29
Box			588	453	389	255	200	136		90		69	50
Finish			880	630		288		196		124			
Casing			489		244		147		96		73	

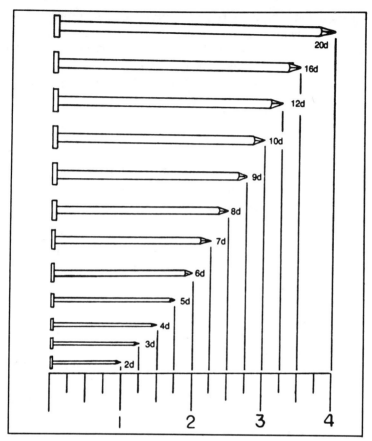

Fig. 5-9. Nail sizes and lengths of common, box, finish and casing nails.

tersunk. A casing nail is used where maximum strength and resistance to head pull-through is required.

■ **Finish.** It is similar to a casing nail but has a smaller, somewhat rounded head and is somewhat thinner in the shank. It is the most widely used nail for finishing purposes.

All of these nails are readily available in sizes ranging from 1 to 4 inches in length. There are a few larger sizes, too. Use "bright" (plain steel) nails for interior work and wherever moisture and rusting is not a problem. The galvanized nails are best for exterior work or wherever moisture and rusting might be a problem. For the most part, you will probably find yourself using only two of these nails, finish nails for all finish work and box nails for general purpose rough work and construction. For nailing plywood, be sure to use the right length nail for the plywood thickness. See Fig. 5-9.

Wire and Brad Nails

For working with thin stock and making up small articles, there are two other kinds of nails that you will need. One is the *brad*, which is merely a pint-sized finish nail. These are sized by specific length in fractions of an inch and by wire gauge (equals shank thickness). The second type is the *wire nail*, which is a pint-sized box nail. These nails are sized in the same manner as brads. Both kinds are usually sold in small packets of individual sizes (assortments can be purchased, too), and generally they are an inch or less in length. In plywood work their primary use lies in nailing ¼-inch stock or in applying thin moldings or trim.

Deformed Shank Nails

There is one other kind of nail worthy of consideration. Where extra resistance against pull-out or head-popping is required, the kind of nail to use is one of the various deformed shank types. Instead of having a straight, smooth shank, these nails have shanks that are cut or manufactured in such a form that they will catch on the wood fibers and lock in place. They may be called *ring-shank, annular, serrated*, etc., but they all perform the same basic function. Once driven, they can be withdrawn only with difficulty. They are widely used for putting down underlayment plywood and can be used wherever their characteristics are desirable.

Noncorrodible Nails

While galvanized nails are all right for general purpose exterior work and are particularly good for interior high-moisture fastening, the galvanized coating can eventually wear away. The result will be rust streaks on the finished surface. The best bet for exterior work is to use a fully noncorrodible kind of nail, either aluminum or stainless steel. One or the other should always be used when fastening redwood, whether for interior or exterior purposes. Stainless steel nails are strongly recommended when building an All Weather Wood Foundation system. And either aluminum or stainless steel nails should always be employed when putting up exterior plywood siding. For the latter purpose, special sinker-head siding nails are excellent. These have a special head configuration that is something of a cross between a casing and a box nail. Table 5-2 shows what is readily available in the way of noncorrodible nails.

Table 5-2. Commonly Available Noncorrodible Nails (courtesy of the California Redwood Association).

SIZE OR EQUIVALENT	STAINLESS STEEL LENGTH (inches)	NAILS PER POUND	ALUMINUM SIDING LENGTH (inches)	NAILS PER POUND	HOT-DIPPED GALV. BOX LENGTH (inches)	NAILS PER POUND	HOT-DIPPED GALV. WOOD SIDING LENGTH (inches)	NAILS PER POUND	HOT-DIPPED GLAV. CASING LENGTH (inches)	NAILS PER POUND
6d	1 7/8	270	1 7/8	566	2	194	2	283	2	212
7d			2 1/8	468	2 1/4	172	2 1/4	248	2 1/4	172
8d	2 3/8	201	2 3/8	319	2 1/2	123	2 1/2	189	2 1/2	131
10d	2 7/8	140	2 7/8	215	3	103	3	153	3	85
6d	3 1/2	70	3 1/4	120	3 1/2	60			3 1/2	67

Nailing Plywood

There are no particular extraordinary tricks for effectively nailing plywood. Drive the nails straight when nailing into a narrow workpiece edge, or at an angle for added strength if you have sufficient material to nail into. With finish nails, guide the nail down until just the head protrudes. Complete the job with a nailset, driving the nail head either exactly flush with the surface or about 1/16th-inch below for filling. Common nails should be driven so that the head surface is flush with the wood surface but not smashed down into the material. A line of nails need not be staggered to avoid splitting, as must be done in many plain woods, because the plywood will not split. Nailing can be done quite close to a plywood edge without splitting. If the nail must be set very close, it is a good idea to drill a pilot hole first. This will avoid splintering and the possibility that a small chunk of an interior edge-grain ply might pop free of the edge, leaving a gap.

There are two points in particular to be aware of when edge-nailing plywood—when nailing down through a plywood surface and into the edge of a second workpiece. The first is to make sure that the nail is parallel to the surfaces of the workpiece that is stood on edge. This is because a nail driven at an angle can easily come through either the face or the back if set at a fore-or-aft angle. The nail tip, especially if blunted or malformed, may catch on a glueline or in the wood grain and bend outward at an angle. It is a good idea, however, to drive the nails at a bit of an angle either to left or to right along the lengthwise direction of the edge. This will afford additional holding strength.

The second point is that a nail driven into end-grain wood does not hold nearly as well as when driven into cross-grain wood. Thus, the portion of the nail that you drive down through the surface of the first workpiece will be gripped quite well by the wood. The portion of the nail driven down into the edge of the second piece may not hold as well. This depends upon whether or not the nail drives down into an end-grain ply or a cross-grain ply. Thus, this type of joint, especially if assembled without glue, may not be as strong as you would wish. Where extra strength in such a joint is necessary, it should be further reinforced. Commonly used possibilities are longer nails, deformed shank nails, glue blocks, metal angle braces, wood screws or some combination of these.

Screws

Screws have much greater holding power than nails do; they also cost a good deal more and require considerably more time to install. However, the final results very often justify the added effort and expense. Screws should always be used where a strong joint and a consequently strong assembly is needed, and also where the article is designed so that it can be later taken apart. (In the latter case, of course, no glue is used to reinforce the joint.) There are many different kinds of screws available for all sorts of purposes, but there is one in particular that you will use more than any other. That is the *flathead wood screw*.

The flathead wood screw has a tapered shank, part of which is threaded especially for use in woods. The head is flat across the top and has chamfered sides to fit neatly into a countersunk hole. The head most commonly used is a standard straight slot (Fig. 5-10) for driving or removing with an ordinary straight-blade screwdriver. *Phillips* or *recessed slots* (Fig. 5-10) are also quite common, and this is the kind of frequently supplied with builders' hardware (hinges, latches, etc.). They are made from several metals; the type you will doubtless use most frequently is plain, bright steel. This is the kind to use for all interior purposes where rusting and moisture is no problem. For exterior work, under conditions of weathering or just high-moisture conditions, you can choose aluminum, brass, bronze, stainless steel or galvanized screws. As with galvanized nails, galvanized screws are not permanently non-

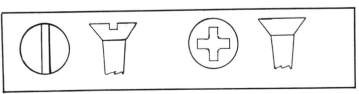

Fig. 5-10. The two most commonly used screw slot configurations are the straight (left) and the Phillips (right).

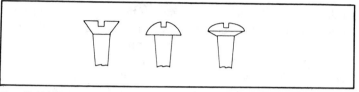

Fig. 5-11. The screw head patterns most used in plywood and other woodworking are the flathead (left), roundhead (center) and the oval head (right).

corrodible while the others are. Plated screws, sometimes touted as being noncorrodible, are not; they will not withstand weathering for more than a short period of time and will soon begin to rust, especially where the screwdriver nicks or abrades away the plating.

There are other head configurations as well (Fig. 5-11), two of which may be of some use to you for particular applications. One is the roundhead wood screw, which has a head that is perfectly flat underneath and fully rounded on the top. This is meant for surface rather than flush-mount or concealed fastening. The other possibility is the *oval head screw*, which can be purchased with a bright chrome-plated finish. This is a finishing screw that is designed to be used with chrome-plated builders' hardware or fixtures, or in conjunction with chrome-plated cup washers for a decorative effect.

These three kinds of wood screws are available in many different standard sizes and can be purchased in small packages or boxes of 100 or 500 pieces. The various lengths and sizes are shown in Table 5-3. The recommended screw sizes and lengths for fastening plywood are shown in Table 5-4. Note that these are minimum lengths; wherever possible, you should use longer ones. The lengths listed are for driving a screw down through the surface of one workpiece and into the edge of the second. Where two pieces are mated surface-to-surface, the length of the screw must be adjusted so that the screw tip does not break through the reverse side of the second workpiece. At the same time it allows threads to grip the maximum amount of material in that piece. For fastening two ¾-inch pieces together, for instance, a 1½-inch #8 screw would be just a tad too long, especially if the head were deeply countersunk. A 1¼-inch screw would have to be used. Sometimes going to a larger screw diameter is helpful, since the larger screw threads have a larger bearing surface; a #10 screw could be used in the above example.

Table 5-3. The Most Readily Available Screw Gauge/Length Combinations and Screw Gauge Diameters.

Screw #	1/4	3/8	1/2	5/8	3/4	7/8	1	1 1/4	1 1/2	1 3/4	2	2 1/4	2 1/2	2 3/4	3	3 1/2	4	Dia.
1	X																	0.073
2	X	X	X															0.086
3	X	X	X	X														0.093
4		X	X	X	X	X	X											0.112
5		X	X	X	X	X	X											0.125
6		X	X	X	X	X	X		X									0.138
7		X	X	X	X	X	X	X	X									0.151
8			X	X	X	X	X	X	X	X	X		X		X			0.164
9				X	X	X	X	X	X	X	X	X	X					0.177
10				X	X	X	X	X	X	X	X	X	X		X	X		0.190
11					X	X	X	X	X	X	X	X						0.203
12						X	X	X	X	X	X	X	X		X	X		0.216
14							X	X	X	X	X	X	X	X	X	X	X	0.242
16								X	X	X	X	X	X	X	X	X	X	0.268
18									X	X	X	X	X	X	X	X	X	0.294
20										X	X	X	X	X	X	X	X	0.320

The usual procedure with flathead wood screws is to countersink them and then fill the hole with a wood filler, sand smooth and paint over to completely conceal the fastener. Often where the heads need not be concealed, they are merely countersunk flush with the workpiece surface and left as is. A common mistake in countersinking wood screws is to simply drive them down hard

Table 5-4. Proper Minimum Wood Screw Size/Length Combinations for Plywood Surface-To-Edge Fastening.

Plywood size	Screw size	Screw length*
1"	10	2"
3/4"	8	1 1/2"
5/8"	8	1 1/4"
1/2"	6	1 1/4"
3/8"	6	1"
1/4"	4	3/4"

Table 5-5. Chart of Clear (Shank) and Pilot Hole Drill Sizes.

Screw#	1	2	3	4	5	6	7	8	9	10	11	12	14	16	18	20
Clear hole - Frac. drill	$\frac{5}{64}$	$\frac{3}{32}$	$\frac{7}{64}$	$\frac{7}{64}$	$\frac{1}{8}$	$\frac{9}{64}$	$\frac{5}{32}$	$\frac{11}{64}$	$\frac{3}{16}$	$\frac{3}{16}$	$\frac{13}{64}$	$\frac{7}{32}$	$\frac{1}{4}$	$\frac{17}{64}$	$\frac{19}{64}$	$\frac{21}{64}$
Clear hole - # drill	49	44	39	33	30	28	24	19	16	11	6	2	-	-	-	-
Pilot hole - Soft wood - Frac. drill	$\frac{1}{32}$	$\frac{1}{32}$	$\frac{3}{64}$	$\frac{3}{64}$	$\frac{1}{16}$	$\frac{1}{16}$	$\frac{1}{16}$	$\frac{5}{64}$	$\frac{5}{64}$	$\frac{3}{32}$	$\frac{3}{32}$	$\frac{7}{64}$	$\frac{7}{64}$	$\frac{9}{64}$	$\frac{9}{64}$	$\frac{11}{64}$
Pilot hole - Soft wood - # drill	68	68	56	56	52	52	52	47	47	42	42	35	35	28	28	17
Pilot hole - Hard wood - Frac. drill	$\frac{1}{32}$	$\frac{3}{64}$	$\frac{1}{16}$	$\frac{1}{16}$	$\frac{5}{64}$	$\frac{5}{64}$	$\frac{3}{32}$	$\frac{3}{32}$	$\frac{7}{64}$	$\frac{7}{64}$	$\frac{1}{8}$	$\frac{1}{8}$	$\frac{9}{64}$	$\frac{5}{32}$	$\frac{3}{16}$	$\frac{13}{64}$
Pilot hole - Hard wood - # drill	68	56	52	52	47	47	42	42	35	35	30	30	28	22	12	6

until they create their own countersink by mashing into the wood. This is not a good idea, particularly with plywood. The screw will grip well in plywood, and there is a good likelihood of twisting the screw head off before it is fully driven. In addition, the wood fibers beneath the screw head will twist and tear. Splintering will probably occur. The wood may also bulge up around the screw head. With plywood especially, always drill a pilot hole and a countersink for each screw.

One easy way is to use a pilot bit, as discussed in the tools chapter. These bits come in standard screw sizes—one each of the #4, #6 and #8 sizes will handle most of your needs. Or you can bore the counterbore and pilot hole with two different twist drill bits, and then countersink the top of the hole. To do this, you need to know the various sizes of bits that are needed. These are listed in Table 5-5. The pilot hole size is the one that will receive the threads of the screw. The clear hole size will pass the shank of the screw. Drill the pilot hole first, to the necessary depth for the screw length. Then drill the clear hole, if necessary, to part depth. Finish off with the countersink, of a type and size that matches the head chamfer of the screws you are using. Where a counterbore is needed so that the screw head can be deeply recessed and then covered with a wood plug or button, drill a counterbore to the necessary depth with a drill bit that matches the size of the plugs that you will use.

Other Fasteners

There are, of course, many other kinds of fasteners. From time to time you will have need of some of them. The best way by far to learn about them is to pay a visit to your local hardware store or lumberyard. Just browse around the hardware store for a while to see what is readily available and can be used for different fastening purposes. Such items as bolts, machine screws, lag screws, washers and nuts, screw anchors, lead shields, hollow-wall fasteners, plastic plug anchors, screw hooks, screweyes and staples will from time to time be necessary in your various projects. There are, however, three particular items that must be mentioned here.

The first item is *sheet metal screws*. Where appearance is not a factor, sheet metal screws can be used to great advantage in plywood. They have great holding power because of their thread pattern, are straight-shanked except for the tip and threaded all the way to the head, and they drive very easily. Several types are available, but the most useful and also the one most commonly stocked in stores is the *Type A panhead*. Flathead sheet metal screws are not made; the panhead has a flat-bottomed head and a rounded top. In many applications it is best used with a small flat washer beneath the head. It can also be recessed in a counter bore and concealed with wood filler. For general purpose, rough-and-ready screw fastening applications, this screw is hard to beat.

Another useful item, especially when working with plywood, is a little device called a *corrugated fastener* or sometimes a *"wiggle nail."* This small piece of heavy corrugated steel is razor sharp on one edge and flat on the other. It comes in various sizes and it used by driving the fastener straight down into the wood, half in one workpiece and half in the other (Fig. 5-12). The fastener is particu-

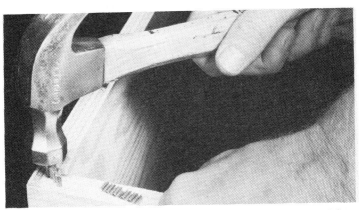

Fig. 5-12. Corrugated fasteners are useful for joining plywood in edge joints, as shown here (courtesy of the American Plywood Association).

Fig. 5-13. Tee-nuts are excellent for blind-fastening two pieces, and can be readily disassembled.

larly useful when edge-joining two pieces of plywood. A series of them can be angled across the joint line, driven in from the reverse side in situations where appearance is of little consequence (they do tear the wood fibers somewhat at times, and they do show). These fasteners also work very well in strengthing miter joints, either flat or edge, by driving them into the workpieces at right angles to the joint line. In fastening an edge-joint corner this way, the fastener must be carefully set so that it does not drive into the outside plies of the workpieces and split or bulge them outward.

The third item is commonly available at hardware stores everywhere, but is often overlooked as a fine method of fastening plywood joints together. This is the *Tee-nut* (Fig. 5-13), which is used where one surface must be solidly bolted to another. After a clearance hole for the machine bolt is drilled, the Tee-nut is inserted from the reverse side of the second workpiece in a hole drilled to accommodate it. The sharp teeth on the Tee-nut bit into the wood and hold it in place. Then the machine screw, which can be flathead and later concealed, is run through the clearance hole in the first workpiece and threaded into the Tee-nut from the front side of the second workpiece. The screw is then taken up and the two pieces are more solidly joined then can be accomplished in practically any other way short of a bolt. When the article is disassembled, the Tee-nut will stay right in place for reassembly.

BUILDER'S HARDWARE

Builder's hardware is a category that includes a tremendous number of highly varied items for both decorative and functional purposes (often both at the same time) that are required in the completion of many projects. These items include hinges of all kinds, locks and latches, catches, knobs and pulls, brackets, drawer slides, door tracks, shelf standards, angle braces and a host of other bits and pieces. There is a vast array of this hardware of both utility and decorative types manufactured today, and exactly what is available varies from locale to locale. There is little point in going into detail in this area, as you can gather up more satisfactory and pertinent information for your own purposes simply by visiting a hardware store or two. Much of this hardware requires no particular instructions or directions for installation. In most cases, the job is so easy and the piece is so simple that what must be done is perfectly obvious. Where instructions are needed, they are almost always included with the hardware; this is particularly true of hinges, locksets and the like. One particular type of hardware, though, does sometimes cause some confusion and installation difficulties. Hinges need a bit of discussion.

Types of Hinges

There are a number of categories of hinges, but those most commonly used by the home mechanic will be *utility hinges, decorative surface hinges* and a specific type called *butt hinges*. Utility hinges are plain, plated steel hinges meant for general-purpose utility work wherever appearance is not important but good strength is. Though they can be mortised in so that they are flush with the workpiece surfaces, they are almost always surface-mounted. They are available in butt type, T and strap configurations (Fig. 5-14), and are simple to install.

Decorative surface hinges are widely used in cabinetry and similar applications. There must be a thousand different styles

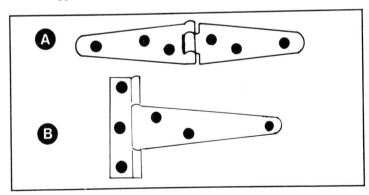

Fig. 5-14. (A) Fixed-pin utility strap hinge. (B) Fixed-pin utility T-hinge.

Fig. 5-15. Various sizes of loose-pin butt hinges.

available, including butts, and most are designed to be attached with screws directly to the surfaces of the workpieces. A few are designed so that one leaf mounts on an outside surface, while the other mounts on an inside surface. These, too, are easy to install. The chief problem is to properly align the leaves on the workpieces and then drive the screws in straight.

Butt hinges (Fig. 5-15) may be either utility or decorative in finish, but they are all approximately the same. They consist of a pair of flat, plain leaves held together by a pin driven through mating curls in the leaves called a barrel. The pin may be non-removable (fixed-pin type) or removable to separate the two leaves (loose-pin type). These hinges can be installed in any one of four different ways, as shown in Fig. 5-16. The surface-mounted portions of these hinges, as well as other surface-mounted types, present few problems provided you learn the trick of properly setting them.

Setting Screws

First, carefully align the hinge leaves, so that the hinge pin is exactly in line with the edges of the two workpieces being hinged. If one hinge is a little bit out of line, none of the hinges in the series will work properly. The door will bind or bow and try to force the hinge into line. Then, holding the hinge firmly positioned, mark the exact centerpoint of each screw hole, using the predrilled holes in the hinge leaves as guides. One trick that works well is to trace a circle around the inside of each hole with a sharp pencil held perpendicular to the hinge. Then remove the hinge and find the centerpoint of the circles. This avoids the possibility of having the hinge slip out of place during the business of working with the scratch awl. Another alternative is to use a centerpunch.

Next, drill pilot holes only—no shank holes—for the screws. In most plywoods, you can undersize the drill bit a few thousandths of an inch from the size listed in the table. The holes must be drilled exactly perpendicular to the workpiece surface, and this is best done on a drill press where possible, or with a portable electric drill fitted in a drilling jig. If the holes are perpendicular, the screws will go in straight. If not, the screws are likely to run in crooked. When the heads contact the hinge leaf, they will pull the leaf out of alignment. Repositioning a hinge-screw hole to get the screw straight is almost an impossibility.

Mortising Problems

The mortising or inletting of hinge leaves, a common practice not only with hinges but also with other kinds of builder's hardware as well, does present a few problems at times, especially when working with plywoods. First, as with surface-mounting, the hinge leaves have to be exactly aligned for proper operation. Then the hinge must be traced around with a scratch awl or some other sharp instrument to form a precise cutting guideline. The problems that are likely to occur during the mortising are fourfold.

First, the mortise must exactly follow the cutting guideline for the sake of appearance. This is no problem when mortising with a router and template, but most mortises are made either with wood

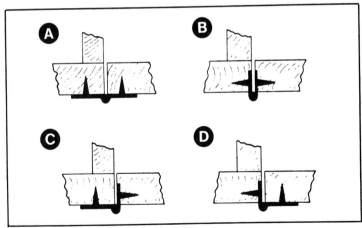

Fig. 5-16. The four principal butt hinge mounting arrangements. (A) Full surface. (B) Full mortise. (C) Half mortise. (D) Half surface.

chisels or a hand-guided router. In this case, the only recourse is to pay close attention to the cutting process as you work, control the cuts carefully, and above all use bits or other cutting edges as sharp as you can get them.

The second problem is splintered edges. These can usually be avoided by using only sharp cutting tools, but sometimes the wood has a tendency to splinter anyway. If this appears to be a possibility, stretch masking tape right along the edge of the cutting guideline and press it down firmly. The surface of the material msut be perfectly clean and dust-free, and the tape should be removed immediately after cutting is done. This trick will often minimize or eliminate splintering, especially of the fine variety. A different kind of splintering can take place when you are working quite close to a plywood edge, or when you are mortising the edge itself. Chipping and splintering, as well as the slivering away of corners and cut edges, are always a good possibility. Use the tape trick here if you can. Be exceptionally cautious when cutting close to the edges of the mortise.

Problem three has to do with depth. It is difficult, especially for beginners, to make a mortise of the correct depth when using hand tools. If the depth happens to be equal to the thickness of a ply, you can use the glueline as an indicator. If not, in most cases you have to go largely by eye. But there are some tricks that you can use. First, measure the actual thickness of the piece to be inletted. You can use a *vernier caliper*, or a pair of inside calipers which then can be placed on a bench rule for a reading. An ordinary rule, however, is not likely to give you an accurate enough measurement.

If you are mortising into an edge, make a depth mark along one surface as a part of the cutting guideline (Fig. 5-17). Take the material off with shallow cuts of the chisel, a little bit at a time, and keep checking the depth with a good bench rule or the brass extension from a carpenter's rule. As you approach the bottom of the cut, keep trying the hinge leaf (or whatever) in the mortise to see how much it lacks of being flush with the workpiece surface. Eventually you will reach the right level. A mortise that is too shallow or too deep will hinder the hinge action, offset one workpiece from the other and thus cause all manner of problems, It certainly will not present a good appearance. Of course, if you are using a router, there should be no problem. Set the depth of cut adjustment to the correct point, after determining the thickness of the piece to be mortised, and then make a trial cut on a piece of

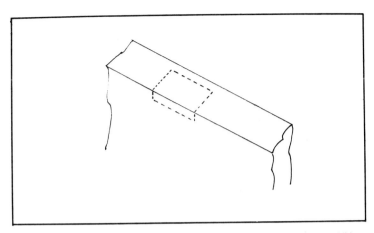

Fig. 5-17. When a mortise is being made at an edge, a depth line in addition to the surface outline is helpful in cutting a correct mortise bottom.

scrap wood. Measure the depth of the cut to ascertain its accuracy, make any readjustment that is necessary and proceed with the mortise.

The fourth problem is allied to the third in that it is difficult to achieve a perfectly flat-bottomed mortise when using hand tools. It may be rough and uneven or humped in the middle or otherwise malformed in some fashion. It is difficult to fasten the hardware into such a mortise. If the screws are cranked down tight, the hardware may well be racked out of line, even to the point where a hinge will refuse to operate. The only way to achieve a flat-bottomed mortise with a chisel is to use a fairly wide blade and make just shaving cuts as you reach the full depth of the mortise. At the same time, keep checking for flatness and planeness with the end of a bench rule, the edge of a carpenter's square, or anything else that will give you some indication of how flat and smooth the bottom of your cut really is. Especially in small mortises, the eye has difficulty in detecting surface irregularities. Material can always be removed, but putting it back is more difficult. Proceed slowly and carefully.

If you get your mortise too deep, smooth the bottom of the cut off and glue a piece of shim stock into the mortise. Any piece of thin wood or veneer, matching if necessary, will do the job. You may be able to cut the mortise to the exact depth required so that when the shim piece is glued in place, the bottom of the mortise will be at exactly the right depth. Otherwise, wait for the glue to cure. Then cut a small amount away from the surface of the shim to complete the proper mortise.

Surface-Mounted and Edge-Mounted Items

As far as the installation of other kinds of builder's hardware is concerned, there should be no particular problems. Mortising is the most difficult job, and the same principles discussed above apply to other kinds of hardware as well. For the rest, it is primarily a matter of proper alignment and positioning, coupled with the drilling of a few holes and the setting of a few screws. Those items that involve any complexity of installation will come complete with directions.

There is just one further point to be noted with respect to mounting builder's hardware on plywood, and that is the disparity between the strength of surface-mounted and edge-mounted items, whether mortised or not. If you mount a piece of hardware on a plywood surface, it will be as rugged as you could possibly hope for, assuming that the mounting job is done correctly. But if you mount a piece of hardware on a plywood edge, it probably will not have as much strength. This is not always true and depends upon the specific kind of plywood. As with nails driven into plywood edges, screws frequently do not have good holding strength, depending upon whether they go into a cross grain or an edge grain. If you can mount to the cross grain of a lumber-core or banded type of plywood, that is all well and good. Sometimes the same can be done in a panel that has a thick veneer core. But with multi-ply panels, especially thin ones, difficulties may arise. The usual recourse is to use considerably longer screws than normal. Be sure that they are driven perfectly straight into the wood, with a properly-sized pilot hole to guide them.

EDGING

Most folks don't find the characteristic stripes around the edges of plywood particularly attractive. In most situations where a transparent or semi-transparent finish is called for, they are particularly objectionable. The stripes often show through even when painted. There are a number of ways to get around this.

One possibility, widely used and quite successful in situations where an opaque finish will later be applied, is to smooth the edge exceptionally well. Sand it well. Fill all the tiny pores, cracks or other defects with spackle. Then sand again, making sure that the edge remains perfectly square. Treat the edge with a sealer. This can be, and usually is, done during the final finishing process. Sand the edge again lightly with very fine paper, or use steel wool and seal again. Repeat this process as many times as necessary to completely hide any vestige of the plies. If properly done, there will be no visible evidence once the edge is finally painted that the material is plywood.

Strips and Moldings

One of the easiest methods of hiding plywood edges is to apply edging strips. You can make up strips yourself, square-edged from a table saw, and then leave them squared or shape their outer corners or surfaces to any form you desire, like a molding. The usual thickness for these strips ranges from about ¼ inch to as much as ½ inch. You can buy a stock pattern of molding in any one of several different woods (clear pine, mahogany and oak are all common) from a lumberyard, in an appropriate width, thickness and configuration.

One of the most commonly used moldings for this purpose is called, not surprisingly, *shelf-edge molding*, and it is made for just this purpose. Both bottom and top edges are nosed or rounded. It is ¾ inch wide and fits exactly on the edge of a ¾-inch plywood panel. The same molding can be trimmed slightly in width so that it can be used for thinner plywood workpieces. These strips are easily cut to size and applied to a squared and smoothed plywood edge with glue and finish nails or brads (Fig. 5-18). Usually the corners of the edge molding are mitered, but butt joints can also be used.

Where greater strength is required, other kinds of molding that inset into a dado in the panel edge or are joined by some other form of joint can also be applied. Though more complicated and

Fig. 5-18. Shelf-edge molding has been applied, miter-cut at corners, to cover the raw edges of this plywood case.

Fig. 5-19. Veneer strip is one method of covering plywood edges (courtesy of the American Plywood Association).

more work, they are also a higher level of quality and somewhat sturdier. A lipped edging can be made by making the edge molding wider than the thickness of the panel. One edge of the molding is then applied flush with one workpiece surface, so that the molding protrudes beyond the opposite workpiece surface. In all cases where edge moldings are used, be sure of the specifics as you lay out the plans for the project, so that the thickness of the moldings can be taken into account as the various workpieces are dimensioned and made up.

Using Veneer Tape

A popular alternative to applying edge moldings to plywood edges is to cover them with *veneer tape* (Fig. 5-19). Several different kinds of this tape, which usually is made of natural woods in several species, are easily obtainable at lumberyards and hardware stores. Some kinds are designed to be glued on with a glue of your choice. Aliphatic resin works well. Others are manufactured with a self-sticking adhesive. All that is necessary is to peel the backing away and press the tape in place. The edge must be perfectly smooth and completely free of dust, or the tape will not stick properly.

Applying Veneer

An alternative to this method is to apply thin strips of veneer to the plywood edges. There are a couple of advantages to using veneer strips. First, they can be obtained from woodworking houses at reasonable cost in several thicknesses and different wood species, including some of the exotics. Second, the veneers can be cut to any desirable width, so that not only edges of any width but also counter lips or other adjoining surfaces (or even a whole project) can be covered. Veneers can be applied with glue or contact cement, either pretrimmed or trimmed after installation. The latter procedure is usually best, whenever possible.

Other Methods

Where it is necessary to hide plywood edges without altering the size of the plywood workpiece, or when it is desirable to use an identical or similar covering material as the workpiece and leave the workpiece surface grain uninterrupted by a border of any sort, there are two other good methods to use (Fig. 5-20). When the appearance of the back face of the workpiece is of little consequence, you can cut the edges of the workpiece to a 45° bevel, slanting in toward the back surface. Then cut a matching triangular strip of plywood or other wood, running with the grain. Glue and nail the strip to the bevel, squaring the edge out again. Where the appearance of both surfaces of the workpiece are important, cut the edge of the workpiece in the form of a notch, with the sides of the notch angled at any appropriate degree. Then cut a matching

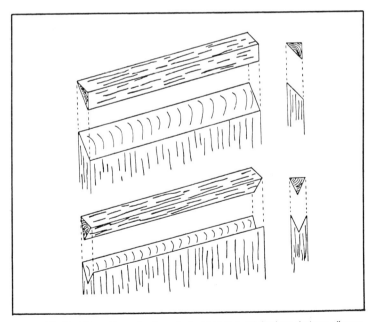

Fig. 5-20. To hide plywood edges but retain the original workpiece dimensions, the edges can be cut away and filled with solid stock.

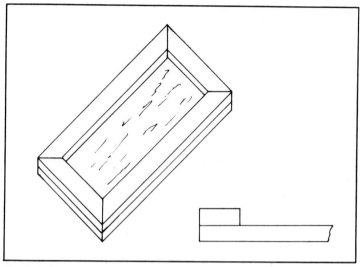

Fig. 5-21. This method of panel framing places the frame on top of the panel; raw plywood edges are visible.

triangular strip that will fit down into the notch and square the edge of the workpiece off again. Glue and nail this piece into place.

FRAMING

Framing a plywood panel may be done to hide the plywood edges, but more often it is a strengthing and stiffing measure for cabinet doors, plaques and similar workpieces made up of thin sheets of plywood. There are several ways to go about the job.

One method is to build the frame on top of the plywood panel. This is a matter of building a "picture frame" that is of the same outside dimensions as the panel, and then setting the frame atop the panel and gluing and fastening it into place (Fig. 5-21). The corner joints of the frame can be end butt, but a miter is more often used and is generally more attractive since this construction shows no end grain. Another possibility, which shows some degree of end grain at each joint corner, is the end lap.

Another method of framing, used where the thickness of the plywood panel is sufficient to allow a decent joint, is to make a frame that fits around the panel, with the pieces joined to one another and also to the edges of the plywood panel (Fig. 5-22). This job is done in much the same fashion as when applying edging strips or molding , except that the frame material is generally considerably wider. Whereas most edge moldings are affixed edgewise to a panel edge, frame stock is laid flat and applied to the panel edges.

There are two variations on this theme that are commonly employed when the plywood panel is substantially thinner than the framing stock. One method is to rabbet the backside of the frame members around the inner perimeter with a groove equal in depth to the thickness of the plywood panel, and of any suitable width (usually ⅜-½ inch). The frame is assembled and laid face down. The panel, also face down, can be dropped into the rabbet and secured. Care must be taken to wipe excess glue from the panel face, should it squeeze out along the rabbet joint. An alternative method is to dado along the longitudinal centerline of the inside edge of each frame member; the dado width should be equal to the thickness of the plywood panel plus just a hair, and a ¼-inch depth is generally sufficient. Then the frame members are fitted around the panel, with the panel edges riding in the slots. Both of these methods are excellent for making panel doors of all kinds.

FINISHING

The last step toward the completion of a project is simply called *finishing*, but actually consists of several different steps that vary considerably with the sort of wood involved, the kind of project or article, and the desired appearance of the finished product. In some cases, of course, there is no finishing to be done. This would be true of a plywood subfloor or roof decking. In other

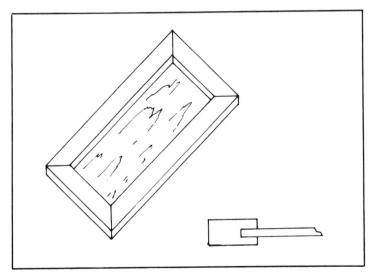

Fig. 5-22. Plywood edges can be hidden by framing around the edges of the panel.

instances, there is only a small amount of finish work to be done where prefinished plywood panels are used, as in paneling a wall. But for most projects, it is necessary once the assembly has been completed to prepare the surface and then apply a finish of some sort. In order to achieve excellent results, great pains should be taken with this process. The finish can make or break a project. If poorly done, it will be the ruination of an otherwise excellent piece of work. Properly done, the finish can make a good fabrication look truly professional, and even an average piece of work very nice. Usually, though, even a superb finish will do little toward distinguishing a poorly-built article.

Finishes and finishing constitutes an extremely complex subject, one that cannot be investigated here to any great extent. Many volumes have been written in this area, and appropriate reference works should be consulted for in-depth information and instructions for various specialized and unusual finishing techniques. The basic information that follows will be sufficient for most kinds of average finishing jobs. Once you have mastered these techniques and assimilated the fundamental principles of finishing, you can easily go forward with more advanced techniques.

SURFACE REPAIRS

No matter how carefully an assembly is put together or how good the material is to begin with, there are always a few surface defects that must be taken care of before any finish can be applied. Nail and screw holes must be filled. There are usually a few scratches, dents, abrasions, cracks, fissures and pinholes. The best way to take care of small dents is not to fill them. Instead, raise the surface grain of the wood until the dent disappears. This is done by laying a quite damp (but not wet) patch over the dent, and then pressing the cloth with the tip of a hot iron. A clothes iron or a soldering iron will work well. Leave the hot iron in place just briefly to create a small pocket of steam that will swell the bent fibers of the wood and allow them to move back into their original position. If the fibers are not torn or gouged, this will work quite effectively.

Other defects, including dents that will not raise by steam, must be filled. The type of filler that you use should be dependent to some extent upon the final finish that will be applied to the surface. Many wood fillers will not take a stain, or they will absorb stain but will turn a different color than the surrounding wood. If the finish will be opaque, you can use practically any wood filler, such as a plastic wood or wood dough. Glazing putty works well in some circumstances, particularly for small nail holes. But it will remain soft for a long period of time and may be attacked by solvents in the finish coatings. Spackle, either in powder form to be mixed with water or as a ready-to-use vinyl paste, makes an excellent filler, especially for small imperfections. Whatever you use, follow the manufacturer's directions for application. Smooth the material into the defect with a putty knife or artist's palette knife. With some fillers it is necessary to overfill the defect, since the putty will shrink slightly upon curing. The idea is to leave enough extra so that the material will not shrink down into the defect, and can be sanded flush without having to fill a second time.

Filling is a bit more tricky if the panel finish is to be natural or semi-transparent. The filler must take the natural finish or the stain to the same degree as the wood surface itself, so that the patches will not be obvious. A few fillers will take stains satisfactorily, and a few will match the surface of the workpiece, depending upon its nature. Some fillers are manufactured ready-to-use in assorted colors to match wood surfaces.

In any case, where a perfect match is desired, it is wise to experiment first with some scrap wood. One common procedure is to mix the filler with pigments or stains until a match is obtained. Another possibility, where the finish will be a stain, is to stain the surface first and match a filler to the final stain color as it appears on the finished piece. Then fill the defects and apply a transparent topcoat over the stained surface. You can also obtain wax or putty sticks in many colors to match various workpiece surface colors and accomplish the same purpose. These fillers are rubbed into the defects or nail holes after the stain coating has been applied, to make an invisible patch.

Another widely used procedure, especially effective for natural finishes, is to save the fine sawdust made when cutting or sanding the material. This should be set aside in a clean envelope so that it will not become contaminated or dirty. Then, when the time comes to fill surface defects, the fine sawdust is mixed with glue to form a paste and pressed into the defects. The finer the sawdust, the more effective the filler will be. It can then be sanded off and looks just like the wood itself. Care must be taken when using this procedure on stained surfaces. Sometimes the stain will not take properly in the sawdust patch because of the glue. The color will probably be different. Again, some experimentation on scrap wood is in order.

Defects, nail holes and screw holes can likewise be filled in workpieces that have a textured surface. Overfill the hole or defect with a suitable filler, and allow it to harden just a bit. Just how much the material should harden depends upon its characteristics. You will have to experiment on scrap to determine the timing involved. At some point before the material has hardened completely, tool the surface of the patch to match the texture of the surrounding surface. You can use almost anything for a tool, and your selection will be governed to some extent by the kind of texturing involved. Brushing the patch surface with an old toothbrush works well for resawn or rough-sawn surfaces, while a wide-bristled wire brush might work better on some other patterns. A blunt-pointed nail could be used on a striated surface.

After all the imperfections are filled, go over all of the exposed joints in the assembly. Check them first for a tight fit. If there are any visible cracks, fill them too. Check also for squeezed-out glue beads, and chip them off with a sharp chisel. If the finish is to be natural or stained, look for glue spots on the wood surface. These should be sanded away with very fine sandpaper, as they may show under a natural finish. Stains will not penetrate them, leaving a light-colored spot. At the same time, examine the entire article to make sure that nothing has been missed, all of the pieces are in place, and no further construction work needs to be done.

SANDING

There are several kinds of sandpaper that you can use, and they are graded according to the size of the grit particles. Cheap papers are simply designated fine, medium and coarse, which tells you very little. They are not good for your kind of work, except perhaps in a pinch. Some papers are also graded by the "ought" system and will be numbered 3/0, 5/0 and so on. The smaller the number, the coarser the paper. Most papers, however, are designated by numbers that indicate the size of the mesh through which the grit particles will pass. The smaller the number, the coarser the paper. A 60-grit or 80-grit paper is very coarse, while 320-grit and 400-grit are very fine.

Grits

There are also several kinds of grits that can be used. Flint paper is the least expensive, has the shortest life and is the least effective type. It is easily spotted by its buff or tan color. Garnet paper is considerably better, though a bit more expensive. It is the usual choice for all-purpose sanding; its color is reddish-brown. Aluminum oxide is an excellent choice for fine finishing and has excellent life. It is brownish-purple in color. Silicon carbide paper is the toughest abrasive and also a good, though expensive, choice. This grit is shiny black. Sandpapers also are graded as to weight. The so-called finishing paper has a lightweight backing, while the cabinet weights are stiffer and heavier. Weights may also be designated by letter, with A being light and C, D and E progressively heavier. There is no B grade.

A variety of grits is also available on cloth or flexible backings. Most craftsmen have their own personal ideas of what constitutes a good sandpaper. They usually settle upon a few grits and types that work well for their purposes. Flint is a good throwaway paper for rough jobs that will clog the paper rapidly and where an expensive sheet is best not used. Garnet may very well become your standby. But the best bet is to try different kinds, determine what you like best and stick with them. If you plan to use much sandpaper, by all means purchase it in bulk quantities from a woodworking supply house. You will save a considerable amount of money this way.

Procedure

The usual procedure when working with solid woods is to start finish sanding as soon as the holes and defects are filled. Start with as coarse a paper as deemed necessary for the particular surface and kind of wood, and progress in stages to an extremely fine grit for a velvety surface. This, however, should not be done with most plywoods. There are two problems, one having to do with the softwood appearance grade plywoods, and the other with hardwood or specialty plywoods. The appearance grade softwood plywoods have already been factory-sanded to a relatively smooth surface by means of specialized machinery. Further sanding of the bare wood surface is not necessary and can cause considerable harm. This is because sanding with small blocks or power sanders normally found in home shops will quickly eat away the soft portions of the wood, while leaving the hard portions raised. The result is a ripply surface that will become obvious when the finish is applied and is very difficult to correct. The problem with the hardwood plywoods is that they too are presanded to certain tolerances. The face veneer is likely to be very thin, and it is too easy to scrub entirely through the face ply in places.

There is no question but that all of the visible surfaces of the article must be thoroughly and completely sanded in order to achieve a fine finish. With all presanded plywoods, the job is simple. First, go over all of the patches that you have put in. Sand them smooth and flush with the surface with a fine sandpaper. Be careful not to create hollow spots by sanding down too far. Check all of the joints and exposed edges and touch them up with a fine sandpaper as necessary, so that they are perfectly smooth. At the same time, ease all of the sharp corners as necessary, again using a fine grit paper.

For touch-sanding patches and any rough spots that might show up, use a sanding block or a small pad sander. Sand only back and forth with the grain. Even slight diagonal movements will cause scratches that are difficult to eradicate. The easiest way to ease edges is to hold a piece of fine-grit, worn sandpaper over the edge, or wrap a small piece around your finger and sand with the grain with a rounding motion. Corners along a surface end-grain can be eased by holding a strip of paper taut between your hands and going across the edge in a shoe-shine motion. Don't slide the paper alone as you do so. Sand one area, lift the paper up and place it on the next area. Once you have completed this process, the entire surface of the article should be quite smooth. The next step is to start the finishing process. There are many different possibilities. In many of them, further sanding (or steel-wooling) is done between coats. Whatever further sanding is necessary will be noted in the following discussion of various kinds of finishes.

GRAIN FILLING

Many woods have a tight, close grain system that makes for a very smooth surface when the wood is finished, exhibiting no pores or openings of any kind. White pine and redwood are typical examples. Other woods, however, have a rough, open grain system. No matter how carefully the surface of such woods is sanded, the pores will remain open and obvious, especially if the surface coating happens to be a glossy one. Oak and mahogany are good examples (Fig. 5-23). In order to achieve a slick, smooth finish on such woods, the grain must be filled. This is done with a special wood-filling paste. These materials are generally cream-colored and can be used as is if an opaque finish is to be applied. They can be tinted with pigments or stains to match the color of the wood for a transparent finish, or the color of the stain for semi-transparent finish. Specific instructions for proper application of the filler are included with each can and should be followed to the letter. Generally, the filler must be diluted somewhat and thoroughly mixed.

Fig. 5-23. A close, tight grain like white pine (left) will sand and finish smooth. An open grain like the mahogany on the right must be sanded as smooth as possible, then filled with a grain filler to achieve a smooth final finish.

Then it is applied to the surface with a stiff brush. After the filler becomes tacky, it must be rubbed across the grain with toweling or burlap to work the paste into the open pores. Then the piece must be rubbed with the grain with a piece of fine cloth. Procedures vary somewhat, and generally a little practice is needed to achieve a perfectly smooth surface.

APPLICATION OF STAIN FINISH

A stain finish is both attractive and easy to apply. There are two different types that are commonly used. One is semi-transparent stain, which tones the wood, emphasizes the grain and figure and has no effect upon texture. The other is solid-bodied or opaque stain, which hides the grain and figure and leaves a coating that is almost completely opaque, but emphasizes the texture of the wood. Though this latter type is normally applied as an exterior stain, it can be used indoors as well. Neither type of stain affords any great protection for the wood surface. Darkening with age, a common characteristic of wood surfaces, will occur beneath a semi-transparent stain and should be taken into account when the stain color is chosen. These stains are available, incidentally, not only in the typical earth or wood tones but also in colors, some of which are fairly brilliant.

Solid-bodied stains are applied to the wood with a brush or roller in the same manner as a paint. For good, solid coverage with no streaking, usually two coats are necessary. For interior use, the stain coating is best covered with a topcoat, especially if it is applied to cabinetry and similar woodwork. One of the best methods is to allow the second coat of stain to dry thoroughly for several days while being careful not to scratch or mar the surface. Then brush on a coat of high-quality varnish. A semi-gloss or low-luster type is usually used for interior work, but a glossy surface can be applied if desired. After this coat has thoroughly cured, rub the entire surface down with fine (3/0 or 4/0) steel wool. Clean the surface thoroughly with a rag or a vacuum cleaner and furniture brush. Then clean it with a *tack rag* (a special, sticky cheesecloth pad, available at any hardware or paint supply store, designed to pick up stray bits of lint and dust), and apply a second coat of the same varnish.

There are two kinds of semi-transparent stains that will work on plywoods, and each has a slightly different application procedure. Water or spirit stains are thin and pigmented relatively lightly. They are simply brushed on to the wood surface and allowed to penetrate. The brushing should be done evenly. Sometimes excess stain must be carefully wiped away with a rag in order to keep the appearance similar over the entire surface. Then the stain is allowed to dry and can be further protected with a topcoat of varnish or wax.

Oil stains are handled a bit differently. The usual procedure is to brush the stain on fairly quickly and liberally, usually onto one full surface or panel of the article at a time. Then the stained surface is immediately rubbed down with a rag, removing excess stain and toning different sections of the surface as desired. With practice, a good deal of control can be exercised in using oil stains to produce whatever effects are desired. Portions of the wood can be toned up or down for both contrasting or similar appearances. As with other kinds of stains, a topcoat can be applied to afford further protection for both the stained finish and the wood itself.

COLOR TONING

Color toning is a method of applying both stain and sealer to the plywood surface at the save time, and makes an attractive and reasonably durable, protective finish. Special nonpenetrating sealers are used with companion stains. The two are separate and must be mixed to suit the conditions. The usual process is to measure out a small specific amount of sealer and then mix the stain in, just a bit at a time. Meanwhile, brush the mix on scrap wood surfaces of the same kind as was used in building the article to be stained, until just the right tone is achieved. It is necessary to keep track of the amount of stain mixed into the sealer. Then a large enough batch is mixed, using the identical proportions, to finish the entire project.

The sealer is especially designed to not penetrate into the wood and to preserve the appearance of the natural surface. The mix can be applied by brush, pad, roller or spray. After it has dried thoroughly, the surfaces should be given a light sanding with a very fine paper or with fine steel wool. Care should be taken not to scrub the toning mix away in spots, which can result in a blotchy appearance. Finally, one or more coats of clear finish can be applied for protection and to give whatever luster is desired. A low-luster or semi-gloss finish is most often used, but a high gloss coating can also be put on. Indeed this is preferable in some situations.

PAINTS

To apply a paint finish, first prepare the surface as discussed earlier. Wipe the surfaces fairly clean of dust, and check them all for knots, sap spots, streaks, glue spots or oil stains. Coat all of these defects with a layer of shellac or sealer before proceeding with painting. Otherwise, the spots may bleed through to the finished coat, or the paint may not adhere properly. *Shellac* is often the preferred sealer for such blemishes, because it is easy to work and dries rapidly. A spray can is by far the easiest course. When the sealer is dry, touch-sand the spots lightly with fine sandpaper and apply a second coat if necessary. Just be sure all of the spots are thoroughly covered.

Priming Coats

The kind of paint that is applied to the plywood surface must be preceded by at least one priming coat. There are several possibilities here. The first and more commonly used arrangement is to apply what ever particular undercoat or primer is recommended by the manufacturer to go along with the particular finish paint being used. In the case of fir softwood plywoods, a primer that is especially designed for use of this kind of plywood works extremely well. This primer penetrates and also hides the rather emphatic grain found on such woods.

Another possibility is to use one or two coats of clear sealer; often this is then followed by a coat of primer as well. On plywoods where the grain is somewhat ripply, you can apply two or more coats of a surfacer composed of 50% denatured alcohol and 50%

four-pound-cut shellac. Apply two coats of the mixture and allow it to dry thoroughly. Then lightly sand the entire surface with a flat sanding block. This will take the shellac off the higher, harder grain areas while leaving the lower, softer areas filled. Continue applying these surfacing coats until the surface is as smooth as you desire.

A third possibility is to use a sander-surfacer, which is a special formulation of primer more often used on auto bodies than wood, but equally effective on both. This surfacer is especially formulated to build up in a fairly heavy layer and is best sprayed on. Then the surface is sanded to take down the high spots and leave the low ones filled. Several coats of this material can be applied to build up a surface that eventually becomes glass-smooth. Then the finish topcoats can be applied.

Regardless of the type of undercoating or primary coat that is applied, it should be lightly sanded with fine sandpaper after it has thoroughly dried, in order to remove dust specks, slight irregularities and high spots. If necessary, a second coat of primer should then be applied and similarly sanded off. It goes without saying that the surface must be thoroughly cleaned, preferably with a vacuum cleaner and a tack rag, before the first coating and between subsequent coats. The quality of the priming coats is extremely important to the appearance of the finished product. Take your time and be as critical of your work as you wish, especially if the article being finished is a piece of cabinetry, furniture or some other item where appearance is of paramount importance.

Oil-Base and Water-Base Paints

The final step is to apply the finish topcoats. In general, there are two possibilities: *oil-base paints* and *water-base paints*. There are numerous formulations of each, with most oil-base paints being a very durable, washable *alkyd resin enamel* with good hiding properties. Thinning and cleanup is done with turpentine or mineral spirits, and drying time is considerably longer than for the water-base paints. Water-base paints are emulsions of resin—acrylic and latex are both common—and are quick-drying, have good color retention and are very easy to apply. Drying time is short. Cleanup is done with water. Thinning is generally neither necessary nor recommended. These paints are available in flat, semi-gloss and high gloss. The particular type that you choose depends upon both the conditions under which the article will be used and its desired final appearance. Specific recommendations should be obtained from your paint supplier, who can furnish you with all the details about the various kinds of paints that he carries.

Once you have made a choice, all that is left to do is apply at least two coats to the primed and sanded surfaces. For brush applications, always paint with the grain direction unless that has been completely obscured by the priming process. Otherwise, always paint in the same direction. Flow the material on evenly and smoothly from a high-quality brush. Try to avoid scrubbing, backbrushing or touch-up as much as possible. Most of these paints can also be applied very successfully with paint pads, sprayers or rollers, using the techniques applicable to each.

VARNISH

Varnish is a traditional natural finish that has always been popular and probably always will be. Many of the old varnish formulations have been superceded by synthetics like *polyurethane*, but various kinds of spar, marine and other varnishes are still readily available and widely used. Varnishes are available in low-luster, semi-gloss and high gloss. They are relatively easy to apply, though the job must be done carefully for best results. It is important that only a top-quality product be used. Otherwise cracking, checking and peeling are likely to occur and refinishing will be required. Perhaps the biggest drawback to varnish is that it is very slow to dry and must be applied in a dust-free environment. When this is not possible you can, however, use one of the various quick-drying varnishes. Varnish also has the advantage that it can be applied over practically any other kind of finish—over stains or sealers for protection, over other kinds of varnishes for refinish, or over paint to change the gloss.

A varnish finish can be started right on the bare wood. After the surface has been properly prepared by filling and touch-sanding, apply the first coat of varnish directly to the wood. No sealing of blemishes is required, but wherever possible they should be removed for the sake of appearance. The varnish for this first coat should be reduced in the ratio of 1 pint of turpentine to 1 gallon of varnish, when is then thoroughly mixed (varnish should never be shaken). Apply this first coat across the grain of the wood in small patches, immediately following up on each patch by brushing the varnish out with the grain direction. Always work from a dry area into a wet one, never the other way around. You must work rapidly so that the edges of the finished areas don't become sticky or tacky by the time you get back to them. Otherwise, brush and lap marks will appear. This coating must be allowed to dry for a long time, especially in high-humidity conditions, until it is thoroughly cured. Then sand the surface lightly with fine sandpaper or go over it with fine steel wool. Clean the surface and apply a second coat, this time of full-strength varnish. Repeat this process for as many coats as are required to provide a fine, deep finish, smoothing the surface between each coat.

The newest kinds of varnishes, which are sometimes actually called varnishes and sometimes not, are the various kinds of clear plastic coatings, of which there are numerous formulations for different specific purposes. These urethane, polyurethane and similar coatings are for the most part quite thin and are very easy to apply. They can be used for a natural finish, often starting with the raw wood, or they can be applied over other finishes to form a hard, protective surface. These finishes are extremely tough, durable, resistant to scratches and abrasions and long-lived. They are also fast-drying and can be quickly and easily flowed onto the surface with a brush. The finishes work very well on plywood surfaces of all kinds; merely follow the manufacturers' application instructions. It should be noted, though, that sometimes the final result is not exactly what the user had in mind. If you have any doubts or if a particular appearance is of great importance to you, buy a small can of whatever coating seems best for your purposes and try it on some scrap pieces of wood.

SHELLAC

A shellac finish has the drawbacks of low water-resistance and not being particularly resistant to spills and stains. Even so, it remains a popular finish for interior woodwork, especially on small pieces of furniture and cabinetry. Shellac is easy to apply, inexpensive and does produce lovely natural-colored results. The process starts with a jar of four-pound-cut shellac and a perfectly clean, dry surface. For the first coat, cut the shellac half and half with alcohol. Apply the shellac smoothly and very quickly with a wide brush. The coating should be even and flowed on with the grain. It will dry very rapidly, and no attempt should be made to go back over any part of the already-coated surface. This coat will almost be completely absorbed into the wood. Then rub the entire surface down with a very fine grit sandpaper or 4/0 steel wool about 4 hours after the application.

Wipe the surface entirely clean once again with a tack rag. If you used steel wool, make a close inspection to ensure that there are no tiny steel wires left embedded in the surface. Then apply a second coat of the shellac. Repeat the smoothing process, and apply a third coat. At least four coats should be applied. Depending upon the nature of the wood, five or six might be needed. After the final coat, rub the surface down with 4/0 steel wool. Then sprinkle extra-fine pumice powder on the surface and hand-rub with an oil-soaked felt pad, with the grain, to achieve perfect, velvety smoothness. As a last step, apply a coating of hard carnauba furniture wax for a protective finish and a fine luster. This finish, incidentally, is especially attractive on close-grained, light-colored hardwood plywoods like white pine, birch or maple.

OILS

Oil finishes of one sort or another have long been used in woodworking. Though for a time they declined in popularity, they are now making a comeback. Several new products and reformulations of old ones that are very effective and easy to apply have helped greatly. The old tried-and-true oil finish is composed of boiled linseed oil mixed with an equal quantity of turpentine. The mix is brushed on liberally, allowed to sink into the wood for a short while, and then excess material is wiped away. This process is repeated, after each coat has dried, for as many as four or five times. After the last coat has dried thoroughly, it is rubbed hard with a soft cloth or with a lamb's-wool polishing pad in an electric polisher until a rich luster appears on the entire surface. The chief difficulty with this process is the length of time required to produce a nice finish, but it does work well. Periodic reapplications, usually necessary only about once a year, can be made to keep the finish looking nice.

The *linseed oil* treatment is particularly effective upon many hardwoods, such as teak, walnut and mahogany, and does much to bring out the richness of the color and grain of such woods. However, there is another equally effective finish that produces a harder and more durable surface and is much easier to apply. This is any one of the special polymerizing oil finishes, such as Watco, that deeply penetrates the wood surface and then hardens the fibers to a tough, wear-resistant surface. These oils are available in clear or in various stain colors; follow the directions included for proper applications.

Another good possibility is one of the most venerable of all oil finishes, *tung oil*, which has recently made a comeback in the finishing scene. Though quite expensive, tung oil can produce a marvelous finish if properly applied to a correctly prepared wood surface. This oil, too, is available in either clear or various stain colors. Perhaps the most readily available and easiest to use formulation is found in the Formby product line, which includes other wood finishing and refinishing materials as well.

EXTERIOR FINISHES

Finishes that will be subjected to constant weathering and high moisture conditions, such as those used on exterior house siding, lawn and patio furniture, house trim and the like, must be applied with special exterior-type coatings made for the purpose. If the finishing job is to be durable and long-lived, the job must be properly done. The finishes that we will discuss are geared particularly to exterior house siding made of plywood panels, but they can be equally well used for other exterior paintwork jobs.

No sanding or smoothing is necessary when preparing exterior plywood siding surfaces for a finish coating (this may not be true of outdoor furniture, though). However, the finish should be applied as soon as possible after the plywood panels are removed from the stack and nailed to the walls. The sooner the job is done, the better will be the performance of the coating. The more time is allowed for the plywood surface to be subjected to sunlight, drying and wetting, the poorer the performance will be.

In addition, all of the panel edges, top, bottom and sides, should be fully sealed with a good sealer before the panels are applied. Any edges that are cut during the fitting process as the panels are put up should be similarly coated. This will minimize moisture pickup through the end grains exposed along the edges and improve the life of the paneling, as well as the performance of the finish.

NATURAL AND STAIN FINISHES

The most natural of the *natural finishes* is to simply leave the wood at it is. Over a period of time that varies from a few months to a few years, the overall tone of the wood will alter, with dark-colored woods generally becoming lighter and light woods becoming darker. To prevent rust-streaking, only noncorrodible fasteners should be used. This "finish" is most economical and requires no maintenance or later refinishing. It is a perfectly acceptable method, especially in arid and semi-arid areas of the country. There is no cause for concern that the wood will deteriorate (if proper construction methods are used). All that will happen is that the surface will weather away at an average rate of about ¼ inch per century.

Where a protective coating is desired that will leave a natural appearance, it is wise not to use coatings like varnish or urethane. These clear coatings will eventually deteriorate from weathering and the effects of sunlight, and will present a scabrous appearance that is impossible to completely correct except by painting with an opaque finish. Instead, a clear, water-repellent solution that contains a wood preservative, such as *pentachlorophenol*, should be applied. This is a particularly popular finish for rough-sawn cedar and redwood siding and will penetrate the wood surface, retard the growth of mildew, reduce warping and provide protection against decay and water staining. Several of these finishes are available from paint supply houses. Perhaps the best of the lot is the so-called FPL finish that was originally formulated and tested by the Forest Products Laboratory. The initial application of this kind of finish is quite short-lived, and will probably have to be reapplied once a year for about the first 3 years. After that, the finish should remain fully protective for as much as 8 to 10 years, depending upon the exposure conditions involved. In the long run, this kind of finish is just as protective (often more so) and is more cost-effective than most other exterior finishes.

Stain finishes are highly recommended for textured exterior plywood siding and are also effective upon sanded plywood. Either semi-transparent or solid-body stains can be used. The former adds color and enhances the wood grain and texture, while the latter adds color and enhances texture but totally obscures the grain and coloration of the wood itself. Only oil-base semi-transparent stains are recommended, as other types do not provide enough protection. With the solid-body stains, either oil-base or latex emulsion stains can be used with equally good results. Both kinds penetrate well, provide a good bond and afford good protection for the wood.

Stains should be applied to clean, dry surfaces and never in strong sunlight or upon a hot surface. Either semi-transparent or solid-body stains should always be applied with a large, fairly stiff-bristled brush and worked thoroughly into the wood texture. Spraying is simpler and faster but gives very poor results, especially on textured woods, and is not a recommended procedure. A long-nap roller can also be employed, but again the results are not as good as with brushing. The process is messy and wasteful.

There are two ways of going about a stain application. One is to put on a single coat, well-brushed and also thoroughly back-brushed. This coating will probably only have a life of 2 to 4 years. Meantime, the surface will have weathered and checked sufficiently to allow excellent adherence of the second coat, which should be applied as soon as the original coat shows definite signs of weathering and possible chalking. The second coat, if properly applied in good weather, should last anywhere from 8 to 10 years.

The second procedure is perhaps the most effective but is also more difficult. This involves applying two initial coats, one right after the other. The drawback is that the second coat must be applied on the same day as the first, before the first coat has a chance to dry. This allows the best penetration of the coatings, and the staining job should have an effective life of approximately 10 years. This method is recommended for highly absorptive rough-sawn or weathered surfaces.

PAINT FINISHES

A great deal of experimentation and study has proven the fact that for an exterior paint finish to be durable and long-lived, the job must be done right and with the proper products. Once the siding surface has been prepared and a reasonable stretch of good weather seems to be in the offing, this is the recommended procedure to follow.

First, treat all wood surfaces, including trim, with a water-repellant preservative solution. This can be done by either brush or spraying. Special care should be taken at all joints, and all cut ends must be sealed. This coating should be allowed to dry for the equivalent of two warm and sunny days.

The next step is to apply a primer, and this is the most important step of all. First, the primer must be one that is compatible with whatever topcoat materials you have chosen. The topcoats can be of a good-quality oil-base, alkyd or latex house paint, and the primer should be one either manufactured or recommended by the producer of the topcoat material. The use of companion products designed to be used together is of extreme importance. For best results, the primer should be a nonporous, oil-base material that is resistant to extractive staining (stains that work up through the paint layers as a result of extractives in the wood itself). The primer should be applied thick enough to obscure the wood grain. This is best done by following the spreading rates recommended by the manufacturer, usually from 350 to 450 square feet per gallon for a good, heavy-bodied primer that is mostly solids by weight. Note that the so-called self-priming acrylic latex paints can indeed be applied without a primer, but this should be done only on woods like pine and fir that are free of colored extractives. Otherwise, stains will bleed through to the surface. Priming should be done thoroughly but rapidly, so as to cover the wood surface as soon as possible after it has been put up.

The final step is to apply two coats of finish paint. This is particularly important for those areas of the house that get the worst weather-beating, so do those first if the job is likely to stretch out over a period of time. The first coat should be applied after allowing a couple of days for the primer to thoroughly cure, but within two weeks at the most. Likewise, the second topcoat should be applied after the first has dried thoroughly, but again within two weeks.

There are some rather emphatic do-nots that should be observed if you expect a top-class paint job. First, the primer should always be applied with a brush, never by any other means. The topcoats should preferably be brushed also, but a roller is an acceptable second-place method. Spraying can be done as well, but it is the least desirable method. Second, never apply oil-base paints on a cool surface that will soon be heated by the sun; this will cause temperature blistering, especially with dark colors. Instead, stay in the shade and follow the sun around the house. Last, do not apply either oil-base or latex paints in the evenings when there is a likelihood of dew forming before the paint has at least surface-dried. Oil-base paints should never be applied at temperatures below 40°, and latex paints must be applied at temperatures of 50° or higher.

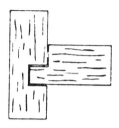

Chapter 6
Constructions and Projects

There are thousands of different ways in which plywood can be used, of course, from common and conventional structural building systems and ordinary casework shop projects to all sorts of unusual and far-out designs. Plywood is such a versatile material that probably all of its applications have never been counted, much less listed, and there are plenty that haven't even been devised yet. The following few constructions and projects are merely representative of some of the more common ways that plywood can be put to good use, and all are well within the scope of the do-it-yourselfer and home mechanic.

CONVENTIONAL PLYWOOD SUBFLOOR

Plywood has largely superseded boards as a subflooring material in houses (and many other kinds of buildings) because it is faster to lay, sturdier and less expensive. In addition, it provides a smooth, stable base for finish floorings of all kinds. The type of plywood most commonly used is C-D INT-APA or equivalent, with exterior glue; the same with intermediate glue can be used where there is unlikely to be any great amount of exposure to moisture during construction. Also, Structural I, II C-C and C-D, and Exterior C-C can be used.

The first step in building a plywood floor is to provide a foundation; the conventional methods are formed and poured concrete walls on a suitable footing, or concrete block on a footing. The foundation may be deep, to enclose a full basement, or shallow

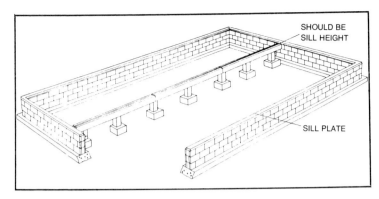

Fig. 6-1. Typical concrete-block crawl-space foundation (courtesy of the American Plywood Association).

for a crawl space. Other alternatives include stone, piers or posts, or the all weather wood foundation (AWWF) system (more about that later). A typical concrete block crawl-space foundation is shown in Fig. 6-1.

Once the foundation is completed, the floor frame must be built. A typical floor frame is shown in Fig. 6-2; construction methods are simple and conventional. The joists are sized accord-

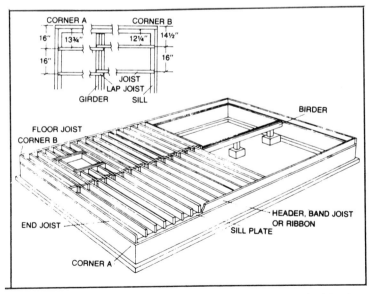

Fig. 6-2. Conventional floor frame assembly, with joist layout and lapping detailed (courtesy of the American Plywood Associations).

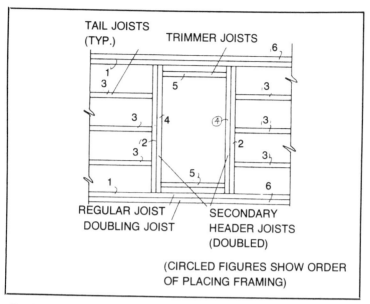

Fig. 6-3. Method of framing openings in floor; numbers indicate sequence of placement of the pieces (courtesy of the American Plywood Association).

ing to applicable span and loading tables that can be found in any residential construction manual. Floor openings are framed in accordance with the pattern given in Fig. 6-3.

With the frame completed, the next step is to determine the type of plywood and the thickness. The latter can be chosen from Table 6-1.; the most common combination is ½-inch or ⅝-inch placed on 16-inch centers. However, thicker panels can always be used, if desired, for extra strength or stiffness. This type of subflooring can be covered with structural tongue-and-groove wood strip or block flooring, or by underlayment for the application of nonstructural floorings like carpeting or sheet vinyl.

To start construction, lay the first panel in place, good face up, at one corner of the floor frame. If you have a corner where the measurement from the outside face of the end joist to the centerline of the first common joist is 16 inches, that's the one to pick. Align the panel with the face grain at right angles to the lie of the joists, and the edges flush with the outside faces of the end and header joists. Leave the panel loose; proceed with the next one in the row, and lay it with a gap of 1/16 inch from the end of the first panel. Continue setting out the panels until you come to one that misses a joist at the far end, by virtue of the gaps left between previous panel

Table 6-1. Plywood Subflooring Listing (courtesy of the American Plywood Association).

Panel Identification Index	Plywood Thickness (inches)	Maximum Span (Inches)
30/12		12*
32/16	1/2, 5/8	16*
36/16	3/4	16**
42/20	5/8, 3/4, 7/8	20**
48/24	3/4, 7/8	24
1-1/8" Groups 1 & 2	1-1/8	48
1-1/4" Groups 3 & 4	1-1/4	48

*May be 16" if 25/32" wood strip flooring is installed at right angles to joists.
**May be 24" wood strip flooring is installed at right angles to joists.

ends. Now, starting at the other end of the row and working backward, adjusting the panels as necessary, you may be able to juggle then about until only one panel will have to be trimmed.

If the pattern doesn't work out, then you will have to trim whatever panels necessary so that all have a good bearing surface underneath for nailing. The end gaps, incidentally, should be 1/8-inch in areas of consistently high humidity, and in any case can be greater than 1/16-inch if that makes adjusting the panels easier. If one panel should happen to miss resting upon a joist by just a small amount, you can install a scab and reset the following panel, or use a small filler strip to close a gap, as shown in Fig. 6-4.

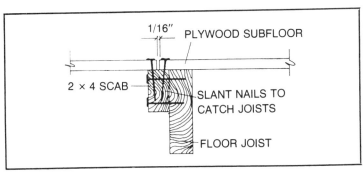

Fig. 6-4. If a subflooring panel misses a joist at one end, a scab can be put in for support (courtesy of the American Plywood Association).

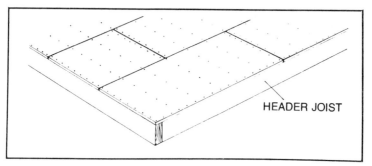

Fig. 6-5. Typical subfloor layout, showing nailing pattern (courtesy of the American Plywood Association).

Once the first row is adjusted, trimmed and set, you can nail the panels in place. The recommended nails are 6d common for ½-inch plywood, or 8d for ⅝ or ¾. However, box nails are also widely used, though they are not as rugged. Deformed shank nails—generally ring shanks—are sometimes used for extra holding power. Set the nails ⅜ inch from the panel edges, and drive them at a slight angle. They should be spaced 6 inches apart all around the perimeter, and 10 inches apart on the interior rows. The dots representing nails in Fig. 6-5 are the correct numbers of nails to use for that spacing. Snapping chalk lines is a great help in keeping the interior nails lined up and correctly driven.

After nailing the first row, start the second row by laying out the panels, just as you did the first. This time, though, the end-joint lines must be offset to produce the staggered joint lines. Generally a half panel is used to start the row, but it could also be a one-third or two-thirds panel. Note that in no case should a piece be laid across just two joists; at least two spans (a joist at each end and one in the middle) must be crosses, with the face grain always at right angles to the joists. Space the second row so that there is a gap of ⅛ inch along the side edges of the first row; double that gap for high humidity conditions. After all the adjustments are made and the trimming is done, nail the row down in the same manner as the first. The third row should be staggered from the second. This is usually done by going back to the original layout pattern, but a third pattern might also be used. Continue in this manner until all of the rows have been laid. Figure 6-6 shows a typical construction.

PLYWOOD UNDERLAYMENT

Plywood makes an excellent *underlayment* and is available in special grades just for the purpose. It can be laid upon a plywood

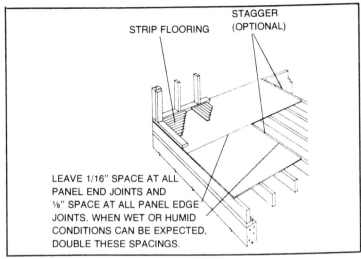

Fig. 6-6. Typical plywood subflooring installation (courtesy of the American Plywood Association).

subfloor of the type just discussed, for later application of tile, carpeting or similar finish floor coverings, either structutal or nonstructural. It can also be used to cover, and add considerable strength to, a new or existing lumber subfloor; it is particularly useful in remodeling and renovation work.

The first step is to choose an appropriate type of plywood underlayment for the job at hand (Table 6-2) and determine the amount of material needed. The panel layout should be calculated

Table 6-2. Plywood Underlayment Listing Association). (courtesy of the American Plywood

Plywood Grades and Species Group	Application	Minimum Plywood Thickness (Inch)
Groups 1, 2, 3, 4, 5 UNDERLAYMENT INT-APA (with interior, intermediate or exterior glue), or UNDERLAYMENT EXT-APA (C-C Plugged)	over plywood subfloor	1/4
	over lumber subfloor or other uneven surfaces	3/8
Same grades as above, but Group 1 only	over lumber floor up to 4" wide. Face grain must be perpendicular to boards.	1/4

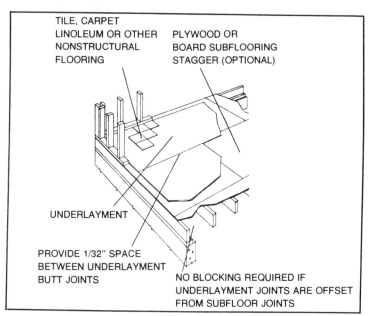

Fig. 6-7. Typical plywood underlayment installation (courtesy of the American Plywood Association).

so that the joints are offset from those in a plywood subfloor; wherever possible, lay the panels in the same direction as the boards in a lumber floor. For greatest stiffness, the face grain of the underlayment should be at right angles to the joists, and the end joints should lie directly above joists for good nailing.

Lay out the underlayment panels in much the same fashion as the plywood subfloor. A fair amount of trimming and cutting may be necessary, not only to cope with existing elements within the floor area (as in a renovation job), but also to make the joints fall properly. Application should take place just prior to laying the finish floor covering, so as to avoid damage to the underlayment. Stagger all panel joints and leave expansion gaps of 1/32 inch at all underalyment joints.

Nail the panels in place with 3d ring-shank nails for up to ½-inch thickness, 4d for thicker panels. You can also use 16-gauge staples in a gun; this method is especially useful for ¼-inch underlayment. Nails should be driven 6 inches apart all around the edges of the panels, and 8 inches apart each way in the interior portions. Staples should be set at 3-inch intervals on panel edges, 6 inches each way in interiors. Figure 6-7 shows the details of a typical installation.

COMBINED SUBFLOOR/UNDERLAYMENT

One of the fastest, most efficient and most effective floor systems combines the subfloor with the underlayment and can be laid on a simple floor frame. This is the APA *Sturd-I-Floor* arrangement, which uses a special plywood. By varying the thickness of the panels to suit, joists can be set on a conventional floor frame with joists on 16, 20, or 24-inch centers, or heavy, 2-inch centers, or heavy, 2-inch thick joists on 32-inch centers. Four-inch girders can be spaced on 48-inch centers. The span and width of all members must be calculated according to the design of the structure.

The plywood panels are made with precision tongue-and-groove joints, and also in a square-edged variety. Any type of finish floor covering can be laid down, whether structural or nonstructural. The panels are laid in the same fashion as ordinary plywood subflooring. End joints should be staggered. Squared edges *must* be blocked underneath for proper support; tongue-and-groove edges need no blocking. An expansion gap of 1/16 inch should be left at all ends and edges. Nail the panels in place preferably with 8d ring-shank nails, or 10d common smooth as a second choice. Nail spacing is 6 inches along the panel edges, and at intermediate supports 6 inches apart for 48-inch spans and 10 inches apart for 32-inch spans. Figure 6-8 details how the Sturd-I-Floor is constructed on a conventional floor frame. Figure 6-9 shows the 48-inch o.c. system.

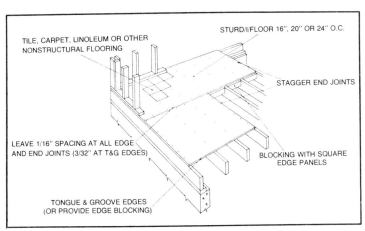

Fig. 6-8. Combined subfloor/underlayment arrangement, using Sturd-I-Floor plywood on a conventional floor frame (courtesy of the American Plywood Association).

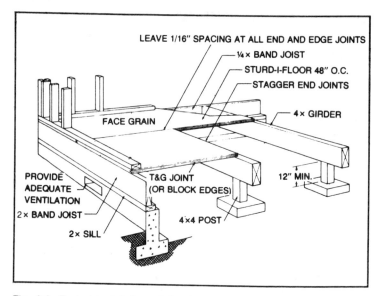

Fig. 6-9. Typical installation of Sturd-I-Floor plywood combined subfloor/underlayment on a beam-type floor frame (courtesy of the American Plywood Association).

GLUED PLYWOOD FLOOR

The APA *Glued Floor System* is economical and provides just about the stiffest, ruggedest and most noise-resistant plywood floor system possible. The standard C-D panels can be glued down to a conventional floor frame to make a two-layer floor, or Sturd-I-Floor can be used for a one-layer system on joist centers up to 48 inches. Any kind of finish flooring can be put down and the floor system will never squeak or bounce; nor will nail heads pop. Construction is fast. Frequently reduced joist size and/or increased spacing is possible because of the system's great strength.

The floor frame parameters must first be coordinated with the design needs of the house. Then the joist size, spacing, spans and the particular kinds of plywood can be chosen. This can be done in combination by consulting Tables 6-3 and 6-4. Note that there are numerous combinations possible. Though the joist span table does not include 32-inch or 48-inch spacing, the glued system can be used with them. The construction details are much the same as for the combined subfloor/underlayment system discussed previously.

Again, laying the panels is similar to the systems just covered. All panels should be spaced 1/16 inch at end joints. The tongue-

Table 6-3. Plywood Recommendations for APA Glued Floor System (courtesy of the American Plywood Association).

Joist Spacing (in.)	Flooring Type	Plywood Grade and Span or Identification Index	Possible Thickness (in.)
16	Resilient Flooring	Sturd-I-Floor 16 o.c.	19/32, 5/8
16	Separate Underlayment or Structural Finish Flooring	C-D 32/16, 42/20, 48/24	1/2, 5/8, 3/4
19.2	Resilient Flooring	Sturd-I-Floor 20 o.c.	19/32, 5/8, 23/32, 3/4
19.2	Separate Underlayment or Structural Finish Flooring	C-D 42/20, 48/24	5/8, 3/4
24	Resilient Flooring	Sturd-I-Floor 24 o.c.	23/32, 3/4, 7/8
24	Separate Underlayment or Structural Finish Flooring	C-D 48/24	3/4
32 or 48	Resilient Flooring	Sturd-I-Floor 48 o.c. (2·4·1)	1-1/8

Table 6-4. Maximum Joist Spans With Sturd-I-Floor Plywood Panels in the APA Glued Floor System (courtesy of the American Plywood Association).

Species and Grade of Joist	Joist Size	Joists 16" o.c. 19/32" Sturd-I-Floor	Joists 16" o.c. 23/32" Sturd-I-Floor	Joists 19.2" o.c. 19/32" Sturd-I-Floor	Joists 19.2" o.c. 23/32" Sturd-I-Floor	Joists 24" o.c. 23/32" Sturd-I-Floor
Douglas fir-Larch No. 1	2x6	11-0	11-4	10-6	10-6	9-5
	2x8	14-3	14-7	13-7	13-10	12-5
	2x10	17-11	18-3	17-0	17-4	15-10
	2x12	21-7	21-11	20-6	20-10	19-3
Douglas fir-Larch No. 2	2x6	10-6	10-6	9-7	9-7	8-7
	2x8	13-10	13-10	12-7	12-7	11-3
	2x10	17-7	17-7	16-1	16-1	14-5
	2x12	21-3	21-5	19-7	19-7	17-6
Douglas fir-South No. 1	2x6	10-4	10-8	9-11	10-2	9-1
	2x8	13-4	13-8	12-8	13-0	12-0
	2x10	16-8	17-0	15-11	16-3	15-4
	2x12	20-1	20-5	19-1	19-5	18-4
Douglas fir-South No. 2	2x6	10-1	10-1	9-3	9-3	8-3
	2x8	13-1	13-4	12-2	12-2	10-10
	2x10	16-4	16-8	15-6	15-6	13-10
	2x12	19-8	20-0	18-8	18-10	16-10
Hem-Fir No. 1	2x6	10-3	10-3	9-5	9-5	8-5
	2x8	13-7	13-7	12-5	12-5	11-1
	2x10	17-0	17-4	15-10	15-10	14-2
	2x12	20-6	20-10	19-3	19-3	17-2
Hem-Fir No. 2	2x6	9-4	9-4	8-6	8-6	7-7
	2x8	12-4	12-4	11-3	11-3	10-0
	2x10	15-8	15-8	14-4	14-4	12-10
	2x12	19-1	19-1	17-5	17-5	15-7
Mountain Hemlock No. 2	2x6	9-6	9-6	8-8	8-8	7-9
	2x8	12-6	12-7	11-6	11-6	10-3
	2x10	15-8	16-0	14-8	14-8	13-1
	2x12	18-9	19-2	17-9	17-9	15-11
Mountain Hemlock-Hem Fir No. 2	2x6	9-4	9-4	8-6	8-6	7-7
	2x8	12-4	12-4	11-3	11-3	10-0
	2x10	15-8	15-8	14-4	14-4	12-10
	2x12	18-9	19-1	17-5	17-5	15-7
Western Hemlock No. 1	2x6	10-8	10-10	9-11	9-11	8-10
	2x8	13-10	14-1	13-0	13-0	11-8
	2x10	17-4	17-8	16-6	16-7	14-10
	2x12	20-11	21-2	19-10	20-2	18-1
Lodgepole Pine No. 2	2x6	8-11	8-11	8-2	8-2	7-3
	2x8	11-9	11-9	10-9	10-9	9-7
	2x10	15-0	15-0	13-8	13-8	12-3
	2x12	18-3	18-3	16-8	16-8	14-11
Western Cedar No. 2	2x6	8-11	8-11	8-2	8-2	7-3
	2x8	11-9	11-9	10-9	10-9	9-7
	2x10	15-0	15-0	13-8	13-8	12-3
	2x12	18-3	18-3	16-8	16-8	14-11
Southern Pine (KD) No. 1	2x6	11-0	11-4	10-6	10-10	9-8
	2x8	14-3	14-7	13-7	13-11	12-9
	2x10	17-11	18-3	17-0	17-4	16-3
	2x12	21-7	21-11	20-6	20-10	19-7
Southern Pine (KD) No. 2	2x6	10-8	10-8	9-9	9-9	8-8
	2x8	13-10	14-0	12-10	12-10	11-6
	2x10	17-4	17-8	16-4	16-4	14-8
	2x12	20-11	21-2	19-10	19-11	17-9
Southern Pine (KD) No. 3	2x6	8-2	8-2	7-5	7-5	6-8
	2x8	10-9	10-9	9-9	9-9	8-9
	2x10	13-8	13-8	12-6	12-6	11-2
	2x12	16-8	16-8	15-2	15-2	13-7

(1) Based on live load of 40 psf, total load of 50 psf, deflection limited to ℓ/360 at 40 psf.
(2) Glue tongue-and-groove joints. If square-edge panels are used, block panel edges and glue between panels and between panels and blocking.

and-groove edges should be spaced 3/32 inch at the faces, except for Sturd-I-Floor 48 o.c. (2•4•1), which should be spaced 1/16 inch. Just before each panel is set in position (lay them out, trim and adjust them at least a row at a time), a heavy bead of special construction glue must be run along the joist tops and also in the tongue-and-groove joints. Whatever glue is used should conform to

Performance Specification AFG-01 to ensure proper performance of the system; an up-to-date listing of the brands and manufacturers can be obtained from the APA. Then the panel is nailed in place with 6d deformed-shank nails spaced 12 inches apart at all of the bearing points. The nailing must be completed within the time period specified by the manufacturer in order to assure proper bonding of the panels to the framework. The floor system is assembled as shown in Fig. 6-10.

CEMENTED-TILE SUBFLOOR

Ceramic tile floors can be laid in either of two ways. They may be cemented to a plywood subfloor or underlayment with an adhesive mastic, or sealted to a concrete subbase with cement-plaster. Of the two methods, the concrete subbase is superior in several respects, especially in heavy-traffic areas. But it is often bypassed in favor of the adhesive method, where the installation is to be made on a conventional wood-frame floor system because of the difficulty in arranging for the poured concrete subbase. There is, however, a simple method of construction, using plywood, that can be employed in either new construction or in a renovation job.

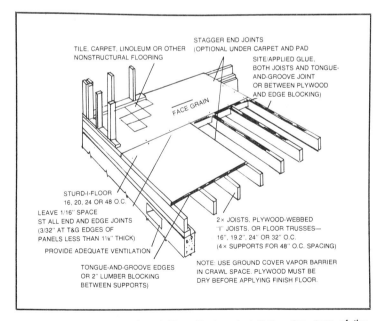

Fig. 6-10. Typical APA Glued Floor System installation (courtesy of the American Plywood Association).

First, the conventional floor frame must be constructed with sufficient strength to bear the weight of a layer of concrete at least 2½ inches thick, plus the weight of the tile. In an existing floor frame, the floor can be beefed up with additional supports of structural members. The joist tops must also be properly aligned with other parts of the floor frame that will receive a different flooring method. Then the top edges of the joists must be chamfered, as shown in Fig. 6-11. This can be done with a radial saw (before the joists are installed) or with a power plane, portable circular saw or router.

The next step is to nail nominal one-by-two ledger plates along both sides of each joist. The tops of the plates should be calculated to lie at least 1½ inches below the joist tops. They can be lower if thicker concrete is desired. Secure the ledgers to the joist sides with plenty of 6d common bright or galvanized nails, angled slightly downward. Then cut suitable strips of ¾-inch exterior type plywood. Use C-C EXT-APA or Structural I or II C-C EXT-APA or equivalent to fit snugly between the joists and atop the ledger plates. Wherever an end joint occurs, nail a two-by-four block on edge between the joists for the joint to rest upon and be secured to. Fasten the blocking with a pair of 16d common or box nails driven through the joist faces and into the ends of the blocks.

The final part of the process involves both concrete work and tile setting, both of which are outside the scope of this book. Basically, however, this consists of laying a sheet of 4-mil polyethylene plastic over the plywood and joist tops, lapped about a foot at all seams, as a waterproofer. Do not staple the plastic in place; the weight of the concrete will force the sheeting tightly against the wood. Then a suitable mix of concrete must be prepared, poured, leveled and floated to make a plane surface for the tile. Wire reinforcing mesh should be placed within the mix during pouring. After proper curing of the concrete, the ceramic tile can be set in a ½-inch bed of cement-plaster, finished and cured.

PLYWOOD CORNER BRACING

In residential construction, there are seeral ways of sheathing the exterior walls of the structure. One of the most common is to apply a sheathing of fiberboard or gypsum, which is then covered with an exterior siding. However, this type of sheathing lacks structural properties and is insufficiently strong to be applied over the entire exterior wall framework. Most building codes will not allow it. But if the corners of the exterior walls—all

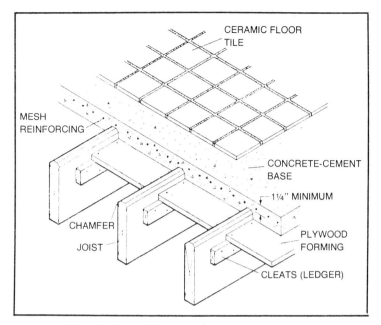

Fig. 6-11. Method of providing a concrete subbase for a ceramic tile installation over a conventional wood floor frame.

of them—are strengthened by some means, the fiberboard or gypsum panels, or similar materials, can be used to cover the remainder of the framing. This can be accomplished by the time-consuming process of installing 45° let-in bracing at all corners, but using plywood is easier, faster, more economical and stronger.

The usual choice of plywood is C-D INT-APA or equivalent. While interior glue is satisfactory, exterior glue will afford greater durability, especially if high moisture conditions are expectable during construction. Exterior types of plywood can also be used. The panel thickness should be ½-inch minimum, both for strength and because this thickness will match that of the standard fiberboard and similar products used for wall sheathing. Full panels should be installed at all outside corners (except as perhaps interrupted by window or door openings) and are often applied at inside corners as well. In addition, the same system can be used at any midwall points where extra structural strength and rigidity are desired, as around large window sections. The panels should be installed with the face grain vertical and the good side facing outward. The nailing schedule is considerably different than that for full plywood sheathing. Use 1½-inch roofing nails, spaced 4

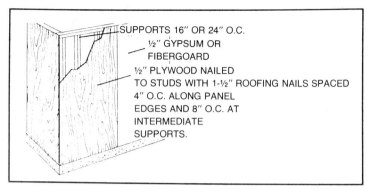

Fig. 6-12. Plywood panels installed as corner-bracing in residential construction (courtesy of the American Plywood Association).

inches apart around the perimeter of the panels, and 8 inches apart at intermediate nailing points. Stud support can be either 16 inches or 24 inches on center. Figure 6-12 shows a typical corner construction.

CONVENTIONAL PLYWOOD WALL SHEATHING

The fastest, strongest and ruggedest method of constructing ordinary wood-framed exterior walls is to sheath them entirely with plywood panels. Corner bracing is unnecessary. The completed construction is much stronger than either a fiberboard-sheathed wall or one covered with diagonal boards. Plywood also makes an excellent nailing base for all types of exterior siding and eliminates the need for additional blocking or nailing strips.

The usual choice for this application is C-D INT-APA or equivalent. Exterior glue will give greater durability, but in many instances interior glue will serve as well, from a practical standpoint. The specifics can be determined from Table 6-5. A ⅜-inch thickness is generally recommended. Panels can be (and most often are) installed with the face grain vertical, but considerable extra stiffness can be gained by placing the panels lengthwise, with the face grain horizontal. Note, however, that in some locales blocking must be installed between the studs at the joint line between the panels (now a bad idea anyway) to satisfy building code requirements.

The panels should be set so that there is a 1/16-inch gap at all end joints, and a ⅛-inch gap at all edges. In areas of consistently high humidity and wet-weather conditions, double the gaps. Trim the panels as necessary (or add filler strips), and fasten them with

6d common smooth, galvanized box, spiral thread or ring-shank nails. Space the nails 6 inches apart around the perimeter and 12 inches apart at intermediate supports. A typical arrangement appears in Fig. 6-13.

EXTERIOR PLYWOOD SIDING

A great many varieties of exterior plywood siding are produced these days. The material makes an excellent exterior covering not only for new houses (and other buildings), but also as a weatherproof, weathertight retrofitted skin for older buildings. Just a few possibilities are shown in Fig. 6-14. The recommendations for siding are given in Table 6-6. Equivalent non-APA materials can also be used.

Regardless of the specific kind of siding used, installation is made in the same way. Panels can be placed either vertically or horizontally. In either case leave a minimum of 6 inches of clearance between the bottoms of the panels and the finished grade line. No building paper or other interface is needed between the siding and plywood sheathing, but it should be placed over board sheathing. Do not use an impermeable, vapor-barrier type of material. Fasten with noncorrodible box, siding or casing nails spaced 6 inches apart along the panel edges and 12 inches apart at interior nailing points. Use 6d nails for thicknesses of ½ inch or l4ss, 8d for thicker panels. When nailing through fiberboard or similar sheathing and into wall studs or blocking, use the next larger size nail.

The trick to making a weathertight, fully sealed exterior wall lies in properly constructing all of the panel joints. The details are

Table 6-5. Listing of Plywood Wall Sheathing
(courtesy of the American Plywood Association).

Panel Idnefification Index	Panel Thickness (Inch) and Construction	Maximum Stud Spacing (inches) Exterior Covering Nailed to:	
		Stud	Sheathing
12/0, 16/0, 20/0	5/16	16	16*
16/0, 20/0, 24/0, 32/16	⅜ and ½ 3-ply	24	16 24*
24/0, 32/16	½ (4 & 5 ply)	24	24
*Apply plywood sheathing with face grain across studs.			

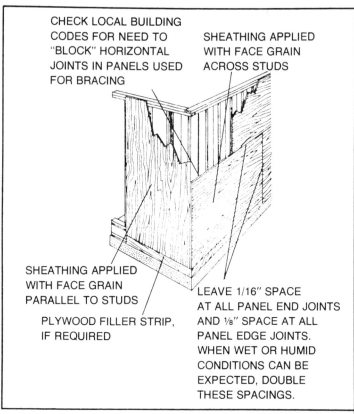

Fig. 6-13. Typical plywood wall sheathing installation; panels can be applied either vertically or horizontally (courtesy of the American Plywood Association).

shown in Fig. 6-15. Apply equally to any construction arrangement that involves the use of exterior plywood siding. A typical siding installation is shown in Fig. 6-16.

SINGLE-LAYER EXTERIOR PLYWOOD WALL

Frequently, even in residential construction, building the conventional two-layer exterior wall (sheathing covered by siding) is not necessary. Most local building codes will accept a single-layer wall, provided proven construction practices are followed. The APA Sturd-I-Wall system is a case in point.

Specification variables for the system are listed in Table 6-7. Application is much the same as for plywood sheathing and/or plywood siding. All joints must fall over a structural member or

Table 6-6. Recommendations for Plywood Panel Siding Over Nailable Plywood or Lumber Sheathing (courtesy of the American Plywood Association).

Description	Nominal Thickness (inch)	Maximum Spacing of Vertical Rows of Nails (inches)		Nail Type & Size	Nail Spacing (inches)	
		Face Grain Vertical	Face Grain Horizontal		Panel Edges	Intermediate Studs
MDO EXT-APA	11/32, 3/8	16	24	6d nonstaining box, siding or casing nails for panels 1/2" thick or less; 8d for thicker panels.	6	12
	1/2 and thicker	24	24			
303-16 o.c. Siding EXT-APA Including T 1-11	5/16 and thicker	16	24			
303-24 o.c. Siding EXT-APA	7/16 and thicker	24	24			

Table 6-7. Recommendations for APA Sturd-I-Wall System (courtesy of the American Plywood Association).

Plywood Panel Siding		Maximum Stud Spacing (inches)		Nail Size* & Type	Nail Spacing (inches)	
Description	Nominal Thickness (inch)	Face Grain Vertical	Face Grain Horizontal		Panel Edges	Intermediate Studs
MDO EXT-APA	11/32, 3/8	16	24	6d nonstaining box, siding, or casing nails for panels 1/2" thick or less; 8d for thicker panels.	6	12
	1/2 and thicker	24	24			
303-16 o.c. Siding EXT-APA Including T1-11	5/16 and thicker	16	24			
303-24 o.c. Siding EXT-APA	7/16 and thicker	24	24			

*Use next regular nail size if applied over sheathing thicker than 1/2 inch.

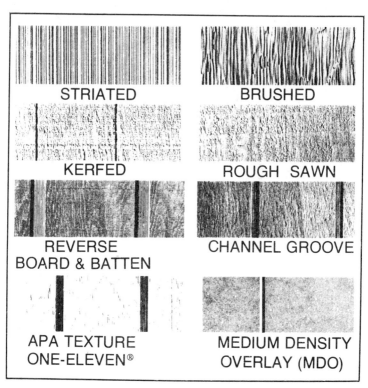

Fig. 6-14. A few of the different kinds of exterior plywood siding panels available (courtesy of the American Plywood Association).

blocking put in for the purpose, and all joints should be backed with building paper. Joints should be treated as in Fig. 6-14, with liberal use made of a high-grade *silicone* or *polyurethane caulk*. A space of 1/16 inch should be left between all panel ends and edges. Figure 6-17 shows typical constructions.

PLYWOOD ROOF DECKING

Roof decking is seldom made up of boards anymore, because plywood panels go down much faster and make a more economical, sturdier roof structure. In addition, the size and/or spacing of the rafters can often be reduced by increasing panel thickness, for a further saving in overall cost. Though one common combination is C-D INT-APA or equivalent on rafters spaced 16 inches on center, there are a good many different possibilities, as can be seen in Table 6-8. The size of the rafters must be determined on the basis of roof design, spans, allowable live loads, etc.

Table 6-8. Plywood Roof Decking Chart (courtesy of the American Plywood Association).

APA Plywood Grade	Panel Ident. Index	Plywood Thickness (inch)	Max. Span (inches)	Unsupported Edge- Max. Length (inches)	Allowable Live Loads (psf) Spacing of Supports Center to Center (inches)										Nail Size & Type	Nail Spacing (inches) Panel Edges	Nail Spacing (inches) Intermediate Supports
					12	16	20	24	30	32	36	42	48	60			
C-D INT-APA	12/0	5/16	12	12	150										6d common smooth, ring-shank or spiral-thread	6 o.c.	12 o.c.
	16/0	5/16, 3/8	16	16	160	75											
C-C EXT-APA	20/0	5/16, 3/8	20	20	190	105	65										
STRUCTURAL I & II C-D INT-APA	24/0	3/8	24	20	250	140	95	50									
		1/2		24													
STRUCTURAL I & II C-C EXT-APA	32/16	1/2, 5/8	32	28	385	215	150	95	50	40					6d (½"), 8d (⅝")		
	42/20	5/8, 3/4, 7/8	42	32		330	230	145	90	75	50	35			8d common smooth, ring-shank or spiral-thread	6 o.c.	12 o.c.***
	48/24	3/4, 7/8	48	36			300	190	120	105	65	45	35				
	48/24*	3/4, 7/8	48	36				225	125	105	75	55	40				
	2-4-1**	1-1/8	72	48				390	245	215	135	100	75	45	8d ring-shank or spiral-thread or 10d common smooth.		
	1-1/8" Grp. 1 & 2	1-1/8	72	48				305	195	170	105	75	55	35			
	1-1/4" Grp. 3 & 4	1-1/4	72	48				355	225	195	125	90	65	40			

*These loads apply only to C-C EXT-APA, STRUCTURAL I C-D INT-APA, and STRUCTURAL I C-C EXT-APA. Check availability before specifying.
**2-4-1 is synonymous with Sturd-I-Floor 48 o.c.
***Where spans are 48 inches or more, space nails 6 inches o.c. along all supports

371

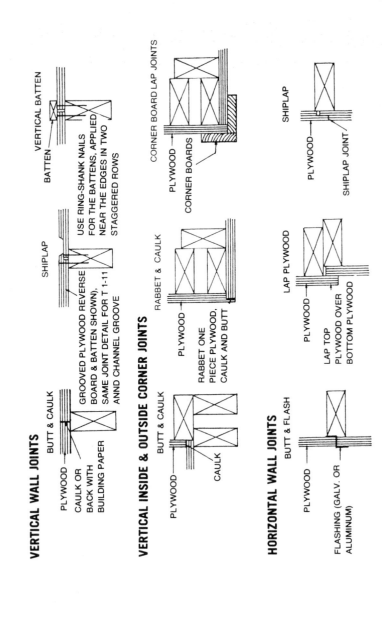

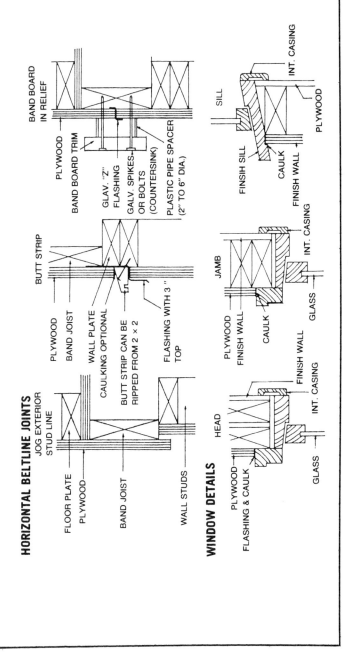

Fig 6-15. Methods of handling various types of joints in exterior plywood siding applications (courtesy of the American Plywood Association).

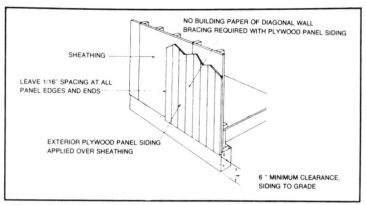

Fig. 6-16. Typical exterior plywood siding installation (courtesy of the American Plywood Association).

Laying a plywood roof decking is done in much the same manner as laying a plywood subfloor. The panels are set out in rows, with slight gaps left at the end joints for expansion. Trim the panels as necessary, and nail as specified in Table 6-8. Panel face grains should run at right angles to the lie of the rafters. At least two spans must be covered by any one piece. When laying the second and subsequent rows of panels, be sure to leave slight expansion gaps along the edges. As with a subfloor, panel end joints should be staggered as shown in Fig. 6-18. Note the placement of exterior type panels if the roof design incorporates open soffits, where the underside of the panels is exposed at the edges (see also next section).

PLYWOOD SOFFITS

In residential construction, roof *soffits* can be of either the *open* or the *closed* variety. In both cases, plywood is an efficient and effective material to use in constructing them. In open soffit construction there actually is no soffit board, but only an open space that exposes the rafters and the underside of the roof decking between them to view. In effect, the bottom on the roof decking becomes the soffit. In closed soffit construction, the rafters are closed in across their bottom edges with a soffit, and across their ends with a *fascia*. The soffit may be attached directly to the rafter bottoms and angled upward to the same degree; or they may be horizontal, attached to a ledger on the wall or to a soft ladder. Open and closed soffits are shown in Fig. 6-19.

Open soffits must be built as a part of the roof decking. But since the underside of an ordinary C-D panel is not especially

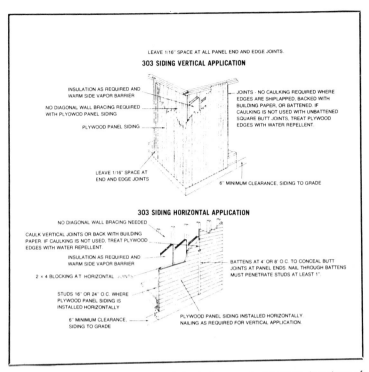

Fig. 6-17. Typical arrangements of single-layer wall sheathing (courtesy of the American Plywood Association).

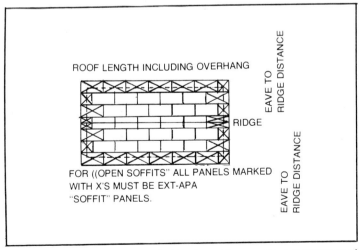

Fig. 6-18. Typical plywood roof decking layout for a gable roof (courtesy of the American Plywood Association).

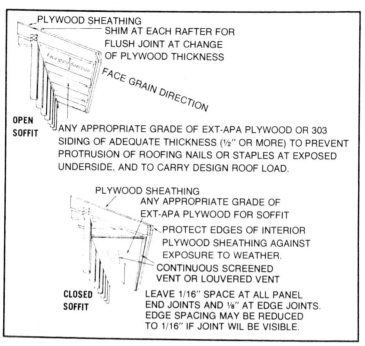

Fig. 6-19. Typical open and closed soffit constructions (courtesy of the American Plywood Association).

attractive, generally a different kind of plywood is chosen for the sake of appearance. Some of those choices are listed in Table 6-9. Note that the same plywoods can also be used as a complete roof decking in a roof design where the underside of the decking is exposed to view in the interior of the structure—the so-called *open-rafter* or *open-beam* arrangement.

Installation of these panels is simple enough. Substitute the panels for the standard roof decking wherever the panel undersides will be visible. The good or finished face should be downward. If the soffit panel thickness differs from that of the roof decking panels, shim at each rafter where the two meet to provide a flush surface. Bits of shim shingle or scrap material will suffice. Fasten the panels according to the details shown in Table 6-9.

For closed soffits, it is first necessary to mount a ledger strip at an appropriate point on the wall, or to build a soffit ladder—a ladder-like framework attached at the eave and the wall—to which the soffit panels can be secured. The soffit material can be chosen from Table 6-10. Fastening details are given there, too. Note the joint spacing and grain direction details.

Table 6-9. Plywood for Open Soffits (courtesy of the American Plywood Association).

Nominal Plywood Thickness (inches) and Panel Descriptions	Group	Maximum Span (inches)	Fastener Size & Type	Fastener Spacing (inches)
7/16 APA 303 Siding	1, 2, 3, 4	16	6d common smooth, ring-shank, or spiral-thread for panels 1/2" or less;	Space at 6 o.c. along panel edges and 12 o.c. along intermediate supports.
1/2 APA Sanded	1, 2, 3, 4			
19/32 APA 303 Siding	1			
1/2 APA Sanded	1, 2, 3			
5/8 APA 303 Siding	1, 2, 3, 4	24		
5/8 APA Sanded	1, 2, 3, 4			
3/4 APA 303 Siding	1, 2, 3, 4		8d for thicker panels.	For 48" support spacing, use nails at 6 o.c. along edges and intermediate supports.
3/4 APA 303 Siding	1	32*		
3/4 APA Sanded	1, 2, 3, 4			
1-1/8 APA Textured	1, 2, 3, 4	48*	8d deformed-shank for 1-1/8" panels; 10d common smooth only if framing is dry.	

*Provide adequate blocking, tongue-and-groove edges, or other supports such as Plyclips.

**Minimum loads are at least 40 psf live load, plus 5 psf dead load, except 1-1/8 inch panels of Groups 2, 3, or 4 species support 35 psf live load, plus 5 psf dead load.

Table 6-10. Plywood for Closed Soffits (courtesy of the American Plywood Association).

Nominal Plywood Thickness (inches) and Panel Descriptions	Group	Maximum Span (inches), All Edges Supported	Fastener Size & Type	Fastener Spacing (inches)	
				Panel Edges	Intermediate Supports
5/16 APA 303 Siding 3/8 APA Sanded	All species groups	24	6d nonstaining box or casing nails for panels 1/2" or less; 8d for thicker panels.	6 o.c.	12 o.c.
7/16 APA 303 Siding 1/2 APA Sanded		32			
19/32 APA 303 Siding or APA Sanded		48			

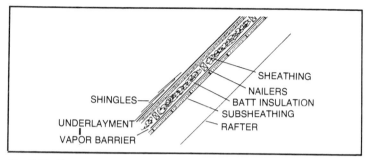

Fig. 6-20. Cross section of built-up thermal roof, with batt or blanket insulation.

BUILT-UP THERMAL ROOF

In these days of high energy costs and supply shortages, a thermally efficient roof is a must. There are many ways of building such a roof. The specific design details are far too complex and variable to investigate here. But a built-up plywood roof filled with thermal insulation is one arrangement that works extremely well. It can be employed in new construction or installed as part of a reroofing job on an existing structure. It is also ideal where the underside of the roof decking is exposed to the interior and thermal insulation cannot be installed in the usual manner (between the rafters and below the decking).

The job starts with a conventional plywood roof deck, as discussed earlier. In an existing building the original decking, if in good condition, can be used whether it is plywood or boards. Then a grid of nominal 1-inch or 2-inch stock must be built, with the members standing on edge, across the entire roof surface to enclose the perimeter of the roof and provide stapling and nailing strips, or *baffles*, for the thermal insulation and the top decking. Insulation is placed within the grid. The type and thickness of the insulation can vary, depending upon the degree of thermal efficiency needed, cost factors and the kind of material that is readily available. The depth of the grid matches the thickness of the insulation. Some of the choices are *vermiculite*, fiber glass or rock wool blanket, reflective sheet, ceramic pour-type or loose rock wool. Another alternative is rigid sheets, in which case a large part of the gridwork might be dispensed with, depending upon the type of finish roof covering contemplated. Rigid insulation (there are several types) should be chosen with care for correct characteristics of resistance to crushing, performance in hot areas (some types deform or melt at fairly low temperatures) and thermal properties.

Once the insulation is emplaced, the next step is to cover the grid with a second deck of plywood, laid in the same fashion as the first (as discussed earlier). After an underlayment of roofing felt or other appropriate material is applied to the top deck, the finish roof covering—roll roofing, asphalt or vinyl shingles, cedar shakes, etc.—can be laid. An exception would be metal roofing sheets, which could be nailed directly to the gridwork under the proper design conditions. If a separate vapor barrier is required, as is often the case, that should be laid across the first roof deck before the gridwork is installed. Figure 6-20 shows a typical built-up thermal roof construction in cross section.

FIRE-RATED ROOF

Even where a fire-rated roof/ceiling (or floor/ceiling) assembly is not required by building codes, its installation is a good idea from the standpoint of added life-safety. The increase in cost is minimal. The actual laying of the plywood panels is no different than for ordinary constructions, but the assembly techniques and material combinations are.

The plywood heavy-timber roof construction is a good one for certain kinds of modern residential designs that feature open ceilings. It can be used with other designs as well. This arrangement is shown in Fig. 6-21. A conventional type of roof can be constructed in the usual manner by substituting fire-retardant treated joists or rafters for plain wood, and covering with a decking of FRT plywood in a minimum thickness of ¾-inch. Note that the spacing may be taken out to as much as 48 inches on center, depending upon roof loading factors and other design parameters (Fig. 6-22). Another

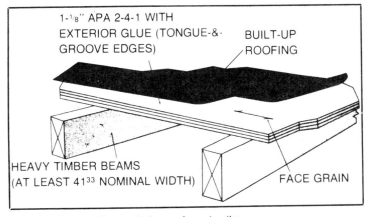

Fig. 6-21. Fire-rated heavy timber roof construction.

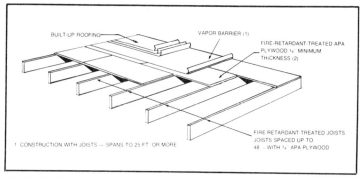

Fig. 6-22. One possibility for fire-rated roof construction with joists (courtesy of the American Plywood Association).

method of building fire-resistant roof/ceiling or floor/ceiling constructions is to use the protected system, which simply involves sheathing the building section with materials of good fire-resistivity. Examples of this method are shown in Fig. 6-23; there are a number of possibilities. For the most part, construction is simple enough. The materials are readily available and not prohibitively expensive.

SOUND CONTROL WITH PLYWOOD

In the interests of privacy and comfort, sound control is often a desirable feature in residential construction, especially in split-level designs or in any situations where activity and noise-producing areas adjoin quiet areas, whether side-by-side or above-and-below. By using the proper design and construction methods, plywood can be employed with other materials to achieve good sound control at reasonable cost. The STC (*Sound Transmission Class*) rating should be 45 or better, and 50 is considered quite good. The IIC (*Impact Insulation Class*) should also be on that order. In either case, the higher the better.

Table 6-11 and Fig. 6-24 show several combinations of materials that can be used with the basic construction of flooring on sleepers to obtain good sound control. Note that the ICC is excellent where carpeting and felt padding are installed. The various methods of construction are explained in the example.

Similar methods can be used to gain sound control with wall assemblies, but achieving high STCs is more difficult. The IIC rating is not used with wall constructions, since impact noise is not a problem. The exterior wall construction shown in Fig. 6-25 is easy to build and affords adequate protection against unwanted

U.L. DESIGN L504

THIS PLYWOOD STRESSED-SKIN CONSTRUCTION IS THE THINNEST ONE-HOUR COMBUSTIBLE ASSEMBLY WHICH HAS PASSED UNDERWRITERS' LABORATORIES TESTS. THE LAYERS OF FIBER INSULATION BOARD AND SPECIAL TYPE X GYPSUMBOARD MAY BE APPLIED IN THE FIELD. 3/8" PLYWOOD MAY BE SUBSTITUTED FOR THE 1/2" INSULATION BOARD.

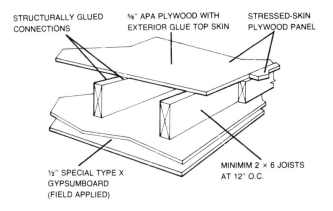

SEE U.L. DESIGN NOS. L501, L502, L503, L512, L514, L515, L519, L521, L522.

5/8" TYPE X GYPSUMBOARD OF 1/2" SPECIAL TYPE X GYPSUMBOARD IS ATTACHED DIRECTLY TO THE BOTTOM OF THE JOISTS. ALTERNATIVELY, THE GYPSUMBOARD MAY BE FASTENED TO RESILIENT METAL FURRING STRIPS FOR IMPROVED ACOUSTICAL PERFORMANCE.

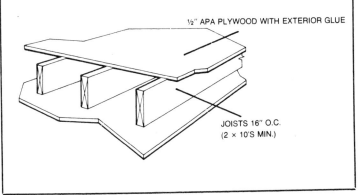

Fig. 6-23. Two roof constructions of one-hour rating using the protected method.

Table 6-11. Various Construction Methods That Can Be Employed to Achieve a High Level of Sound Control in Floor/Ceiling Assemblies (courtesy of the American Plywood Association).

Test number & sponsor	Finish Floor	Deck	Gypsum board ceiling	Insulation	STC	IIC	Weight (lbs./sq. ft.)
Case 1 KAL-224-7 & 8 APA	44 oz. carpet and 40 oz. pad on 19/32" T&G Sturd-I-Floor	19/32" T&G Sturd-I-Floor nailed to 2x3 sleepers which in turn were glued halfway between joists to 1/2" insulation board stapled to 1/2" plywood subfloor on 2x joists at 16" o.c.	5/8" screwed to channels	3" glass fiber	52	78	11.7
Case 2 KAL-224-9 & 10 APA	25/32" wood-strip flooring nailed to wood sleepers	Sleepers glued, halfway between joists, to 1/2" insulation board stapled to 1/2" plywood subfloor on 2x joists at 16" o.c.	5/8" screwed to channels	3" glass fiber	53	51	13.0
Case 3 KAL-736-8 & 9 ISU	25/32" wood-strip flooring nailed to wood sleepers	Sleepers glued to 3" wide strips of 1/2" sound board glued, halfway between joists, to 1/2" plywood subfloor on 2x joists at 16" o.c.	5/8" screwed to channels	3" batts between joists; 1½" blanket between sleepers	55	51	11.3
Case 4 R-TL 70-61 R-IN 70-9 W	Vinyl flooring glued to 1/2" plywood underlayment	Underlayment over 1x3 furring strips halfway between joists, on top of 1/2" sound board over 5/8" plywood subfloor on 2x joists at 16" o.c.	5/8" screwed to channels	3" glass fiber	57	56	11.6
Case 5 R-TL 71-279 R-IN 71-19 WWPA	.070 vinyl glued to 19/32" T&G Sturd-I-Floor	19/32" I&G Sturd-I-Floor stapled over 2x2 sleepers glued halfway between joists, to 1/2" plywood subfloor glued to 2x joists at 16" o.c.	5/8" screwed to channels	3" glass fiber between joists; 1½" sand between sleepers	59	56	20.2

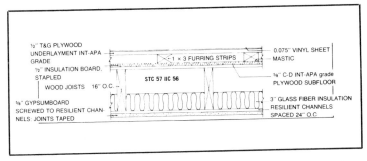

Fig. 6-24. Material with an excellent IIC rating (courtesy of the American Plywood Association).

sound, especially in exurban and rural areas where outside sound levels are low anyway. To gain a high level of resistivity against noise within the house, a construction like the party wall shown in Fig. 6-26 works well.

ALL WEATHER WOOD FOUNDATION

This foundation system is a proven one that has been thoroughly tested and in use for many years. It is made up of pressure-preservative treated plywood panels, lumber, polyethylene sheeting and special sealers. It is put together with stainless steel nails. The basic construction is the same as for any stud wall. No concrete is required for crawl-space designs, though it is pouted for full-basement floors. The AWWF system can be used beneath a light-frame type of building and is strong, long-lived and economical.

Though construction is quite simple, the foundation must be properly designed for each individual construction job. Attention must be paid to a number of details not common to ordinary stud-wall fabrication. These points are variable. A typical full-basement AWWF construction is shown in Fig. 6-27. Full details can be obtained from the American Plywood Association.

INTERIOR WALL PANELING

One of the most popular home-remodeling projects involves putting up decorative plywood wall paneling. Though fairly time-consuming if the job is to be done right, paneling is not a difficult task and makes an economical, attractive way to entirely change the appearance and character of a room. Hardwood paneling is often installed in new construction as well. The Georgia-Pacific Corporation is one of the leading producers of hardwood plywood wall

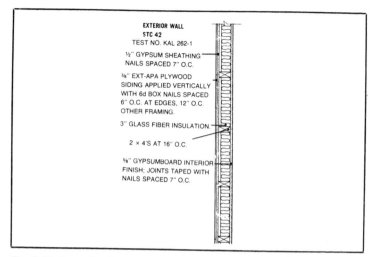

Fig. 6-25. A construction method for a sound resistant exterior wall (courtesy of the American Plywood Association).

paneling. They have put together an excellent series of drawings and instructions of how to put up wall paneling. We can do no better than to quote those instructions in their entirety; the drawings appear in Figs. 6-28 through 6-33.

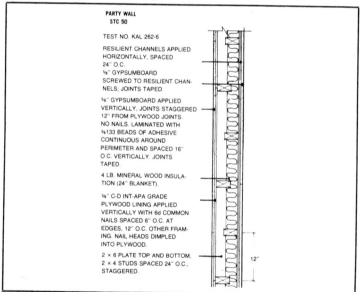

Fig. 6-26. A party wall construction such as this one will afford excellent sound control between rooms within a house (courtesy of the American Plywood Association).

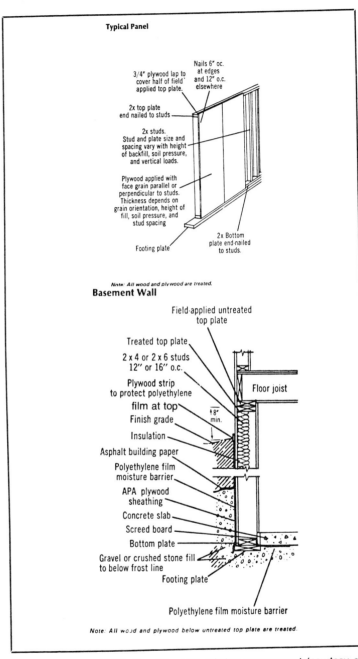

Fig. 6-27. Typical All Weather Wood Foundation arrangement (courtesy of the American Plywood Association).

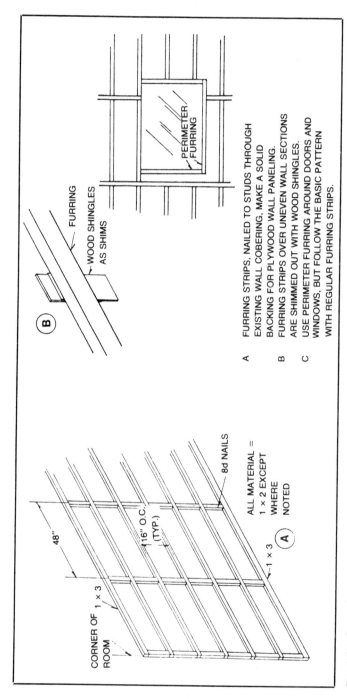

Fig. 6-28. General arrangement of furring for a wall paneling installation and details of furring-strip application (courtesy of Georgia-Pacific Corporation).

How Many Panels?

The room's perimeter will tell you. Assume a room measures 15' × 18'. Two (15') plus two (18') equals 66'. Panels are 4' wide, so 66' divided by 4' comes out as 16½ panels. Door, window and fireplace areas are deductible; about half a panel for each usually works out. Of course, these are variables, so measuring for the deductions is wise. You don't buy half panels. If your arithmetic says 14½ panels, but 15. Be generous—it's better than being short. Extra paneling can be useful for matching cornice or a shield for concealed lighting. Many people use the same paneling as a veneer on existing doors.

Application

If walls are sound and true, paneling may be applied directly with nails or glue. Most times when remodeling, it's been found that covering the wall with a gridwork of furring strips solves all problems. Usually 1 × 2 sound, dry lumber is used, but plywood is a good candidate. Plywood furring can be as thin as ⅜", as wide as 2". "Scrap" pieces of ¾" plywood ripped into 2" strips make excellent furring.

In any event, the furring is applied as shown in Fig. 6-28—verticals, 48" on centers (over studs); horizontals, 16" on centers and nailed at each stud crossing. The horizontal pieces will be easy to install if you make a special gauge—a piece of wood that is 16"

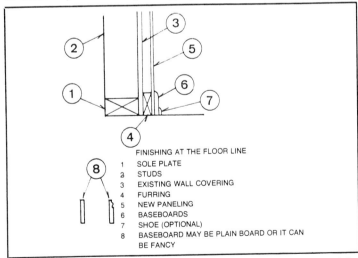

FINISHING AT THE FLOOR LINE
1. SOLE PLATE
2. STUDS
3. EXISTING WALL COVERING
4. FURRING
5. NEW PANELING
6. BASEBOARDS
7. SHOE (OPTIONAL)
8. BASEBOARD MAY BE PLAIN BOARD OR IT CAN BE FANCY

Fig. 6-29. Details of finishing wall panels at the floor line (courtesy of the Georgia-Pacific Corporation).

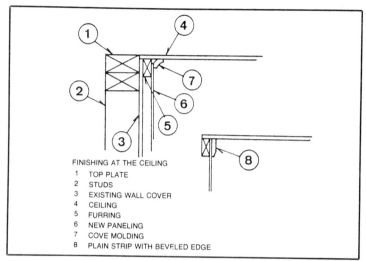

Fig. 6-30. Details of finishing wall panels at the ceiling line (courtesy of the Georgia-Pacific Corporation).

long, less the width of one furring piece. Thus, if furring is 1½" wide, the spacing gauge will be 14½" long.

The joint between horizontal and vertical strips must not be tight. Also, leave a gap of about ¼" between the top strip and the ceiling and between the bottom strip and the floor to allow for breather space.

Positioning the Panels

First, spread the panels around the room so you can judge the best placement in relation to tone and grain pattern. Number the panels on the back so you'll know where each will go.

The first panel goes in a corner, placed so its free edge hits the center of a vertical furring strip. Use a level to be sure the panel is perfectly vertical. You may have to reshape the corner edge of the panel if the room corner is not plumb. If the panel abuts an irregular wall—a brick or a stone surface—you'll have to shape the panel's edge accordingly. The best way is to hold the panel in place and then work with a compass, using the irregular wall as a guide so you can transfer the cut-line to the panel. If the cut-line is straight, saw with a crosscut saw. If irregular, use a coping saw or saber saw. With a coping saw or crosscut saw, keep the good side of the panel up. Do the reverse with a saber saw.

Once the first panel is up, others will install rather quickly. Simply butt edges, glue, and nail in place. Use 3d finishing nails,

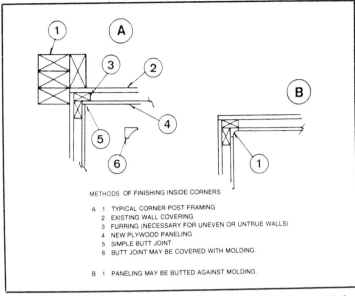

Fig. 6-31. Methods of finishing inside corners in a wall paneling installation (courtesy of the Georgia-Pacific Corporation).

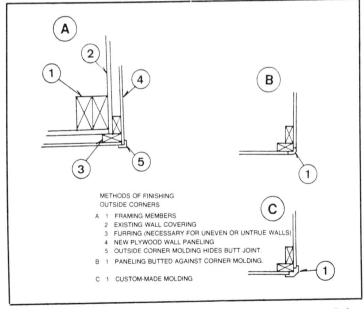

Fig. 6-32. Methods of finishing outside corners in a wall paneling installation (courtesy of the Georgia-Pacific Corporation).

spacing them 8" along all edges and 16" along horizontal furring. Set nail heads about 1/32" deep, then conceal them with a putty stick. The colored sticks are available to match the tone of the paneling.

Figures 6-29 through 6-33 show how to finish professionally in corners, around doors and windows, and at floor and ceiling joints. Matching moldings for the paneling you have installed are available. Unfinished moldings in a wide variety are also available should you decide to do your own thing.

Points To Consider

Existing mouldings, if carefully removed, may be reusable. Existing baseboard, if of correct thickness and not saved for reuse, can serve as the bottom furring strip. All edges of each panel must be attached to furring. Add extra strips if necessary. Use a block plane or something similar to slightly bevel the back edges of panels that abut. Leave a ¼" gap at top, bottom and ends of installations so panels can expand if necessary. Moldings hide the gaps. It's a good idea to store the panels in the room for a few days before installation. This is to allow moisture content to equalize.

Application of paneling is not difficult, but accurate measuring and careful sawing and fitting are the keys to a successful project. This is one situation where it pays to use an hour to do a half-hour job. The paneling will be nice to look at for a long time.

WALL-HUNG UTILITY DESK

This plywood utility desk can be easily built in the shop and then hung on a wall at any convenient point as a mini-office or handy work-storage unit (Figs. 6-34 and 6-35). It could even be set up as a modelmaking shop. Cut the sides and ends of the case. Miter the ends of the four pieces to form the corners, and rabbet the back edge of each to receive the back panel. Then assemble the case with aliphatic resin glue and 4d finish nails, taking care to maintain squareness. Apply glue to the rabbet, drop the back panel in place, and secure with brads or wire nails about an inch long. Cut the divider and shelves to size. Make alignment marks on the case, and check the pieces for proper fit in their respective positions. Install the upper shelf first, then the divider, followed by the remaining two shelves. Secure them with 6d finish nails and glue.

Then cut the door to size, trim and fit it, and install it with flap hinges and lid supports. If this kind of hardware is unavailable (it usually can be ordered), substitute a strip of ¾-inch piano hinge and a lid support arm. In this case, the height of the door and the width

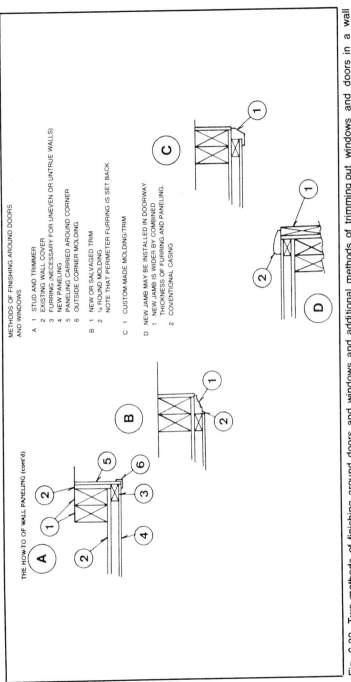

Fig. 6-33. Two methods of finishing around doors and windows and additional methods of trimming out windows and doors in a wall paneling installation (courtesy of the Georgia-Pacific Corporation).

Fig. 6-34. This wall-hung utility desk is easy to make and can provide a convenient cork-center (courtesy of the Georgia-Pacific Corporation).

of the case bottom piece will have to be reduced by ¾ inch. Install a single magnetic catch near the top center of the door, or one at each upper corner, to keep the door closed in the upright position. A pull, or locking latch can be substituted for the half-circle finger-pullout. Cover all of the raw edges of plywood that are exposed, and finish to suit (for prefinished hardwood plywood, trim the edges with matching veneer tape)

CUBE STORAGE

Plywood cubes are easy to make and can be turned to any number of purposes—storage beneath a bed, stacked as book cases, record storage or a music wall, stacked with a shelf added to make a desk, doorless and face up as planters, in vertical pairs as nightstands or end tables, etc. The basic design shown here is a simple case (Fig. 6-36). There are a number of construction options. The door can be omitted, for instance. If a number of cubes are to be stacked, they could be made with butt joint. Mitered joints are more attractive and also hide the raw plywood edges. This is the best bet for hardwood plywoods that will receive a natural finish (or are prefinished). For greater strength (as for a seat), use splined miter joints. The back panel is intended in this design to be inset, but could be made slightly larger and set in a case rabbet. Shelves can be set with plain butt joints, but a dado makes for greater strength and a better construction.

Many types of plywood can be used for this project, either softwood or hardwood. A ¾-inch thickness is best; use an Exterior type for a deck or patio furnishings. The raw edges at the face of the cube should be concealed with veneer, molding or edging. Secure all of the joints with aliphatic glue (resorcinol for outdoor cubes) and 4d finish nails at mitered joints, 6d at butt joints. Squareness is very important. Corner clamps are helpful during assembly. Shevles should be centered so that the cubes can turned on their sides, making the shelves into dividers. Doors can be mounted on mortised butt hinges, concealed or pivot hinges, or small surface-

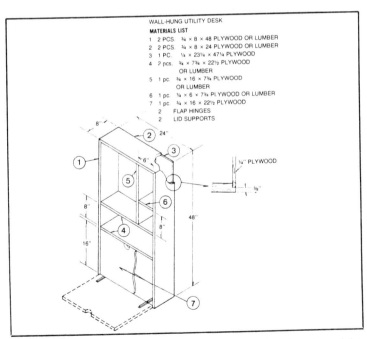

Fig. 6-35. Construction details and materials for the desk (courtesy of the Georgia-Pacific Corporation).

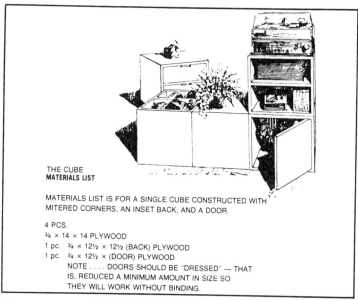

Fig. 6-36. Easy-to-make plywood cubes can be put to all sorts of uses as furnishings (courtesy of the Georgia-Pacific Corporation).

mounted decorative hinges. Likewise, there are dozens of possibilities for pulls or knobs—just a simple drilled finger-hole will do the job. Magnetic catches can be used to hold the doors closed. Details of the cube system are shown in Fig. 6-37.

FIRESIDE BENCH

This fireside bench is a typical casework-and-faceframe construction that can be modified in the form of kitchen cabinets, shelving and similar items (Fig. 6-38). Softwood plywood can be used for the bench with either a natural or a painted finish. You can use hardwood plywood for a furniture finish. In the case of a natural finish, the faceframe material should be matched as closely as possible to the appearance of the plywood, but this is not necessary for a painted finish.

Cut and fit the case parts first. Though butt joints are shown, the piece dimensions can be altered slightly for dado joints at the top and dividers. This would allow substantially greater strength and rigidity. Assemble the case carefully, keeping it fully in square. Fasten the parts with aliphatic resin glue and 8d finish nails for butt joints, 6d for dadoed joints. The order of assembly is: back to sides, bottom to back and sides, top to back and sides, dividers to top, bottom and back.

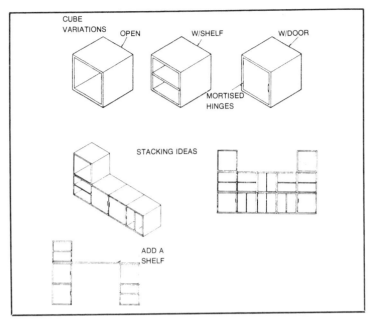

Fig. 6-37. Details of the cube system (courtesy of the Georgia-Pacific Corporation).

Fig. 6-38. The fireside bench is an excellent beginning casework project (courtesy of the Georgia-Pacific Corporation).

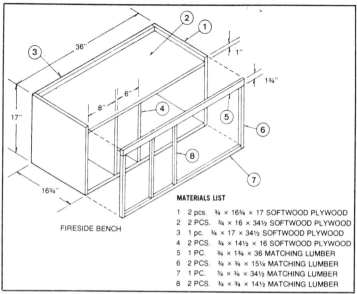

MATERIALS LIST

FIRESIDE BENCH

1 2 pcs. ¾ × 16¾ × 17 SOFTWOOD PLYWOOD
2 2 PCS. ¾ × 16 × 34½ SOFTWOOD PLYWOOD
3 1 pc.   ¾ × 17 × 34½ SOFTWOOD PLYWOOD
4 2 PCS. ¾ × 14½ × 16 SOFTWOOD PLYWOOD
5 1 PC.   ¾ × 1¾ × 36 MATCHING LUMBER
6 2 PCS. ¾ × ¾ × 15¼ MATCHING LUMBER
7 1 PC.   ¾ × ¾ × 34½ MATCHING LUMBER
8 2 PCS. ¾ × ¾ × 14½ MATCHING LUMBER

Fig. 6-39. Construction details and materials for the fireside bench (courtesy of the Georgia-Pacific Corporation).

The parts for the faceframe should be cut individually to fit in each location. Ripping and trimming should be done with fine-toothed blades. The edges and ends should be carefully squared and smoothed. The frame can be made up on the bench as a unit by carefully transferring measurements, but it is generally easier to install the parts piece by piece on the case itself. Secure the faceframe to the case with glue and 3d or 4d nails. To avoid the visible end grain at the top corners of the faceframe, the pieces can be cut in a modified offset miter.

Sand the cabinet and fill holes or defects as necessary. Apply a finish of your choice. Then cover a 2-inch thick foam upholstery pad, approximately 16 inches wide by 34½ inches long, with a suitable material and fit into place. Two or three small dots of Velcro will keep it from slipping forward. Construction details appear in Fig. 6-39.

DOGHOUSE

Here is a doghouse that is attractive, easy to clean, inexpensive and fun to build. It will afford a snug and comfortable outside shelter for your dog, complete with observation deck. The size can be modified if necessary, to accommodate different breeds of dogs (Figs. 6-40 and 6-41).

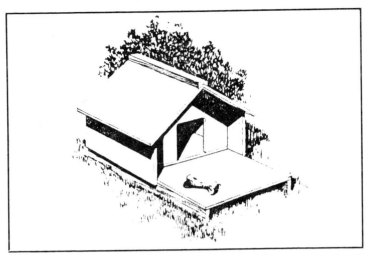

Fig. 6-40. This fine doghouse design can be made bigger or smaller to exactly accommodate Rover, and is easy to construct (courtesy of the Georgia-Pacific Corporation).

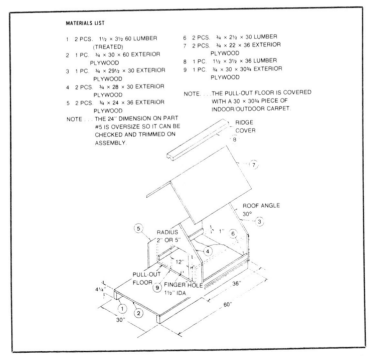

Fig. 6-41. Construction details and materials for the doghouse (courtesy of the Georgia-Pacific Corporation).

LARGE PLANTER

MATERIALS LIST

1. 4 PCS. 3½ × 3½ × 18 LUMBER
2. 2 PCS. ⅝ × 18 × 24 EXTERIOR PLYWOOD
3. 2 PCS. ⅝ × 16 × 18 EXTERIOR PLYWOOD
4. 1½ × 16 × 22¾ LUMBER
   (MAKE FROM TWO PIECES OF 2 × 10 STOCK)
5. 2 PCS. ⅝ × 2½ × 27¾ LUMBER
6. 2 PCS. ⅝ × 2½ × 21 LUMBER

NOTES CUT PARTS #5 AND 6 LOGNER THAN NECESSARY, TRIM TO SIZE.
⅝" MATERIAL = EXTERIOR PLYWOOD SIDING.
ATTACH FOUR SWIVEL TYPE PLATE CASTERS WITH 1½ ROUND HEAD WOOD SCREWS.

Fig. 6-42. This large planter is simple to construct (courtesy of the Georgia-Pacific Corporation).

Start by putting the floor platform together; attach the support rails to the floor with resorcinol glue and 8d galvanized nails. Then cut out the front and back pieces. The height of the front from bottom to peak is 28 inches, and the same measurement for the back is 29½ inches. Cut the angle at 30°. Make the door cutout in the front piece and thoroughly ease all of the cut edges. Then glue and nail the back piece in place. Use 8d galvanized nails and align the bottom of the back piece flush with the bottom surface of the floor.

Then cut two sides to a length of 36 inches. Determine their correct width by putting them in place against the back and floor. The top edge of each side must be bevel-cut to match the roof pitch. Next, cut two runners (#6) and fasten them to the inside surfaces of the sides with glue and 4d galvanized nails. They should be positioned 1 ¾ inches up from the bottom of the sides, and ¾ inch in from the back edge. Then attach the sides to the floor and back, using glue and 6d galvanized nails. The bottom edges should be flush with the underside of the floor, and the back end should be flush with the outside surface of the back piece. Secure the front next, with its bottom edge flush with the bottom edges of the runners.

Cut the roof pieces, align and fit them, and secure them with glue and 6d galvanized finish nails. Cut a length of two-by-four to match the roof length. Cut a notch the length of it to match the pitch

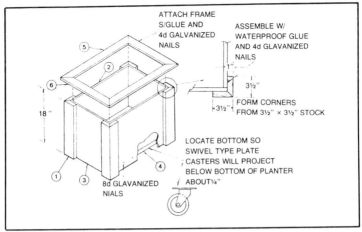

Fig. 6-43. Construction details for the large planter (courtesy of the Georgia-Pacific Corporation).

angle of the roof, and glue it in place (no nails used here). Then cut the inside floor piece to size and drill the finger-hole. Check the piece for easy sliding in and out, and trim or adjust as necessary. When you've gotten a good fit, slightly round all four corners and ease all of the bottom edges. Cut a piece of indoor/outdoor carpeting to fit the floor. Secure it with double-sided carpet tape so that it

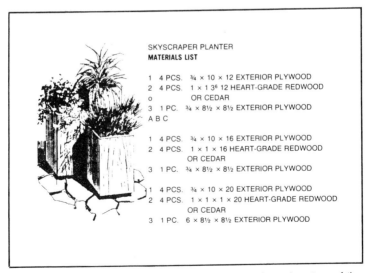

Fig. 6-44. The skyscraper planter looks attractive outdoors (courtesy of the Georgia-Pacific Corporation).

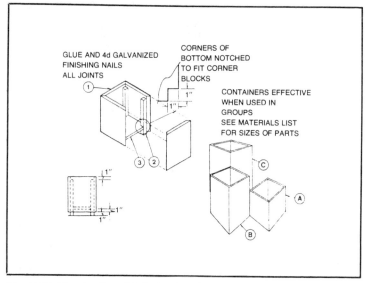

Fig. 6-45. Construction details for the skyscraper planter (courtesy of the Georgia-Pacific Corporation).

can be easily removed for washing or replacement. The interior of the doghouse should be left unfinished, but an applied finish on the outside will add to the appearance (unless you have used an exterior siding plywood that already has a finish). Stain or paint are two good possibilities. Make sure that whatever you use is nontoxic.

PLANTERS

Planters are easy and fun to build. They make attractive furnishings not only for decks and patios, but for indoor accents as well. The designs shown here are only two of dozens—maybe hundreds—of possibilities. You can use appearance grade Exterior type softwood plywood (or even engineered grade if you wish), or a plywood siding material of redwood, cedar or some other kind. Even an exterior grade hardwood plywood could be used.

Use miter joints wherever appearance is important. Fasten the parts with resorcinol glue and galvanized or other noncorrodible nails of suitable length. Finish nails are fine, but casing nails will afford a bit more holding power. For a natural finish that is virtually maintenance-free, coat the planters (on the outside) with an oil finish that is formulated to be weather-residtant. (See Figs. 6-42 through 6-45.)

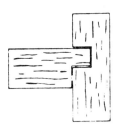

Index

A
Acoustical properties — 121
Aeronautical, plywood uses — 17
All weather wood foundation — 383
American Plywood Association Plywood — 65
American plywood industry — 15
Automotive, plywood uses — 17

B
Bandsaw cuts — 260
Barrel-top trunks — 14
Bending — 51
Beveling — 277
Bits — 187-192
Black walnut — 27
Block flooring — 140
Bobs — 152
Builder's hardware — 324

C
Cemented-tile subfloor — 363
Chamfering — 274
Chemical resistance — 129
Chisels — 178
Clamps — 192
Cold press method — 43
Color toning — 342
Commercial uses — 13
Coverage — 49
Cube storage — 392
Curves, cutting — 270
Cutters — 187-192

D
Debarking logs — 28
Decay resistance — 128
Dimensional stability — 48
Distributive strength — 44
Doghouse — 396
Doors, plywood — 14
  plywood uses — 17
Drilling — 281-288
Drills — 182-187

E
Edge finishing — 272
Edging — 330
Electrical properties — 128
Equilibrium
  moisture content — 124
Expense — 59
Exterior finishes — 348
Exterior plywood siding — 367

F
Fasteners — 314-324
Filing — 266
Finishes — 347-351
Finishing — 335
Finishing tools — 195-200
Fire protection
  types of construction — 131
Fire retardant
  hardwood plywood panels,
  producing — 134
Fire retardant paints — 133
Fire-rated roof — 379
Fireside bench — 394
Flame spread rating — 130
Flitches — 35
Forming tools — 195-200
Framing — 334
FRP plywood — 136
Furniture, plywood uses — 17

G
Glued plywood floor — 360
Glues — 18, 306-314
  new developments — 21
Gluing plywood — 306
Grade trademarks — 91
Grain figures,
  demand for — 26
Grain filling — 340
Grain patterns,
  demand for — 26
Green clipper — 33

H
Hammers — 175
Handling difficulty — 58
Hardware — 324
Hardwood plywoods — 97

manufacturing standards — 99
panel constructions — 108
types — 111
varieties — 98
veneer grades — 101
wood species — 99
Hardwoods — 63
HDO-MDO plywood — 135
Hinges — 325

I
Insect resistance — 128
Interior wall paneling — 383-390

J
Jointing — 269
Joints — 291-306

L
Layout tools — 147
Levels — 153
Lines — 152

M
Machinery — 11
Maneuvering space — 217
Matching — 116
Mayo, John K. — 12
Mayo's patents — 12
Measuring tools — 147
Miscellaneous tools — 205-208
Molding — 51
Moldings — 331
Mortising — 288
  problems — 327

N
Nails — 315
Natural finishes — 348
1950's, new developments — 22
1960's, new developments — 22

O
Oil finishes — 347

P
Pacific Coats Plywood Manufacturers, Inc. — 20

401

Paine Lumber Company	16	
Paint finishes	350	
Paints	343-345	
Patterning tools	154	
Peeler blocks	28	
heating	29	
Peeling lathes	31	
Permeability	127	
Personal protection gear	202	
Phenol-formaldehyde resin glue	21	
Planes	180	
Planing	264	
Planters	400	
Plywood, added strength	46	
advantages	44	
benefits for the do-it-yourselfer	52-57	
buying	236	
conservation	52	
conventional subfloor	352	
conventional wall sheathing	366	
cutting with handsaw	245	
cutting with portable circular saw	246-252	
cutting with radial arm saw	258	
cutting with sabersaw	252	
cutting with table saw	254	
disadvantages	57-60	
engineered grades	83	
FRP	136	
gluing	306	
handling	232	
HDO-MDO	135	
how it's made	24	
mill controlled	67	
miscellaneous properties	121	
origins	9	
post WWI changes	19	
pressure-preserved	137	
selecting	236	
siding face grades	86	
specialty	135-143	
storage	232	
understanding characteristics	61	
with sound control	380	
Plywood corner bracing	364	
Plywood floor glued	360	
Plywood identification index	88	
Plywood industry, today	23	
Plywood roof decking	370	
Plywood siding, exterior	367	
Plywood soffits	374	

Plywood subfloor/underlayment, combined	359	
Plywood underlayment	356	
Plywood workshop	213	
Portland Manufacturing Company	16	
Pressure preserved plywood	137	

R

Redwood plywood grades	86	
Resins	306-314	
Roofs	370	
Rounding	278	
Router	200	
Rules	147	

S

Sanding	268, 338	
Sawing	263	
Saws	158-175	
Screwdrivers	177	
Screws	319, 326	
Scrolls, cutting	270	
Shaping	280	
Shellac	346	
Single-layer exterior plywood wall	368	
Softwood plywoods	65	
exterior	70	
interior	70	
manufacturing standards	94	
panel construction	69	
sizes	67	
species group classification	73	
thicknesses	67	
Softwoods	63	
Southern Plywood Manufacturers Association	22	
Span index	90	
Speciality manufacturing	43	
Speciality plywoods	135-143	
Split resistance	46	
Squares	150	
Squaring	262	
Stain finish, application	341	
Stain finishes	348	
Stay log cutting	42	
Stiffness	45	
Storage space	222	
Straightedges	148	
Strength-to-weight ratio	49	
Strips	331	
Subfloor/underlayment, combined	359	

Surface repairs	336	

T

Thermal properties	122	
Thermal roof, built-up	378	
Three lathe-to-clipper process	33	
Toolbox, starting	208	
Tools, costs	146	
finishing	195-200	
forming	195-200	
keeping sharp	209	
layout	147	
measuring	147	
miscellaneous	205-208	
patterning	154	
selecting	145	
Tree farming	25	

U

Utility desk, wall-hung	390	

V

Varnish	345	
Veneer, applying	332	
drying process	36	
hot press	40	
layup process	38	
slicing	34	
sorting	35	
stacking method	35	
Veneer panels, grading	41	
inspecting	41	
sanding	40	
trimming	40	
Veneer appearance grades	79	
Veneer grade letters	78	
Veneer grades	76	
Veneering, a new name	18	
early uses	10	

W

Warp resistance	47	
Water exposure	126	
Woods, increased availability	50	
Work surfaces	219	
Working characteristics	59	
Workpiece layout	238	
Workshop, layout	231	
planing	231	
plywood	213	
where to put it	214	
Workshop space	216	
Workshop utilities	224-231	